中国科学院中国孢子植物志编辑委员会　编辑

中　国　真　菌　志

第七十卷

金孢属及其相关属

韩燕峰　主编

国家自然科学基金重大项目
中国科学院前沿科学重点研究项目
（国家自然科学基金委员会　中国科学院　资助）

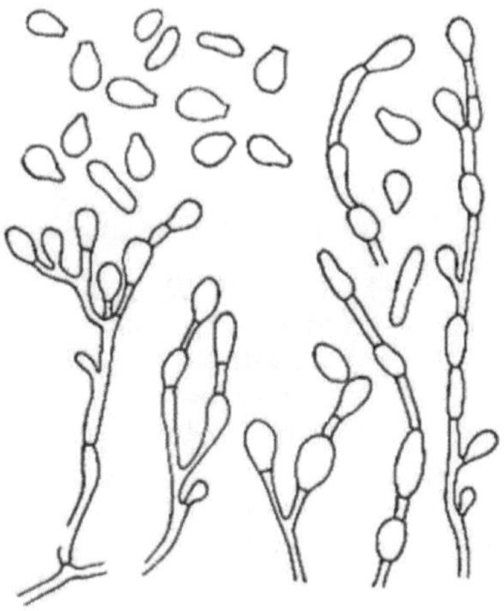

科 学 出 版 社
北　京

内 容 简 介

本卷记载了我国金孢属及嗜角蛋白真菌相关属共15属136种，包括金孢属26种、栉霉属5种、地丝霉属4种、角蛋白癣菌属11种、冠癣菌属1种、小孢子菌属5种、毁丝霉属1种、奈尼兹皮菌属5种、帕拉癣菌属2种、瓶霉属20种、假裸囊菌属14种、赛多孢属9种、帚霉属18种、嗜热疣霉属5种和发癣菌属10种。文中提供了种的形态描述、图示、培养性状等信息。对未观察的种，提供了相关的参考文献及扼要介绍。此类真菌大多嗜角蛋白，部分种能引发人类和动物的皮肤癣菌病，因此本卷对嗜角蛋白真菌和皮肤癣真菌的研究进行了扼要概述，并介绍了各属真菌的分类历史、研究状况和应用价值。这些真菌与人类生活和健康具有密切联系，许多种在国民经济中具有潜在应用价值。

本书可供大专院校菌物学、中药学、植物保护学和生物制药等有关专业师生，以及从事真菌资源研究和开发的相关人员参考。

图书在版编目 (CIP) 数据

中国真菌志. 第七十卷, 金孢属及其相关属 / 韩燕峰主编. --北京 : 科学出版社, 2025. 6. -- ISBN 978-7-03-081146-2

Ⅰ. Q949. 32

中国国家版本馆 CIP 数据核字第 2025NS1529 号

责任编辑：刘新新 赵小林/责任校对：杨 赛

责任印制：肖 兴/封面设计：刘新新

科学出版社出版

北京东黄城根北街 16 号

邮政编码：100717

http://www.sciencep.com

北京建宏印刷有限公司印刷

科学出版社发行 各地新华书店经销

*

2025 年 6 月第 一 版 开本：787×1092 1/16

2025 年 6 月第一次印刷 印张：16 3/4 插页：1

字数：430 000

定价：268.00 元

（如有印装质量问题，我社负责调换）

CONSILIO FLORARUM CRYPTOGAMARUM SINICARUM
ACADEMIAE SINICAE EDITA

FLORA FUNGORUM SINICORUM

VOL. 70

CHRYSOSPORIUM AND ITS RELATED GENERA

REDACTOR PRINCIPALIS

Han Yanfeng

A Major Project of the National Natural Science Foundation of China
A Project of the Key Research Program of Frontier Sciences of the Chinese Academy of Sciences
(Supported by the National Natural Science Foundation of China and the Chinese Academy of Sciences)

Science Press
Beijing

金孢属及其相关属

本 卷 著 者

韩燕峰　陈万浩　张延威　张芝元　梁建东　梁宗琦

（贵州大学生命科学学院真菌资源研究所）

CHRYSOSPORIUM AND ITS RELATED GENERA

AUCTORES

Han Yanfeng　Chen Wanhao　Zhang Yanwei
Zhang Zhiyuan　Liang Jiandong　Liang Zongqi

(*Institute of Fungus Resources, College of Life Science, Guizhou University*)

序

中国孢子植物志是非维管束孢子植物志，分《中国海藻志》、《中国淡水藻志》、《中国真菌志》、《中国地衣志》及《中国苔藓志》五部分。中国孢子植物志是在系统生物学原理与方法的指导下对中国孢子植物进行考察、收集和分类的研究成果；是生物物种多样性研究的主要内容；是物种保护的重要依据，与人类活动及环境甚至全球变化都有不可分割的联系。

中国孢子植物志是我国孢子植物物种数量、形态特征、生理生化性状、地理分布及其与人类关系等方面的综合信息库；是我国生物资源开发利用、科学研究与教学的重要参考文献。

我国气候条件复杂，山河纵横，湖泊星布，海域辽阔，陆生和水生孢子植物资源极其丰富。中国孢子植物分类工作的发展和中国孢子植物志的陆续出版，必将为我国开发利用孢子植物资源和促进学科发展发挥积极作用。

随着科学技术的进步，我国孢子植物分类工作在广度和深度方面将有更大的发展，这部著作也将不断被补充、修订和提高。

中国科学院中国孢子植物志编辑委员会

1984年10月 • 北京

中国孢子植物志总序

中国孢子植物志是由《中国海藻志》、《中国淡水藻志》、《中国真菌志》、《中国地衣志》及《中国苔藓志》所组成。至于维管束孢子植物蕨类未被包括在中国孢子植物志之内，是因为它早先已被纳入《中国植物志》计划之内。为了将上述未被纳入《中国植物志》计划之内的藻类、真菌、地衣及苔藓植物纳入中国生物志计划之内，出席1972 年中国科学院计划工作会议的孢子植物学工作者提出筹建“中国孢子植物志编辑委员会”的倡议。该倡议经中国科学院领导批准后，“中国孢子植物志编辑委员会”的筹建工作随之启动，并于 1973 年在广州召开的《中国植物志》、《中国动物志》和中国孢子植物志工作会议上正式成立。自那时起，中国孢子植物志一直在“中国孢子植物志编辑委员会”统一主持下编辑出版。

孢子植物在系统演化上虽然并非单一的自然类群，但是，这并不妨碍在全国统一组织和协调下进行孢子植物志的编写和出版。

随着科学技术的飞速发展，在人们对真菌知识的了解日益深入的今天，黏菌与卵菌已从真菌界中分出，分别归隶于原生动物界和管毛生物界。但是，长期以来，由于它们一直被当作真菌由国内外真菌学家进行研究，而且，在“中国孢子植物志编辑委员会”成立时已将黏菌与卵菌纳入中国孢子植物志之一的《中国真菌志》计划之内，因此，沿用包括黏菌与卵菌在内的《中国真菌志》广义名称是必要的。

自“中国孢子植物志编辑委员会”于 1973 年成立以后，作为“三志”的组成部分，中国孢子植物志的编研工作由中国科学院资助；自 1982 年起，国家自然科学基金委员会参与部分资助；自 1993 年以来，作为国家自然科学基金委员会重大项目，在国家基金委资助下，中国科学院及科技部参与部分资助，中国孢子植物志的编辑出版工作不断取得重要进展。

中国孢子植物志是记述我国孢子植物物种的形态、解剖、生态、地理分布及其与人类关系等方面的大型系列著作，是我国孢子植物物种多样性的重要研究成果，是我国孢子植物资源的综合信息库，是我国生物资源开发利用、科学研究与教学的重要参考文献。

我国气候条件复杂，山河纵横，湖泊星布，海域辽阔，陆生与水生孢子植物物种多样性极其丰富。中国孢子植物志的陆续出版，必将为我国孢子植物资源的开发利用，为我国孢子植物科学的发展发挥积极作用。

中国科学院中国孢子植物志编辑委员会

主编　曾呈奎

2000 年 3 月　北京

Foreword of the Cryptogamic Flora of China

Cryptogamic Flora of China is composed of *Flora Algarum Marinarum Sinicarum*, *Flora Algarum Sinicarum Aquae Dulcis*, *Flora Fungorum Sinicorum*, *Flora Lichenum Sinicorum*, and *Flora Bryophytorum Sinicorum*, edited and published under the direction of the Editorial Committee of the Cryptogamic Flora of China, Chinese Academy of Sciences (CAS). It also serves as a comprehensive information bank of Chinese cryptogamic resources.

Cryptogams are not a single natural group from a phylogenetic point of view which, however, does not present an obstacle to the editing and publication of the Cryptogamic Flora of China by a coordinated, nationwide organization. The Cryptogamic Flora of China is restricted to non-vascular cryptogams including the bryophytes, algae, fungi, and lichens. The ferns, a group of vascular cryptogams, were earlier included in the plan of *Flora of China*, and are not taken into consideration here. In order to bring the above groups into the plan of Fauna and Flora of China, some leading scientists on cryptogams, who were attending a working meeting of CAS in Beijing in July 1972, proposed to establish the Editorial Committee of the Cryptogamic Flora of China. The proposal was approved later by the CAS. The committee was formally established in the working conference of Fauna and Flora of China, including cryptogams, held by CAS in Guangzhou in March 1973.

Although myxomycetes and oomycetes do not belong to the Kingdom of Fungi in modern treatments, they have long been studied by mycologists. *Flora Fungorum Sinicorum* volumes including myxomycetes and oomycetes have been published, retaining for *Flora Fungorum Sinicorum* the traditional meaning of the term fungi.

Since the establishment of the editorial committee in 1973, compilation of Cryptogamic Flora of China and related studies have been supported financially by the CAS. The National Natural Science Foundation of China has taken an important part of the financial support since 1982. Under the direction of the committee, progress has been made in compilation and study of Cryptogamic Flora of China by organizing and coordinating the main research institutions and universities all over the country. Since 1993, study and compilation of the Chinese fauna, flora, and cryptogamic flora have become one of the key state projects of the National Natural Science Foundation with the combined support of the CAS and the National Science and Technology Ministry.

Cryptogamic Flora of China derives its results from the investigations, collections, and classification of Chinese cryptogams by using theories and methods of systematic and evolutionary biology as its guide. It is the summary of study on species diversity of cryptogams and provides important data for species protection. It is closely connected with human activities, environmental changes and even global changes. Cryptogamic Flora of

China is a comprehensive information bank concerning morphology, anatomy, physiology, biochemistry, ecology, and phytogeographical distribution. It includes a series of special monographs for using the biological resources in China, for scientific research, and for teaching.

China has complicated weather conditions, with a crisscross network of mountains and rivers, lakes of all sizes, and an extensive sea area. China is rich in terrestrial and aquatic cryptogamic resources. The development of taxonomic studies of cryptogams and the publication of Cryptogamic Flora of China in concert will play an active role in exploration and utilization of the cryptogamic resources of China and in promoting the development of cryptogamic studies in China.

C.K. Tseng
Editor-in-Chief
The Editorial Committee of the Cryptogamic Flora of China
Chinese Academy of Sciences
March, 2000 in Beijing

《中国真菌志》序

《中国真菌志》是在系统生物学原理和方法指导下，对中国真菌，即真菌界的子囊菌、担子菌、壶菌及接合菌四个门以及不属于真菌界的卵菌等三个门和黏菌及其类似的菌类生物进行搜集、考察和研究的成果。本志所谓“真菌”系广义概念，涵盖上述三大菌类生物(地衣型真菌除外)，即当今所称“菌物”。

中国先民认识并利用真菌作为生活、生产资料，历史悠久，经验丰富，诸如酒、醋、酱、红曲、豆豉、豆腐乳、豆瓣酱等的酿制，蘑菇、木耳、茭白作食用，茯苓、虫草、灵芝等作药用，在制革、纺织、造纸工业中应用真菌进行发酵，以及利用具有抗癌作用和促进碳素循环的真菌，充分显示其经济价值和生态效益。此外，真菌又是多种植物和人畜病害的病原菌，危害甚大。因此，对真菌物种的形态特征、多样性、生理生化、亲缘关系、区系组成、地理分布、生态环境以及经济价值等进行研究和描述，非常必要。这是一项重要的基础科学研究，也是利用益菌、控制害菌、化害为利、变废为宝的应用科学的源泉和先导。

中国是具有悠久历史的文明古国，古代科学技术一直处于世界前沿，真菌学也不例外。酒是真菌的代谢产物，中国酒文化博大精深、源远流长，有几千年历史。约在公元300 年的晋代，江统在其《酒诰》诗中说：“酒之所兴，肇自上皇。或云仪狄，一曰杜康。有饭不尽，委余空桑。郁积成味，久蓄气芳。本出于此，不由奇方。”作者精辟地总结了我国酿酒历史和自然发酵方法，比意大利学者雷蒂(Radi，1860)提出微生物自然发酵法的学说约早 1500 年。在仰韶文化时期(5000～3000 B. C.)，我国先民已懂得采食蘑菇。中国历代古籍中均有食用菇蕈的记载，如宋代陈仁玉在其《菌谱》(1245)中记述浙江台州产鹅膏菌、松蕈等 11 种，并对其形态、生态、品级和食用方法等作了论述和分类，是中国第一部地方性食用蕈菌志。先民用真菌作药材也是一大创造，中国最早的药典《神农本草经》(成书于 102～200 A. D.)所载 365 种药物中，有茯苓、雷丸、桑耳等 10 余种药用真菌的形态、色泽、性味和疗效的叙述。明代李时珍在《本草纲目》(1578)中，记载“三菌”、“五蕈”、“六芝”、“七耳”以及羊肚菜、桑黄、鸡㙡、雪蚕等30 多种药用真菌。李时珍将菌、蕈、芝、耳集为一类论述，在当时尚无显微镜帮助的情况下，其认识颇为精深。该籍的真菌学知识，足可代表中国古代真菌学水平，堪与同时代欧洲人(如 C. Clusius，1529～1609)的水平比拟而无逊色。

15 世纪以后，居世界领先地位的中国科学技术逐渐落后。从 18 世纪中叶到 20 世纪 40 年代，外国传教士、旅行家、科学工作者、外交官、军官、教师以及负有特殊任务者，纷纷来华考察，搜集资料，采集标本，研究鉴定，发表论文或专辑。如法国传教士西博特(P.M. Cibot)1759 年首先来到中国，一住就是 25 年，写过不少关于中国植物(含真菌)的文章，1775 年他发表的五棱散尾菌(*Lysurus mokusin*)，是用现代科学方法研究发表的第一个中国真菌。继而，俄国的波塔宁(G.N. Potanin，1876)、意大利的吉拉迪(P. Giraldii，1890)、奥地利的汉德尔-马泽蒂(H. Handel-Mazzetti，1913)、美国的梅里尔(E.D. Merrill，1916)、瑞典的史密斯(H. Smith，1921)等共 27 人次来我国采集标本。研究发表中国真菌论著 114 篇册，作者多达 60 余人次，报道中国真菌 2040 种，其中含

10 新属、361 新种。东邻日本自 1894 年以来，特别是 1937 年以后，大批人员涌到中国，调查真菌资源及植物病害，采集标本，鉴定发表。据初步统计，发表论著 172 篇册，作者 67 人次以上，共报道中国真菌约 6000 种(有重复)，其中含 17 新属、1130 新种。其代表人物在华北有三宅市郎(1908)，东北有三浦道哉(1918)，台湾有泽田兼吉(1912)；此外，还有斋藤贤道、伊藤诚哉、平冢直秀、山本和太郎、逸见武雄等数十人。

国人用现代科学方法研究中国真菌始于 20 世纪初，最初工作多侧重于植物病害和工业发酵，纯真菌学研究较少。在一二十年代便有不少研究报告和学术论文发表在中外各种刊物上，如胡先骕 1915 年的“菌类鉴别法”，章祖纯 1916 年的“北京附近发生最盛之植物病害调查表”以及钱穟孙(1918)、邹钟琳(1919)、戴芳澜(1920)、李寅恭(1921)、朱凤美(1924)、孙豫寿(1925)、俞大绂(1926)、魏喦寿(1928)等的论文。三四十年代有陈鸿康、邓叔群、魏景超、凌立、周宗璜、欧世璜、方心芳、王云章、裘维蕃等发表的论文，为数甚多。他们中有的人终生或大半生都从事中国真菌学的科教工作，如戴芳澜(1893～1973)著“江苏真菌名录”(1927)、“中国真菌杂录”(1932～1939)、《中国已知真菌名录》(1936，1937)、《中国真菌总汇》(1979)和《真菌的形态和分类》(1987)等，他发表的“三角枫上白粉病菌之一新种”(1930)，是国人用现代科学方法研究、发表的第一个中国真菌新种。邓叔群(1902～1970)著“南京真菌之记载”(1932～1933)、“中国真菌续志”(1936～1938)、《中国高等真菌》(1939)和《中国的真菌》(1963)等，堪称《中国真菌志》的先导。上述学者以及其他许多真菌学工作者，为《中国真菌志》研编的起步奠定了基础。

在 20 世纪后半叶，特别是改革开放以来的 20 多年，中国真菌学有了迅猛的发展，如各类真菌学课程的开设，各级学位研究生的招收和培养，专业机构和学会的建立，专业刊物的创办和出版，地区真菌志的问世等，使真菌学人才辈出，为《中国真菌志》的研编输送了新鲜血液。1973 年中国科学院广州“三志”会议决定，《中国真菌志》的研编正式启动，1987 年由郑儒永、余永年等编撰的《中国真菌志》第 1 卷《白粉菌目》出版，至 2000 年《中国真菌志》已出版 14 卷。《中国真菌志》自第 2 卷开始实行主编负责制，2.《银耳目和花耳目》(刘波，1992)；3.《多孔菌科》(赵继鼎，1998)；4.《小煤炱目Ⅰ》(胡炎兴，1996)；5.《曲霉属及其相关有性型》(齐祖同，1997)；6.《霜霉目》(余永年，1998)；7.《层腹菌目 黑腹菌目 高腹菌目》(刘波，1998)；8.《核盘菌科 地舌菌科》(庄文颖，1998)；9.《假尾孢属》(刘锡琎、郭英兰，1998)；10.《锈菌目(一)》(王云章、庄剑云，1998)；11.《小煤炱目Ⅱ》(胡炎兴，1999)；12.《黑粉菌科》(郭林，2000)；13.《虫霉目》(李增智，2000)；14.《灵芝科》(赵继鼎、张小青，2000)。盛世出巨著，在国家“科教兴国”英明政策的指引下，《中国真菌志》的研编和出版，定将为中华灿烂文化做出新贡献。

余永年
庄文颖 谨识

中国科学院微生物研究所

中国・北京・中关村

2002 年 9 月 15 日

Foreword of Flora Fungorum Sinicorum

Flora Fungorum Sinicorum summarizes the achievements of Chinese mycologists based on principles and methods of systematic biology in intensive studies on the organisms studied by mycologists, which include non-lichenized fungi of the Kingdom Fungi, some organisms of the Chromista, such as oomycetes etc., and some of the Protozoa, such as slime molds. In this series of volumes, results from extensive collections, field investigations, and taxonomic treatments reveal the fungal diversity of China.

Our Chinese ancestors were very experienced in the application of fungi in their daily life and production. Fungi have long been used in China as food, such as edible mushrooms, including jelly fungi, and the hypertrophic stems of water bamboo infected with *Ustilago esculenta*; as medicines, like *Cordyceps sinensis* (caterpillar fungus), *Poria cocos* (China root), and *Ganoderma* spp. (lingzhi); and in the fermentation industry, for example, manufacturing liquors, vinegar, soy-sauce, *Monascus*, fermented soya beans, fermented bean curd, and thick broad-bean sauce. Fungal fermentation is also applied in the tannery, paperma-king, and textile industries. The anti-cancer compounds produced by fungi and functions of saprophytic fungi in accelerating the carbon-cycle in nature are of economic value and ecological benefits to human beings. On the other hand, fungal pathogens of plants, animals and human cause a huge amount of damage each year. In order to utilize the beneficial fungi and to control the harmful ones, to turn the harmfulness into advantage, and to convert wastes into valuables, it is necessary to understand the morphology, diversity, physiology, biochemistry, relationship, geographical distribution, ecological environment, and economic value of different groups of fungi.

China is a country with an ancient civilization of long standing. In ancient times, her science and technology as well as knowledge of fungi stood in the leading position of the world. Wine is a metabolite of fungi. The Wine Culture history in China goes back to thousands of years ago, which has a distant source and a long stream of extensive knowledge and profound scholarship. In the Jin Dynasty (*ca.* 300 A.D.), JIANG Tong, the famous writer, gave a vivid account of the Chinese fermentation history and methods of wine processing in one of his poems entitled *Drinking Games* (Jiu Gao), 1500 years earlier than the theory of microbial fermentation in natural conditions raised by the Italian scholar, Radi (1860). During the period of the Yangshao Culture (5000—3000 B.C.), our Chinese ancestors knew how to eat mushrooms. There were a great number of records of edible mushrooms in Chinese ancient books. For example, back to the Song Dynasty, CHEN Ren-Yu (1245) published the *Mushroom Menu* (Jun Pu) in which he listed 11 species of edible fungi including *Amanita* sp. and *Tricholoma matsutake* from Taizhou, Zhejiang Province, and described in detail their morphology, habitats, taxonomy, taste, and way of cooking. This was the first local flora of the Chinese edible mushrooms. Fungi used as medicines originated in ancient China. The earliest Chinese pharmacopocia, *Shen-Nong Materia Medica* (Shen Nong Ben Cao Jing), was published in 102—200 A.D. Among the 365 medicines recorded, more than 10 fungi, such as *Poria cocos* and *Polyporus mylittae*, were included. Their fruitbody shape, color, taste, and medical functions were provided. The great pharmacist of Ming Dynasty, LI Shi-Zhen published his eminent work *Compendium Materia Medica* (Ben Cao Gang Mu) (1578) in which more than thirty fungal species were accepted as medicines, including *Aecidium mori*, *Cordyceps sinensis*, *Morchella* spp., *Termitomyces* sp., etc. Before the invention of microscope, he managed to bring fungi of different classes together, which demonstrated his intelligence and profound knowledge of biology.

After the 15th century, development of science and technology in China slowed down. From middle of the 18th century to the 1940's, foreign missionaries, tourists, scientists, diplomats, officers, and other professional workers visited China. They collected specimens of plants and fungi, carried out taxonomic studies, and published papers, exsi ccatae, and monographs based on Chinese materials. The French missionary, P.M. Cibot, came to China in 1759 and stayed for 25 years to investigate plants including fungi in different regions of China. Many papers were written by him. *Lysurus mokusin*, identified with modern techniques and published in 1775, was probably the first Chinese fungal record by these visitors. Subsequently, around 27 man-times of foreigners attended field excursions in China, such as G.N. Potanin from Russia in 1876, P. Giraldii from Italy in 1890, H. Handel-Mazzetti from Austria in 1913, E.D. Merrill from the United States in 1916, and H. Smith from Sweden in 1921. Based on examinations of the Chinese collections obtained, 2040 species including 10 new genera and 361 new species were reported or described in 114 papers and books. Since 1894, especially after 1937, many Japanese entered China. They investigated the fungal resources and plant diseases, collected specimens, and published their identification results. According to incomplete information, some 6000 fungal names (with synonyms) including 17 new genera and 1130 new species appeared in 172 publications. The main workers were I. Miyake (1908) in the Northern China, M. Miura (1918) in the Northeast, K. Sawada (1912) in Taiwan, as well as K. Saito, S. Ito, N. Hiratsuka, W. Yamamoto, T. Hemmi, etc.

Research by Chinese mycologists started at the turn of the 20th century when plant diseases and fungal fermentation were emphasized with very little systematic work. Scientific papers or experimental reports were published in domestic and international journals during the 1910's to 1920's. The best-known are "Identification of the fungi" by H.H. Hu in 1915, "Plant disease report from Peking and the adjacent regions" by C.S. Chang in 1916, and papers by S.S. Chian (1918), C.L. Chou (1919), F.L. Tai (1920), Y.G. Li (1921), V.M. Chu (1924), Y.S. Sun (1925), T.F. Yu (1926), and N.S. Wei (1928). Mycologists who were active at the 1930's to 1940's are H.K. Chen, S.C. Teng, C.T. Wei, L. Ling, C.H. Chow, S.H. Ou, S.F. Fang, Y.C. Wang, W.F. Chiu, and others. Some of them dedicated their lifetime to research and teaching in mycology. Prof. F.L. Tai (1893—1973) is one of them, whose representative works were "List of fungi from Jiangsu"(1927), "Notes on Chinese fungi"(1932—1939), *A List of Fungi Hitherto Known from China* (1936, 1937), *Sylloge Fungorum Sinicorum* (1979), *Morphology and Taxonomy of the Fungi* (1987), etc. His paper entitled "A new species of *Uncinula* on *Acer trifidum* Hook. & Arn." (1930) was the first new species described by a Chinese mycologist. Prof. S.C. Teng (1902—1970) is also an eminent teacher. He published "Notes on fungi from Nanking" in 1932—1933, "Notes on Chinese fungi" in 1936—1938, *A Contribution to Our Knowledge of the Higher Fungi of China* in 1939, and *Fungi of China* in 1963. Work done by the above-mentioned scholars lays a foundation for our current project on *Flora Fungorum Sinicorum*.

Significant progress has been made in development of Chinese mycology since 1978. Many mycological institutions were founded in different areas of the country. The Mycological Society of China was established, the journals *Acta Mycological Sinica* and *Mycosystema* were published as well as local floras of the economically important fungi. A young generation in field of mycology grew up through postgraduate training programs in the graduate schools. In 1973, an important meeting organized by the Chinese Academy of Sciences was held in Guangzhou (Canton) and a decision was made, uniting the related scientists from all over China to initiate the long term project "Fauna, Flora, and Cryptogamic Flora of China". Work on *Flora Fungorum Sinicorum* thus started. The first volume of Chinese Mycoflora on the Erysiphales (edited by R.Y. Zheng & Y.N. Yu, 1987) appeared. Up to now, 14 volumes have been published: Tremellales and Dacrymycetales edited by B. Liu (1992), Polyporaceae by J.D. Zhao (1998), Meliolales Part I (Y.X. Hu, 1996), *Aspergillus* and its related teleomorphs (Z.T. Qi, 1997), Peronosporales (Y.N. Yu, 1998), Hymenogastrales, Melanogastrales and Gautieriales (B. Liu, 1998), Sclerotiniaceae and Geoglossaceae (W.Y. Zhuang, 1998), *Pseudocercospora* (X.J. Liu & Y.L. Guo, 1998), Uredinales Part I (Y.C. Wang & J.Y. Zhuang, 1998), Meliolales Part II (Y.X. Hu, 1999), Ustilaginaceae (L. Guo, 2000), Entomophthorales (Z.Z. Li, 2000), and Ganodermataceae (J.D. Zhao & X.Q. Zhang, 2000). We eagerly await the coming volumes and expect the completion of Flora *Fungorum Sinicorum* which will reflect the flourishing of Chinese culture.

Y.N. Yu and W.Y. Zhuang
Institute of Microbiology, CAS, Beijing
September 15, 2002

[illegible] research and teaching in mycology. [illegible] (1893—1973) is one of the [illegible] "[illegible]" [illegible] Chinese [illegible]

Significant progress has been made in the development of Chinese mycology [illegible]. After [illegible] institutions were founded [illegible] the country. The Mycological Society of China was established [illegible] as well as [illegible]. The young generation of mycologists [illegible] postgraduate training programs [illegible] an important [illegible] Chinese Academy of Sciences [illegible] scientists from all over China to initiate the long-term project [illegible]. The first volume of Chinese [illegible]

[illegible]

致　谢

本卷研编工作先后得到国家自然科学基金委员会、中国科学院微生物研究所、华南农业大学、贵州省科技厅和贵州大学经费的支持。

特对下列单位和个人给予的大力支持表示衷心的感谢：中国科学院微生物研究所庄文颖院士；华南农业大学姜子德教授、张炼辉教授；福建师范大学黄建忠教授；遵义医科大学肖建辉教授。感谢下列单位和个人在标本采集、照片摄影、文献资料馈赠方面的帮助：贵州大学邹晓教授、安徽农业大学陈名君教授、江西农业大学马建教授、福建农林大学邱君志教授、福建师范大学张怀东博士和李芹博士。感谢以下作者在引用其论著中图片方面的支持：中国科学院微生物研究所蔡磊研究员和王雪薇副研究员、山东农业大学张天宇教授及其所带研究生、山东农业大学张修国教授、华南农业大学姜子德教授、汕头大学李伟教授、贵州大学姜于兰教授和王勇教授等。感谢对本卷提出很多宝贵意见的中国科学院微生物研究所庄文颖院士和郑焕娣博士、山东农业大学张修国教授及中国科学院微生物研究所郭英兰研究员。感谢中国普通微生物菌种保藏管理中心（CGMCC）、中国科学院微生物研究所菌物标本馆（HMAS）和贵州大学标本馆（GZAC）给予的支持和帮助。

感谢先后参与标本采集、描述鉴定、绘图和制版、论文撰写，以及资源前期开发利用研究工作的研究生：王宝林、王垚、王玉荣、王成、罗韵、郑欢、杨娟、沈鑫、胡胜兰、董醇波、任玉连、邵秋雨、李欣、王海燕和彭兰等。

最后谨以此书深切缅怀我国著名真菌学家梁宗琦先生，恩师对科学的热爱和执着精神永远激励着学生，难忘师恩！

目　录

绪 论

一、概 述

目前，全球已知真菌种数约 150 000 种（Hawksworth，2012）。土壤作为真菌重要的栖息场所，生存繁衍着大量各种各样的真菌。我国的菌物资源非常丰富，特别是 21 世纪后，基于形态和分子系统学发现与报道的菌物数量逐年上升（戴玉成和庄剑云，2010；Dai *et al.*，2015）。真菌学家长期积累的真菌形态学知识和材料，对建立真菌自然分类系统是不可缺少的元素，也是确保获得真菌分子数据后进一步研究的前提条件（Gams，2016）。人类利用真菌的历史，在中国已有 6000 年，而在西方至少也有 3500 年（邵力平等，1983）。除我们熟知的直接食用、发酵、酿酒之外，真菌在药物生产、医学保健、酶制剂生产、生物防治、生态恢复等方面有着重要的不可替代的作用和价值。有一类能够以角蛋白基质为碳源和氮源而生长的真菌，在降解畜禽羽毛方面具有潜在利用价值。例如，金孢属 *Chrysosporium* Corda 是一类全球分布的真菌，能够生存于角蛋白腐质、粪便、土壤和富角蛋白基质等环境中，而且部分菌株还是脊椎动物的机会性病原体（梁建东等，2007a）。研究表明，印度角蛋白癣菌 *Keratinophyton indicum* (Garg) Kandemir & de Hoog（=印度金孢 *Chrysosporium indicum* Garg）、嗜毛金孢 *C. keratinophilum* D. Frey ex J.W. Carmich.、棉毛栉霉 *Ctenomyces vellereus* (Sacc. & Speg.) P.M. Kirk（=刺柄金孢 *C. asperatum* J.W. Carmich.）和热带金孢 *C. tropicum* J.W. Carmich. 能以人的毛发或禽类羽毛作为底物基质生长（李欣等，2022）；乔治金孢 *C. georgiae* (Varsavsky & Ajello) Oorschot 不能在牛、羊及人的毛发上生长，但其能以鸡毛为碳氮源进行生长繁殖。本卷所涉及的属种很多被报道为嗜角蛋白真菌，其中能引起人类/动物皮肤癣菌病的一些种类被认为是皮肤病原真菌，常见种类 20～30 种（林燕荣，2010），因此在公共卫生安全上需要引起重视；也有很多类群能产生有用的代谢物，在医药、农业、饲料和皮革方面具有潜在应用价值（李欣等，2022），值得深入研究并开发利用。

（一）嗜角蛋白真菌的研究现状

近年来，土生角蛋白降解真菌受到了全球许多真菌学者的关注。究其原因有以下两点：一方面是这类真菌能高效降解结构十分稳定的富角蛋白基质；另一方面是嗜角蛋白腐生真菌是潜在的病原真菌（Vidyasagar *et al.*，2005）。

嗜角蛋白真菌的界定长期以来一直与角蛋白降解真菌混淆，很多时候它们被认为是同义词，但二者具有细微差别。角蛋白降解真菌常指具有角蛋白降解能力的真菌类群，而嗜角蛋白真菌则泛指所有能够以含有角蛋白基质作为天然栖息地的真菌，与角蛋白的亲和力弱于角蛋白降解真菌（Dominik & Majchrowicz，1970）。Błyskal（2009）则认为角蛋白降解真菌是能够产生角蛋白酶而降解 α-角蛋白的一类微生物，而嗜角蛋白真菌是指仅能利用角蛋白的部分易降解组分及相关物质的真菌。近年来，嗜角蛋白真菌和角蛋白降解真菌的几种定义一直在沿用，甚至被认为是同义词。结合作者前期工作及前人的相关研究，现将嗜角蛋白真菌界定为一类偏好各种角质基质，并能降解利用角蛋白作为唯一碳氮源的真菌类群（沈鑫等，2018；Calin *et al.*，2016；李欣等，2022）。因此，人们可以利用角蛋白基质如毛发、指甲等富集并获得更多种类的嗜角蛋白真菌（Mahariya & Sharma，2018）。

许多丝状真菌都能降解羽毛角蛋白，最典型的是发癣菌属 *Trichophyton* Malmsten。Nannizzi（1927）报道皮肤癣菌可水解角蛋白；Yu 等（1968）研究发现，须癣发癣菌 *T. mentagrophytes* (C.P. Robin) Sabour.具有降解角蛋白的能力，通过离子交换层析纯化获得的角蛋白酶的分子质量约为 48 kDa；Jain 和 Agrawal（1980）研究发现，轮枝菌属 *Verticillium* Nees 真菌和马发癣菌 *Trichophyton equinum* Gedoelst 有较高的羽毛降解能力；罗文等（1993）分离到一株能降解羽毛粉角蛋白的真菌，初步鉴定为一种串孢属菌 *Catenularia* sp.；Singh（1997）报道了猴发癣菌 *Trichophyton simii* (Pinoy) Stockdale, D.W.R. Mack. & Austwick HN50 的角蛋白降解活性及其所分泌的角蛋白酶特性；Tawfik 和 Adnan（2000）从股癣患者的皮肤中分离到须癣发癣菌，其所产的角蛋白酶的分子质量为 38 kDa，对豚鼠毛和纤维蛋白也具有高水解活力。曹军等（2001）也从土壤中分离到一株产角蛋白酶的栖土曲霉 *Aspergillus terricola* Marchal & É.J. Marchal。谭盈盈等（2005）从养殖场污水处理系统中分离到羽毛角蛋白降解菌 F1、A1、Z1，经初步鉴定属于拟青霉属 *Paecilomyces* Bainier。梁建东等（2007b）利用羽毛粉或人体头发作碳氮源对 9 个耐热金孢属真菌进行筛选，经鉴定发现，嗜热毁丝霉 *Myceliophthora thermophila* (Apinis) Oorschot H49 菌株对角蛋白降解能力最强。很多皮肤癣菌也有降解角蛋白和角质层的能力，从而引起人和动物角蛋白组织（皮肤、头发和指甲）的感染（Deshmukh，2008），被称为皮肤癣菌病（Hedayati *et al.*，2004）。能分泌角蛋白酶的皮肤癣菌有红色发癣菌 *Trichophyton rubrum* (Castell.) Sabour.、短帚霉 *Scopulariopsis brevicaulis* (Sacc.) Bainier、石膏样奈尼兹皮菌 *Nannizzia gypsea* (Nann.) Stockdale［=石膏样小孢子菌 *Microsporum gypseum* (E. Bodin) Guiart & Grigoraki］、犬小孢子菌 *M. canis* E. Bodin ex Guég.、白色念珠菌 *Candida albicans* (C.P. Robin) Berkhout、热带念珠菌 *C. tropicalis* (Castell.) Berkhout、近平滑念珠菌 *C. parapsilosis* (Ashford) Langeron & Talice 和克柔念珠菌 *C. krusei* (Castell.) Berkhout。因此，嗜角蛋白真菌主要属于丝孢类菌，包括皮肤癣菌和许多腐生丝状真菌（Jain & Sharma，2012）。

嗜角蛋白真菌广布全球，但也受人和/或动物等多种因素影响（Deshmukh & Verekar，2006）。该类真菌栖息于各种生境中，但当出现在与人类密切相关的环境中时，能对部分人群构成潜在威胁（Aquino *et al.*，2012）。近年来，许多以前被认为是非致病性的真菌引起人类感染的现象频繁发生，当这些真菌在自然界的密度增大时，常与感染密切相关。例如，有学者报道其常感染癌症患者、HIV/AIDS 患者和接受免疫抑制治疗等存在免疫缺陷的患者（Perdelli *et al.*，2006；Carmo *et al.*，2007）。医院作为一种特殊的公共场合，具有人员流动性大、病原微生物种类多和患者免疫力较低等特点，常常存在对人类具有潜在感染风险的嗜角蛋白真菌类群，国外对该场所的真菌多样性的研究已有报道（Singh *et al.*，2009；Abbas Soleymani *et al.*，2015）。同时研究发现，许多医院的空气、粉尘、水及盆栽的土壤中均存在许多嗜角蛋白病原真菌和机会性病原真菌（Hedayati *et al.*，2004；Patel *et al.*，2016；Tong *et al.*，2017）。张芝元等（2017a）对我国川、黔、渝的几所医院土中的该类真菌进行了初步调查和多样性分析，发现了链格孢属 *Alternaria* Nees、曲霉属 *Aspergillus* P. Micheli ex Haller、毛壳属 *Chaetomium* Kunze、芽枝霉属 *Cladosporium* Link、镰刀菌属 *Fusarium* Link、地丝霉属 *Geomyces* Traaen、地霉属 *Geotrichum* Link、小孢子菌属 *Microsporum* Gruby、青霉属 *Penicillium* Link、紫霉属 *Purpureocillium* Luangsa-ard, Hywel-Jones, Houbraken & Samson、篮状菌属 *Talaromyces* C.R. Benj.、木霉属 *Trichoderma* Pers.和发癣菌属共 13 个属。宠物受到全球越来越多人的喜爱，其活动性强，又与人类接触亲密，其皮肤上是否携带嗜角蛋白真菌，是否是导致人类皮肤癣菌病的诱因，也受到了关注。有研究表明，宠物狗体表存在较为丰富的该类真菌，还包含有一批潜在的非皮肤癣菌类致病真菌，如交链格孢 *Alternaria alternata* (Fr.) Keissl.、暗孢节菱孢 *Arthrinium phaeospermum* (Corda) M.B. Ellis、黄曲霉 *Aspergillus flavus* Link、球毛壳菌 *Chaetomium globosum* Kunze、球孢芽枝霉 *Cladosporium sphaerospermum* Penz.、芽枝状枝孢霉 *Cladosporium cladosporioides* (Fresen.) G.A. de Vries、细小枝孢霉 *Cladosporium tenuissimum* Cooke、拟粗壮弯孢 *Curvularia pseudorobusta* Meng Zhang & T.Y. Zhang、黑附球菌 *Epicoccum nigrum* Link（张芝元等，2017b）。动物园也是富含角蛋白基质如毛发等的一个生境，其嗜角蛋白真菌也非常丰富，有研究表明，利用毛发诱饵法从动物园土壤中分离获得 5000 多株嗜角蛋白真菌，其中还包含了很多新种（Li *et al.*，2022a，2022b，2022c）。上述研究结果表明，嗜角蛋白真菌广布于人类生存的环境中，其种类和分布值得重视。

角蛋白广布于动物的外胚层细胞，结构稳定，其在自然界中的含量仅次于纤维素和几丁质（Lange *et al.*，2016），为人和脊椎动物体被主要的结构组成成分（Mckittrick *et al.*，2012）。全球每年会产生大量富含角蛋白的废物，如家禽养殖场、屠宰场、皮革行业等（Brandelli，2008；Khardenavis *et al.*，2009；Fakhfakh-Zouari *et al.*，2010）。富含角蛋白的废物若处理不当不仅是一种资源浪费，也会带来严重的生态环境问题。嗜角蛋白真菌因其能分解利用角蛋白，加速此类物质的降解，维持生态平衡，故具有重要的生态作用。此外，一些角蛋白降解真菌是引起人类和动物感染的病原真菌（Anane *et al.*，2015），而一些非致病的角蛋白降解真菌与某些病原真菌的协同作用，则促进了病原真菌对宿主的感染（de Hoog *et al.*，2000a）。

通过总结以往研究发现，嗜角蛋白真菌类群所涉及的应用领域极其广泛。例如，肉座菌目 Hypocreales 镰刀菌属真菌菌株 1A 能产生对动物皮肤具有良好活性的角蛋白酶（Calin *et al.*，2019），因而在皮革行业展现出较好的应用潜力。木霉属真菌中棘孢木霉 *Trichoderma asperellum* Samuels, Lieckf. & Nirenberg 对植物种子的萌发及幼苗的生长有促进作用（Calin *et al.*，2019）；哈茨木霉 *T. harzianum* Rifai 已广泛用于生物防治（Sundram *et al.*，2008）；也有将该属真菌应用于皮革行业的报道，表明木霉属真菌在农业、工业上都有较大的应用潜力（Ismail *et al.*，2012）。爪甲团囊菌目 Onygenales 金孢属真菌代谢产物表现出极高的应用价值，如嗜毛金孢的分泌物能够有效抑制皮肤癣菌的活性，可用于生物制药；热带金孢代谢物能有效杀死致倦库蚊幼虫，因此可用于生物防治（Priyanka & Prakash，2003；梁建东等，2007a）。散囊菌目 Eurotiales 曲霉属真菌应用最为广泛，在饲料业、皮革业、医药、农业、生物修复和功能性食品方面都具有应用价值（李欣等，2022）。例如，米曲霉 *Aspergillus oryzae* (Ahlb.) Cohn 对羽毛进行生物转化后，其产生的粗蛋白中含有动物体中最缺乏的必需氨基酸——赖氨酸（Ali *et al.*，2011）。溜曲霉 *A. tamarii* Kita 去除山羊皮毛的同时提高了皮革的强度性能（Dayanandan *et al.*，2003）。黑曲霉 *A. niger* Tiegh.的羽毛水解产物对植物的产量和生长发育也表现出显著影响，还可防治植株疾病（Adetunji *et al.*，2012）。一些其他嗜角蛋白真菌还可作为生物指标（Lydolph *et al.*，2005）。嗜角蛋白真菌在各方面的应用中均表现出环境友好性及应用的广泛性，证明了嗜角蛋白真菌具有极高的实用价值。

（二）皮肤病原真菌的研究现状

皮肤癣菌病（dermatophytosis）是一种常见的皮肤真菌感染病，病理学上，这些皮肤病原真菌的入侵可引起永久性的脱发进而导致秃头。亲角蛋白性决定其侵犯体表角化组织（皮肤、毛发、甲），从而引起皮肤癣菌病。按病变部位分为头癣、体癣、股癣、手癣、足癣及甲真菌病（朱敬先，2017）。

皮肤癣菌（dermatophyte）又称皮肤丝状真菌，主要有发癣菌属、小孢子菌属和表皮癣菌属 *Epidermophyton* Sabour.（冉玉平等，2017）。许兰氏发癣菌 *Trichophyton schoenleinii* (Lebert) Langeron & Miloch. ex Nann.是一种常见的皮肤真菌。研究发现，光滑皮肤和指甲感染以红色发癣菌和须癣发癣菌为主，毛发感染以犬小孢子菌、须癣发癣菌、断发发癣菌 *T. tonsurans* Malmsten、紫色发癣菌 *T. violaceum* Sabour. ex E. Bodin 及铁锈色小孢子菌 *Microsporum ferrugineum* M. Ota、粉奈尼兹皮菌 *Nannizzia fulva* Stockdale（=粉小孢子菌 *M. fulvum* Uriburu）为主（Havlickova *et al.*，2008；Fuller，2009；郭志丽，2007），而许兰氏发癣菌、奥杜盎小孢子菌 *M. audouinii* Gruby 及絮状表皮癣菌 *Epidermophyton floccosum* (Harz) Langeron & Miloch.正逐渐消失，仅限某些欠发达国家或地区存在（Zhan & Liu，2017），由于人类生活环境和个人卫生的不断改善，上述三种菌引起的皮肤癣菌病的流行逐年降低，仅在中国、尼日利亚和伊朗曾被报道。其余皮肤真菌如石膏样奈尼兹皮菌、犬小孢子菌、紫色发癣菌和疣状发癣菌 *Trichophyton verrucosum* E. Bodin 等可引起人类头皮癣和脱皮（Ilkit，2010；Krunic *et al.*，2007）。陈立良等（2017）对惠州地区浅部真菌病进行了调查，发现红色发癣菌占比最大，石膏样奈尼兹皮菌次之，须癣发癣菌第三，这些真菌主要引起足癣、股癣和体癣等疾病。皮肤癣菌如石膏样奈尼兹皮菌和土生发癣菌 *T. terrestre* Durie & D. Frey 均为复合种，广布全球，主要存在于土壤和人及动物皮肤病病灶部位，可引起人和动物的皮肤感染（Hoarau *et al.*，2016；Moon *et al.*，2007；张芝元，2018）。据报道，皮肤癣菌是特定的角蛋白亲和物，用角蛋白材料富集土壤微生物技术，可能会刺激土壤中休眠的土传繁殖体活跃生长，进而分离获得该类真菌（Kushwaha，2000）。由于该类群真菌的致病性和降解角蛋白特性，其分类研究备受关注。

随着分子生物学技术的迅猛发展，本卷所记载类群的分类系统主要考虑其系统发育关系，同时按照最新的《国际藻类、菌物和植物命名法规》（ICN）中的“一种真菌一个名称”（Turland *et al.*，2018）和优先权原则。本卷册涉及描述鉴定的 15 个真菌属，即金孢属 *Chrysosporium* Corda、栉霉属 *Ctenomyces* Eidam、地丝霉属 *Geomyces* Traaen、角蛋白癣菌属 *Keratinophyton* H.S. Randhawa & R.S. Sandhu、冠癣菌属 *Lophophyton*

Matr. & Dassonv.、小孢子菌属 *Microsporum* Gruby、毁丝霉属 *Myceliophthora* Costantin、奈尼兹皮菌属 *Nannizzia* Stockdale、帕拉癣菌属 *Paraphyton* Y. Gräser, Dukik & de Hoog、瓶霉属 *Phialophora* Medlar、假裸囊菌属 *Pseudogymnoascus* Raillo、赛多孢属 *Scedosporium* Sacc. ex Castell. & Chalm.、帚霉属 *Scopulariopsis* Bainier、嗜热疣霉属 *Thermothelomyces* Y. Marín, Stchigel, Guarro & Cano 和发癣菌属 *Trichophyton* Malmsten，均采用了传统的形态学分类标准和现代分子系统学研究结果（Barnett & Barry，1998；Kiffer & Morelet，1999；Taylor *et al.*，2000；Kirk *et al.*，2001）。

在撰写本卷的过程中，作者发现同一个定名人在发表论文时存在不一致的情况，为避免被误认为是不同的定名人，本卷中核查后将其修改一致，但这就会与原发表时的定名人写法有所不同，请读者在引用种名和其定名人时查阅发表新种时的定名人，特此说明。此外，本卷中出现的标本馆和菌种保藏中心的全称和缩写如下：中国科学院微生物研究所菌物标本馆（HMAS）、中国普通微生物菌种保藏管理中心（CGMCC）、贵州大学标本馆（GZAC 或 GZU）和贵州大学真菌资源研究所标本馆（GZUIFR 或 GZDXIFR），在此一并说明。

二、各属研究概述

本节介绍研究记载较多的 15 个属的分类地位、分类历史、分类特征、经济价值等概况，有助于全面了解这些属的研究进展。

（一）金孢属

1. 金孢属的研究历史

金孢属隶属于子囊菌门 Ascomycota 子囊菌纲 Ascomycetes 散囊菌亚纲 Eurotiomycetidae 爪甲团囊菌目 Onygenales 爪甲团囊菌科 Onygenaceae（Kirk *et al.*，2001）；该属是以革质金孢 *Chrysosporium corii* Corda 为模式种建立（Corda，1833），由于革质金孢是粪生金孢的同物异名，因此现该属的模式种为粪生金孢。Vidal 等（2000）则认为金孢属的成员，有性型大多属于爪甲团囊菌目，少数种的有性型则属于粪壳菌目 Sordariales Chadef. ex D. Hawksw. & O.E. Erikss。Wijayawardene 等（2020）将该属全部归入爪甲团囊菌科。Kandemir 等（2022）基于多基因系统学研究发现，金孢属中成员隶属于爪甲团囊菌目爪甲团囊菌科中的角蛋白癣菌属和隐囊菌属 *Aphanoascus* Zukal 及未定地位类群中。

金孢属自建立以后，此属分类地位先后被不同学者予以修正（van Oorschot，1980）。Saccardo（1901）废除了此属的分类地位，将其视为侧孢属 *Sporotrichum* Link 的异名属。Hughes（1958）重新评价了金孢属，将具色泽、无间生孢子的种归入侧孢属，并将这两个特征视为二属间的区别依据。Carmichael（1962）将地丝霉属、毁丝霉属、芽酵母属 *Blastomyces* Costantin & Rolland 和伊蒙菌属 *Emmonsia* Cif. & Montemart.视为金孢属的异名，使此属成为一个形态异型属。von Arx（1971）观察侧孢属模式种金黄侧孢 *Sporotrichum aureum* Link 的菌株时发现了锁状联合现象，之后 Donk（1974）认为此菌有性型与 *Poria metamorphosa* (Fuckel) Sacc.有关，才明确侧孢属为担子菌的属级阶元。

Dominik（1967）发表了金孢属专论，基于 Carmichael 的分类观点对该属概念进行了拓展，同时囊括了瘤孢属 *Sepedonium* Link 内的一些种。van Oorschot（1980）基于前人研究基础对金孢属做了较大的修订，对此属的分类鉴定做出了重要贡献，明确了金孢属及其相关属的主要区别特征，并对相关菌株进行了重新评价和鉴定，确定了金孢属有 15 个有效种，即裂叶金孢 *Chrysosporium lobatum* Scharapov、嗜毛金孢、热带金孢、毡生金孢 *C. pannicola* (Corda) Oorschot & Stalpers、印度金孢、卡氏金孢 *C. carmichaelii* Oorschot、同步金孢 *C. synchronum* Oorschot、硫磺金孢 *C. sulphureum* (Fiedl.) Oorschot & Samson、乔治金孢、弱金孢 *C. inops* J.W. Carmich.、拟粪生金孢 *C. pseudomerdarium* Oorschot、嗜旱金孢 *C. xerophilum* Pitt、粉生金孢 *C. farinicola* (Burnside) Skou、粪生金孢 *C. merdarium* (Ehrenb.) J.W. Carmich.和昆士兰金孢 *C. queenslandicum* Apinis & R.G. Rees。

20 世纪 80 年代，除了 Oorschot 的工作，Sigler 等（1982，1986）、van Oorschot 和 Piontelli（1985）、Al-Musallam 和 Tan（1989）、Kushwaha 和 Shrivastava（1989）先后发表了 6 个新种，分别为：线状金孢 *Chrysosporium filiforme* Sigler, J.W. Carmich. & H.S. Whitney、巴耶纳尔金孢 *C. vallenarense* Oorschot & Piont.、欧洲金孢 *C. europae* Sigler, Guarro & Punsola、异味金孢 *C. mephiticum* Sigler、喜土金孢 *C. geophilum* Kushwaha & Shrivast 和轮带金孢 *C. zonatum* Al-Musallam & C.S. Tan。

20 世纪 90 年代，此属的分类研究有了较快的发展，梁宗琦（1991）、Skou（1992）、Jain 等（1993）、Gené 等（1994）、Sharma 等（1994）、Cano 和 Guarro（1994）、Matsushima 和 Matsushima（1996）及 Vidal 等（1996，1999）先后报道了 18 个新种，分别为：中国金孢 *Chrysosporium sinense* Z.Q. Liang、葡串金孢 *C. botryoides* Skou、球形金孢 *C. globiferum* Skou、球形金孢节状变种 *C. globiferum* var. *articulatum* Skou、球形金孢雪白变种 *C. globiferum* var. *niveum* Skou、西班牙金孢 *C. hispanicum* Skou、霍姆氏金孢 *C. holmii* Skou、居中金孢 *C. medium* Skou、居中金孢厚壁变种 *C. medium* var. *spissescens* Skou、小金孢 *C. minor* Skou、梨形金孢 *C. pyriformis* Skou、戈氏金孢 *C. gourii* P.C. Jain, Deshmukh & S.C. Agarwal、丝毛金孢 *C. pilosum* Gené, Guarro & Ulfig、序状金孢 *C. racemosus* K.D. Sharma, S. Bhaitacharjee & Bhadauria、西格氏金孢 *C. siglerae* Cano & Guarro、发酵金孢 *C. fermentotritici* K. Matsush. & Matsush.、蝙蝠金孢 *C. vespertilium* Guarro, P. Vidal & De Vroey 和曲丝金孢 *C. undulatum* P. Vidal, Guarro & Ulfig。

2000 年以来，Vidal 等（2000，2002）、Beguin 等（2005）、Tripathi 和 Kushwaha（2005）、Abarca 等（2010）、Nováková 和 Kolařík（2010）、Han 等（2013）、Zhang 等（2017a，2017b，2020）等先后发表了 22 个新种，即河泥金孢 *Chrysosporium fluviale* P. Vidal & Guarro、微孢金孢 *C. minutisporosum* P. Vidal & Guarro、水生金孢 *C. submersum* P. Vidal & Guarro、翼手金孢 *C. chiropterorum* Beguin & Larcher、基督金孢 *C. christchurchicum* Tripathi & Kushwaha、蛇生金孢 *C. ophiicola* Guarro, Deanna A. Sutton, Wickes & Rajeev、临汾金孢 *C. linfenense* Z.Q. Liang, J.D. Liang & Y.F. Han、瓜罗金孢 *C. guarroi* J. Cabañes & Abarca、洞穴金孢 *C. speluncarum* A. Nováková & M. Kolařík、长孢金孢 *C. longisporum* Stchigel, Deanna A. Sutton, Cano & Guarro、大孢金孢 *C. magnasporum* Stchigel, Cano, Mac Cormack & Guarro、海岩金孢 *C. oceanitesii* Stchigel, Cano, Archuby & Guarro、青海金孢 *C. qinghaiense* Y.F. Han, J.D. Liang & Z.Q. Liang、三

亚金孢 *C. sanyaense* Y.W. Zhang, Y.F. Han, J.D. Liang & Z.Q. Liang、具刺金孢 *C. echinulatum* Hubka, Mallátová, Čmoková & M. Kolařík、贵州金孢 *C. guizhouense* Y.W. Zhang, Y.F. Han & Z.Q. Liang、湖北金孢 *C. hubeiense* Yan W. Zhang, Y.F. Han & Z.Q. Liang、山西金孢 *C. shanxiense* Y.W. Zhang, W.H. Chen, X. Zou, Y.F. Han & Z.Q. Liang、荆州金孢 *C. jingzhouense* Y.W. Zhang, Y.F. Han & Z.Q. Liang、棒孢金孢 *C. clavisporum* Yan W. Zhang, Y.F. Han & Z.Q. Liang、雷公山金孢 *C. leigongshanense* Z. Li, G.P. Zeng, J. Ren, X. Zou & Y.F. Han、阿尔维亚金孢 *C. alvearium* F. Liu & L. Cai。其间，Pitt 等（2013）将嗜旱金孢作为模式种建立了嗜旱金孢属 *Xerochrysium* Pitt。

近年，作者团队鉴定报道了一批金孢新种（Li *et al.*，2019；Han *et al.*，2022），分别为：开阳金孢 *Chrysosporium kaiyangense* Y.F. Han, W.H. Chen, J.D. Liang & Z.Q. Liang、侧孢金孢 *C. laterisporum* Z. Li, Y.W. Zhang, W.H. Chen & Y.F. Han、梭形金孢 *C. fusiforme* Y.F. Han, W.H. Chen, J.D. Liang & Z.Q. Liang、甘肃金孢 *C. gansuense* Y.F. Han, W.H. Chen, J.D. Liang & Z.Q. Liang、广西金孢 *C. guangxiense* Y.F. Han, W.H. Chen, J.D. Liang & Z.Q. Liang、不规则金孢 *C. irregularum* Y.F. Han, W.H. Chen, J.D. Liang & Z.Q. Liang、江苏金孢 *C. jiangsuense* Y.F. Han, W.H. Chen, J.D. Liang & Z.Q. Liang、多形金孢 *C. multiforme* Y.F. Han, W.H. Chen, J.D. Liang & Z.Q. Liang、卵孢金孢 *C. ovalisporum* Z. Li, Y.W. Zhang, W.H. Chen & Y.F. Han、四川金孢 *C. sichuanense* Y.F. Han, W.H. Chen, J.D. Liang & Z.Q. Liang、绒毛金孢 *C. villiforme* Y.F. Han, W.H. Chen, J.D. Liang & Z.Q. Liang。Labuda 等（2021）基于 ITS rDNA 和 LSU rDNA 对金孢属和角蛋白癣菌属进行了系统发育研究，将金孢属的 9 个种移入了角蛋白癣菌属中，分别是棒孢金孢[现为棒孢角蛋白癣菌 *Keratinophyton clavisporum* (Yan W. Zhang, Y.F. Han & Z.Q. Liang) Labuda & Bernreiter]、具刺金孢[现为具刺角蛋白癣菌 *Keratinophyton echinulatum* (Hubka, Mallátová, Čmoková & M. Kolařík) Labuda & Bernreiter]、河泥金孢[现为河泥角蛋白癣菌 *Keratinophyton fluviale* (P. Vidal & Guarro) Labuda & Bernreiter]、湖北金孢[现为湖北角蛋白癣菌 *Keratinophyton hubeiense* (Yan W. Zhang, Y.F. Han & Z.Q. Liang) Labuda & Bernreiter]、临汾金孢[现为临汾角蛋白癣菌 *Keratinophyton linfenense* (Z.Q. Liang, J.D. Liang & Y.F. Han) Labuda & Bernreiter]、微孢金孢[现为微孢角蛋白癣菌 *Keratinophyton minutisporosum* (P. Vidal & Guarro) Labuda & Bernreiter]、西格氏金孢[现为西格氏角蛋白癣菌 *Keratinophyton siglerae* (Cano & Guarro) Labuda & Bernreiter]、水生金孢[现为水生角蛋白癣菌 *Keratinophyton submersum* (P. Vidal & Guarro) Labuda & Bernreiter]和青海金孢[现为青海角蛋白癣菌 *Keratinophyton qinghaiense* (Y.F. Han, J.D. Liang & Z.Q. Liang) Labuda & Bernreiter]。毡状金孢 *C. pannorum* (Link) S. Hughes[现为毡状假裸囊菌 *Pseudogymnoascus pannorum* (Link) Minnis & D.L. Lindner]移入了假裸囊菌属中，随后阿尔维亚金孢[现为阿尔维亚角蛋白癣菌 *Keratinophyton alvearium* (F. Liu & L. Cai) Kandemir & de Hoog]也被移入角蛋白癣菌属中（Kandemir *et al.*，2022）。还有一些金孢与其有性型隐囊菌属中的种合并。这些研究使金孢属内保留的种逐渐减少。根据 Index Fungorum（2022）数据库的统计，全世界目前报道记载的金孢属种名为 112 个。经作者整理，迄今全球金孢属的有效种为 67 种。

金孢属的中国分类研究概况：中国最早开展金孢属的分类研究是自冬虫夏草上发现

的一新种——中国金孢（梁宗琦，1991），随后国内学者开始关注此属，并陆续报道一批新种和新记录种，排除已移出和同物异名的种，迄今国内报道金孢属 26 个种（表 1）。

表 1　中国已报道的金孢属物种名录

序号	名称
1	节状金孢 *C. articulatum* Scharapov
2	法斯特金孢 *C. fastidium* Pitt
3	梭形金孢 *C. fusiforme* Y.F. Han, W.H. Chen, J.D. Liang & Z.Q. Liang
4	甘肃金孢 *C. gansuense* Y.F. Han, W.H. Chen, J.D. Liang & Z.Q. Liang
5	乔治金孢 *C. georgiae* (Varsavsky & Ajello) Oorschot
6	广西金孢 *C. guangxiense* Y.F. Han, W.H. Chen, J.D. Liang & Z.Q. Liang
7	贵州金孢 *C. guizhouense* Y.W. Zhang, Y.F. Han & Z.Q. Liang
8	不规则金孢 *C. irregularum* Y.F. Han, W.H. Chen, J.D. Liang & Z.Q. Liang
9	江苏金孢 *C. jiangsuense* Y.F. Han, W.H. Chen, J.D. Liang & Z.Q. Liang
10	荆州金孢 *C. jingzhouense* Y.W. Zhang, Y.F. Han & Z.Q. Liang
11	开阳金孢 *C. kaiyangense* Y.F. Han, W.H. Chen, J.D. Liang & Z.Q. Liang
12	嗜毛金孢 *C. keratinophilum* D. Frey ex J.W. Carmich.
13	侧孢金孢 *C. laterisporum* Z. Li, Y.W. Zhang, W.H. Chen & Y.F. Han
14	雷公山金孢 *C. leigongshanense* Z. Li, G.P. Zeng, J. Ren, X. Zou & Y.F. Han
15	裂叶金孢 *C. lobatum* Scharapov
16	粪生金孢 *C. merdarium* (Ehrenb.) J.W. Carmich.
17	多形金孢 *C. multiforme* Y.F. Han, W.H. Chen, J.D. Liang & Z.Q. Liang
18	卵孢金孢 *C. ovalisporum* Z. Li, Y.W. Zhang, W.H. Chen & Y.F. Han
19	苍白金孢 *C. pallidum* Z.F. Zhang & L. Cai
20	三亚金孢 *C. sanyaense* Y.W. Zhang, Y.F. Han, J.D. Liang & Z.Q. Liang
21	山西金孢 *C. shanxiense* Y.W. Zhang, W.H. Chen, X. Zou, Y.F. Han & Z.Q. Liang
22	四川金孢 *C. sichuanense* Y.F. Han, W.H. Chen, J.D. Liang & Z.Q. Liang
23	中国金孢 *C. sinense* Z.Q. Liang
24	热带金孢 *C. tropicum* J.W. Carmich.
25	绒毛金孢 *C. villiforme* Y.F. Han, W.H. Chen, J.D. Liang & Z.Q. Liang
26	轮带金孢 *C. zonatum* Al-Musallam & C.S. Tan

金孢属的生境和主要形态特征研究概况：长期以来，金孢属种级分类研究主要依据菌落和分生孢子等形态学特征，有时也结合对温度的耐受性、降解角蛋白的能力和对皮肤的感染症状等特征。

（1）生境：金孢属真菌栖息基质广泛，较耐干旱，多分离自土壤（韩燕峰等，2017a），如印度金孢、临汾金孢、雷公山金孢、裂叶金孢、毡状金孢、热带金孢、青海金孢、三亚金孢、勒克瑙金孢 *C. lucknowense* Garg 和乔治金孢等；部分种分离自人体或动物组织，如自人脚皮肤癣分离的具刺金孢，从啮齿类动物肺部分离的拟粪生金孢，从人耳皮肤上分离的卡氏金孢，从鸡冠中分离到的嗜毛金孢，以及从人皮肤上分离的弱金孢等；有的可从粪便中获得，如粪生金孢和嗜毛金孢等；有些可从富含皮肤、头发或指甲等角蛋白的场所如沙滩、河流沉积物、鸟巢或动物栖居地分离获得，如山西金孢、河泥金孢、巴耶纳尔金孢、裂叶金孢和粉生金孢等（Liang *et al.*，2009a；Li *et al.*，2017a；韩燕峰等，2013；Zhang *et al.*，2013；张延威等，2016；Vidal *et al.*，2002；van Oorschot，1980）。

（2）菌落颜色：金孢属真菌在大豆胨酵母浸粉琼脂（phytone yeast extract agar，PYE）、马铃薯葡萄糖琼脂（potato dextrose agar，PDA）及查氏培养基（Czapek agar，CA）等培养基上生长的菌落颜色多样，多呈白色，如卡氏金孢、昆士兰金孢、热带金孢、河泥金孢和喜土金孢等；有时呈淡黄色至黄色，如粪生金孢、拟粪生金孢和弱金孢等；有时呈黄褐色至褐色，如欧洲金孢、丝毛金孢和序状金孢（van Oorschot，1980；Kushwaha & Shrivastava，1989；Sharma *et al.*，1994；Sigler *et al.*，1986；Cano & Guarro，1994）。

（3）球拍状菌丝：球拍状菌丝的有无是金孢属种级分类鉴定的主要依据之一。具球拍状菌丝的种较多，如常见的卡氏金孢、昆士兰金孢、毡状金孢、贵州金孢、湖北金孢、山西金孢和三亚金孢等。无球拍状菌丝的种有粪生金孢、拟粪生金孢、热带金孢、裂叶金孢、青海金孢和雷公山金孢等（van Oorschot，1980；Zhang *et al.*，2013，2016；韩燕峰等，2013；Li *et al.*，2017a）。

（4）顶侧生孢子：顶侧生孢子的着生方式、形状、大小和表面纹饰等也是金孢属重要的鉴别特征之一。顶侧生孢子大多着生于短突起或短侧分枝上，分生孢子形状多样：球形至近球形的，如球形金孢、葡串金孢和燕形金孢 *Chrysosporium hirundinis* Sharapov 等；卵形至拟卵形的，如卡氏金孢、昆士兰金孢、热带金孢、弱金孢及同步金孢等；少数棒状的，如棒孢金孢；梨形的如梨形金孢。孢子表面大多光滑，少数粗糙、具细刺或细瘤，如粪生金孢、卡氏金孢、嗜旱金孢和拟粪生金孢等（Skou，1992；van Oorschot，1980；Zhang *et al.*，2017a）。

（5）间生孢子：研究发现金孢属内的 48 个种具有间生孢子，如嗜毛金孢、弱金孢、裂叶金孢、印度金孢、居中金孢和中国金孢等；其形状主要以柱状和桶状为主，表面光滑或粗糙（van Oorschot，1980；Skou，1992；梁宗琦，1991）。

（6）其他特征：除以上鉴别特征，有些独特的结构也可作为金孢属的分种依据，如山西金孢的产孢细胞与分生孢子连接处具有膨大的“领状”结构（张延威等，2016）；轮带金孢的菌落具有白色、浅黄色至淡褐色的轮带特征（Kushwaha & Shrivastava，1989）；序状金孢可见顶侧生孢子在菌丝隔膜处聚集成簇（Sharma *et al.*，1994）。最适温度也可作为鉴别依据（van Oorschot，1980），如卡氏金孢和拟粪生金孢的最适温度为 25℃，而热带金孢的最适温度则为 30℃。

（7）有性阶段：由于目前报道的金孢属真菌仅见无性产孢结构，依据 Kandemir 等（2022）的系统发育研究结果，它们能与爪甲团囊菌目（爪甲团囊菌科 Onygenaceae、裸被囊菌科 Gymnoascaceae 和裸囊菌科 Arthrodermataceae）和球囊霉目 Ascosphaerales

的球囊霉科中的部分产有性结构的成员聚在一起，其对应关系仍需继续研究。

金孢属分子系统学研究概况：随着现代科学的发展和各学科知识相互渗透，真菌分类学从单一的形态特征鉴别发展到以形态特征为主，其他特征并重，并朝着包括其他类型特征（生理学、生物化学、遗传学、分子生物学等）发展。其中，分子生物学的发展对分类学研究起了极大的促进作用，形成了新的学科——分子系统学。关于金孢属的分子系统学研究起步较晚，20 世纪 90 年代才陆续有少量报道，但这些研究并不侧重于真菌系统分类研究，而是为了证明某些种能引起某种疾病（Bowman & Taylor，1993；Bowman *et al.*，1996；Harmsen *et al.*，1995；Peterson & Sigler，1998）。直到 1999 年，Sugiyama 等（1999）为了进一步厘清爪甲团囊菌目的分类和系统发育，采用 SSU rDNA 序列分析，对爪甲团囊菌目的主要成员进行了系统学的研究，但这个研究中仅涉及金孢属的几个种。Vidal 等（2000）对金孢属进行了较为全面的分子系统学研究，分析了该属 57 个分类单元的 ITS 序列和 5.8S rRNA 基因序列，认为金孢属的成员应隶属于爪甲团囊菌目和粪壳菌目。近年来，分子生物学技术的迅猛发展，对金孢属分类单位的准确鉴定起到了极大的推动作用。21 世纪以来，金孢属的分类研究多综合形态学和分子系统学数据进行鉴定，如合格发表的曲丝金孢（Vidal *et al.*，1999）、河泥金孢（Vidal *et al.*，2000）、水生金孢（Vidal *et al.*，2002）、青海金孢（韩燕峰等，2013）、三亚金孢（Zhang *et al.*，2013）、湖北金孢（Zhang *et al.*，2016）等。Labuda 等（2021）基于 ITS 和 LSU rDNA 对部分金孢及其有性型进行了系统发育研究，把金孢属中一些种移入到角蛋白癣菌属和隐囊菌属，并提议了 4 个角蛋白癣菌新种。Li 等（2022a）在调查嗜角蛋白真菌资源时，基于 ITS 和 LSU rDNA 发表了 2 个角蛋白癣菌属新种。由于角蛋白癣菌属是基于多基因系统发育研究而将部分种从金孢属中移出建立的新属，其形态特征与金孢属非常相似，因此在后面进行该属信息介绍时不再赘述。根据最新《国际藻类、菌物及植物命名法规》中“一种真菌一个名称”的原则（Turland *et al.*，2018），无性型金孢属与其有性型的命名联系在一起，部分新种发表时采用了其有性型属名，如粗糙裸囊菌 *Gymnoascus exasperatus* Z.F. Zhang, F. Liu & L. Cai（Zhang *et al.*，2017b）、土星状隐囊菌 *Aphanoascus saturnoideus* Cano & Guarro（Labuda *et al.*，2021）等。为方便科研同行阅读，本卷仅列出中国金孢属保留的种，未包括上述无性型为金孢属结构的有性型名录和已被移走的金孢名录。

2. 金孢属真菌的经济意义

金孢属真菌分布广泛，适应性强，从赤道到极地的不同土壤、空气、海洋、河床污泥、植物材料、家畜粪便、动物和鸟类的体表或体内、低温岩洞、高湿鸟巢、烘烤食品、荒漠、核辐射地区、盐池及污水等材料和生境中均有报道（Kushwaha，2000；Moore，2001；Gock *et al.*，2003；Deshmukh，2004；Khizhnyak *et al.*，2003；Da Silva *et al.*，2003）。而这种强的环境适应能力与其产生不同的、特殊的酶及其他代谢产物有极大的相关性，表明了其独特的潜在应用价值。现已发现金孢属真菌可以产生角蛋白酶、纤维素酶、脂肪酶、菊粉酶和 α-半乳糖苷酶等 19 种以上酶类物质和一些次生代谢产物，如抗癌、抗真菌、抗寄生虫等抗生素类物质及其他活性物质。此外，金孢属是一个喜角蛋白的类群，在目前承认的 67 个种中，能降解角蛋白的种占一半以上（Kushwaha，2000）。

微生物来源的角蛋白酶是目前酶领域研究的热点之一。这类酶在医药、饲料、肥料、皮革、食品、纺织、日化等工业生产及环境治理方面都有广阔的应用前景（曾毅等，2004）。角蛋白酶不可替代的重要作用已成为业界的共识，金孢属作为喜角蛋白的真菌类群，在角蛋白酶生产中的潜在价值备受关注，有望从其中找到对环境适应能力强的理想产角蛋白酶菌株，以期对其研发并用于工业生产。该属在医药卫生研究开发中的意义及利用前景主要集中于以下三方面。

1）天然新药物筛选

生物来源的新药研发，一直是生物技术领域研究的热点之一。金孢属在此领域具有巨大的开发潜力。由于癌症治疗中手术和放疗的局限性，药物治疗被寄予了厚望。近年来，深部真菌感染的高死亡率和治疗上的困难越来越受到人们关注，寻找高效、低毒、用于治疗真菌感染的天然药物成了医学科学工作者当前的紧迫任务之一。金孢在新药物的开发中将是一类新的候选对象。目前，金孢中已发现许多含有抗癌、抗真菌、抗寄生虫等抗生素类物质。Kohno 等（2001）报道了一株金孢菌可产生 TMC-69 抗生素，它的加氢衍生物对鼠 P388 和 B16 黑色素瘤具很强的抗癌活性。程永庆等（1992）从一株疣金孢 *Chrysosporium verrucosum* Tubaki（现为毡状假裸囊菌 *Pseudogymnoascus pannorum* (Link) Minnis & D.L. Lindner）的发酵液中分离到两个抗生素组分，这两个活性成分具有很强的抑制肿瘤细胞核苷转运和小鼠脾淋巴细胞摄取核苷的作用。此外，分别从一金孢菌株和裂叶金孢中分离到一个能诱导癌细胞凋亡的新抗生素 adenopeptin 和能抑制癌基因活性蛋白形成的物质（Hayakawa *et al.*，1998；Vander *et al.*，1995）。Yamashita 等（1984）发现毡状金孢能产生一种新的抗真菌抗生素 chryscandin，这种抗生素是第一个天然来源的、分子中具有一个 3-氨基呋喃核糖苷酸（3-aminoribofuranuronic acid）的核苷类抗生素，他们测定了此抗生素的结构并进行了全合成。抗菌实验表明，此抗生素对白色念珠菌和革兰氏阳性菌具有明显抑制作用，毒性实验表明它对老鼠的急毒性非常低。有实验表明，昆士兰金孢和热带金孢等几个种所分泌的代谢物具有强烈的抗黑曲霉活性（Bahuguna & Kushwaha，1992），且昆士兰金孢还产生一种新的抗真菌化合物 queenslandon 和两个新的抗革兰氏阳性菌的抗菌混合物（Hoshino *et al.*，2002；Ivanova *et al.*，2002）。此外，嗜毛金孢能产生抗多种表皮寄生真菌的热稳定活性物质（Gokulshankar *et al.*，2005）。一些念珠状菌株则能被来源于某金孢培养液的抗真菌物质硫酸固醇明显抑制（Yang *et al.*，2003）。粪生金孢能产生抗寄生虫的生物活性物质 sporandol，这个物质与其他联二萘类物质相比，对哺乳动物的毒性更低。除此物质外，从粪生金孢培养液中还分离到一种抗 HIV-1 的蛋白酶和 EGF-R 酪蛋白激酶的抑制剂 semicochliodinol A 和 semicochliodinol B（Tsipouras *et al.*，1997；Fredenhagen *et al.*，1997）。

2）媒介昆虫的生物控制

媒介昆虫可以传播许多严重的疾病，如疟疾、登革热、痢疾、伤寒、肝炎、鼠疫、地方性斑疹伤寒、森林脑炎、流行性出血热等。采用生物控制，减少化学防治对环境造成的不良影响，一直是害虫防治研究的热点之一，金孢在此领域有明显的应用潜力。Mohanty 和 Prakash（2002）的研究表明，裂叶金孢对传播疟疾的斯氏按蚊的 3 龄幼虫有良好的杀灭作用。在实验室条件下，热带金孢的代谢物对五带淡色库蚊的低龄幼虫有较高的致死率（Priyanka & Prakash，2003）。

3）值得重视的兼性病原

真菌对人或动物的感染由于其治疗的顽固性，以及临床易被误诊而忽视的现象越来越受到人们的重视。人们试图通过对感染人畜真菌的种类、特性及感染症状、特点和机制的研究找到应对措施。已知金孢属的一些种是人或动物的潜在致病菌或机会感染菌。其中，少数种会对人或动物造成严重的感染。对此菌的研究将有助于人们对其危害进行防控。目前观察到金孢感染人类的主要靶组织是肺、皮肤及角膜。例如，Gan 等（2002）报道了一株金孢感染了一个白血病患者的肺、皮肤，并且可能伤及其神经系统。Hayashi 等（2002）和 Roilides 等（1999）都报道了一种由轮带金孢引起的非扩散的肺病。Brandt 等（2005）则在肺炎患者的支气管中发现了瑞氏拟奈尼兹皮菌 *Nannizziopsis vriesii* (Apinis) Currah 的金孢无性型，并认为它与肺炎的发生有关。毡状金孢是人和某些动物皮肤病的致病菌，此菌能引起人的指甲、手指和前臂的皮肤感染。值得关注的是，小金孢和某些金孢能引起人的角膜感染。此外也有人报道，金孢属的一些种是无色菌丝霉菌病的病原之一（Zelenková，2006；Wagoner *et al.*，1999；Qiu *et al.*，2005；Hocquette *et al.*，2005）。金孢属真菌也是某些动物的病原真菌之一，它们主要侵染动物的皮肤和肺。Thomas 等（2002）报道瑞氏拟奈尼兹皮菌的金孢无性型能引起变色龙和棕树蛇的皮肤感染，也与咸水鳄鱼的皮肤感染和死亡相关。Paré 等（2006）用高冠变色龙 *Chamaeleo calyptratus* 作为实验动物对瑞氏拟奈尼兹皮菌金孢无性型的致病性进行了分组研究，结果表明所有实验组均被感染。这些观察研究也印证了 Mitra（1998）的观点，即某些金孢与动物的皮肤传染病相关。昆士兰金孢能侵染蛇的皮肤、肺和肝；小金孢会引起欧洲海狸的肺部感染；而某些金孢则侵染伶鼬 *Mustela nivalis* 的肺部，在感染部位它的不育大孢子周围长满了肉芽（Vissiennon，1999；Mörner *et al.*，1999；Laakkonen *et al.*，1998）。

金孢属在造纸和纺织生物技术中的应用价值也受到关注。20 世纪以来，生物技术在造纸和纺织工业中的应用日趋广泛，其中生物来源的纤维素酶和半纤维素酶在造纸的生物制浆、生物漂白、废纸生物脱墨、改善浆料性能和树脂的生物控制，以及纺织过程中的棉花精炼、棉织物的整理等一系列处理中均有重要应用。Buckley 和 Dobson（1998）所做的试验表明，在固定化培养条件下，木质金孢 *Chrysosporium lignorum* O. Bergman & T. Nilsson 产生的胞外纤维素酶能使聚合染料脱色，此研究结果在木纤维的生物降解分析中，可用来代替放射性同位素标记的木纤维素底物，其应用将为实验室安全和环保做出贡献。Gusakov 等（2001，2005）和 Bukhtojarov 等（2004）经突变育种获得一株产纤维二糖水解酶的勒克瑙金孢。这种酶对热稳定，可在 50℃条件下保持活性 24 h 或在 60℃条件下 7 h 后仍有大于 90%的活力。此菌产生的纤维素酶在牛仔布水洗中已显示出了较大的应用价值。此外，这个菌株还产生一种特殊的不同于 β-1,4-葡聚糖酶的木聚糖酶。

金孢属在食品、饲料、日用化工等工业及环保中也具有重要的利用价值。香精油、脂肪酶、菊粉酶和 α-半乳糖苷酶在保健、化妆品、生物化工、洗涤剂工业、皮革工业、功能食品的生产及饲料工业等方面都有较广泛的应用。Kinderlerer 等（1988）报道了嗜旱金孢在以椰子作原料发酵时可产生含有脂肪链取代基甲基酮的香精油。Venkateshwarlu 和 Reddy（1993）研究发现，硫磺金孢可产生高温碱性脂肪酶。而菊粉酶和 α-半乳糖苷酶也可由某些金孢产生（周帼萍等，2002）。目前，已有学者对金孢的一些种在环保中潜在的应用进行了初步研究。在实验条件下，毡状金孢对毒性大、难降解的2,4-二氯代酚有良好的降解作用（Vroumsia *et al.*，2005）。嗜角蛋白金孢对污染物和盐度具有极大的抵抗力，可以作为水污染的指示剂（Ulfig *et al.*，1997）。另外，有报道认为某些金孢对镉亦有很好的抵抗力（Płaza *et al.*，1998）。

近年来，欧美等发达国家和印度等发展中国家对这类真菌资源的生态分布、分类鉴定、嗜角蛋白特性、对人类感染致病的潜在危险（特别是在新的抗真菌抗生素及抗癌药物筛选）、环境治理，以及在食品、饲料、日用化工等工业中的应用做了不少研究，已展现出了诱人的应用前景。但是，这些工作也仅涉及此属已报道67个种中的15个种左右，不足总数的25%，大量的资源尚待开发。我国幅员辽阔，生态环境多样，必然蕴含此属丰富特异的物种资源，但国内对此类重要资源的研究、开发相对滞后。重视和加强此类独特真菌资源的研究具有重要的资源学意义，就像Link于1809年建立青霉属时未能料到百年后弗莱明会发现青霉素一样，随着人们对金孢属研究应用的深入，一些新奇资源包括它们的代谢产物将会不断地被发掘而服务于人类（梁建东等，2007a）。

金孢属种的形态描述方法：参照Carmichael（1962）和van Oorschot（1980）的文献，金孢属的分类研究主要根据宏观的培养性状（菌落特征），微观的个体形态特征及一些生物学性状进行描述鉴定，同时对一些形态上不易区分的疑难种结合分子系统学的原理及方法进行分类鉴定。

培养基：查氏培养基和PDA培养基。

方法：将固体培养基制成平板，以点种法接种，即用接种针尖蘸取少量孢子点植在平板的中心位置，然后于15℃、25℃、35℃培养7天或14天进行观察（个别菌株在5℃、10℃、15℃、20℃、25℃、30℃、35℃、40℃、45℃、50℃、53℃及60℃的某些温度下进行培养）。

观察的要点如下。

（1）大小：以菌落的直径（mm）表示。

（2）颜色：包括菌落表面和背面的颜色，以及色素是否渗入培养基。

（3）菌落表面特征：如皱纹、辐射沟纹、同心环、整个菌落致密或疏松等。

（4）菌落质地：毡状、毯状、绒毛状、棉絮状、粉粒状、革质状等。

（5）菌落高度：扁平、突起或隆起，以及菌落中心部分状况等。

（6）菌落边缘：全缘、锯齿状、树枝状等。

（7）渗出物：菌落表面有无液滴及液滴的颜色等。

（8）菌丝特征：表面特征（粗糙或光滑）、宽度等。

（9）球拍状菌丝：存在与否及其大小。

（10）分生孢子着生方式：直接生于菌丝、生于短突或短柄、生于侧分枝等。

（11）间生孢子：存在与否及其大小、形状和表面特征等。

（12）顶侧生分生孢子形态：形状、大小、表面状况等。

（二）栉霉属

栉霉属隶属于爪甲团囊菌目裸囊菌科，其模式种为锯齿栉霉 *Ctenomyces serratus* Eidam（Eidam，1880；Wijayawardene *et al.*，2017）。Langeron和Milochevitch（1930a）

将乳色毛癣菌 *Trichophyton lacticolor* Sabour.和须癣小孢子菌 *Microsporum mentagrophytes* C.P. Robin 移入栉霉属，分别命名为乳色栉霉 *Ctenomyces lacticolor* (Sabour.) Langeron & Miloch.和须癣栉霉 *C. mentagrophytes* (C.P. Robin) Langeron & Miloch.。随后，10 个物种（齿状发癣菌 *Trichophyton denticulatum* Sabour.、马发癣菌 *T. equinum* Gedoelst、羊毛发癣菌 *T. eriotrephon* Papeg.、多粉发癣菌 *T. farinulentum* Sabour.、猫发癣菌 *T. felineum* R. Blanch.、浅灰发癣菌 *T. griseum* F. Fisch.、桃色发癣菌 *T. persicolor* Sabour.、浅灰发癣菌放射褶变种 *T. gypseum* var. *radioplicatum* W. Fisch.、维安发癣菌 *T. viannai* Mello 和石膏表皮癣菌 *Epidermophyton gypseum* L. MacCarthy）也被移入了栉霉属（Nannizzi，1934）。Milochevitch（1935）和 Szathmáry（1960）分别描述了新种博萨栉霉 *Ctenomyces bossae* Miloch.和发癣栉霉 *C. trichophyticus* Szathmáry。后来的研究表明，博萨栉霉是错误命名；发癣栉霉是无效发表；猫栉霉 *C. felineus* (R. Blanch.) Nann.是锯齿栉霉的异名；桃色栉霉 *C. persicolor* (Sabour.) Nann.被转移至奈尼兹皮菌属，即桃色奈尼兹皮菌；须癣栉霉、马栉霉 *C. equinus* (Gedoelst) Nann.和羊毛栉霉 *C. eriotrephon* (Papeg.) Nann.都被移入了发癣菌属，分别命名为：须癣发癣菌、马发癣菌和羊毛发癣菌；乳色栉霉和锯齿栉霉被移回了发癣菌属，并被命名为须癣发癣菌；其余的 4 个种：多质栉霉 *C. farinulentus* (Sabour.) Nann.、放射褶栉霉 *C. radioplicatus* (W. Fisch.) Nann.和维安栉霉 *C. viannai* (Mello) Nann.被移到发癣菌属，石膏样栉霉 *C. gypseus* (L. MacCarthy) Nann.被移到表皮癣菌属，它们最终被视为无效种或可疑种（Orr *et al.*，1963；de Hoog *et al.*，2017，2000a）。因此，栉霉属仅保留了一个有效种，即模式种锯齿栉霉。

Zhang 等（2019）在调查中国土壤中的嗜角蛋白真菌时报道了白色栉霉 *Ctenomyces albus* Y.F. Han, Z.Q. Liang & Zhi Y. Zhang、倒卵形栉霉 *C. obovatus* Y.F. Han, Z.Q. Liang & Zhi Y. Zhang 和青灰栉霉 *C. peltricolor* Y.F. Han, Z.Q. Liang & Zhi Y. Zhang 等 3 个新种，并承认了棉毛栉霉 *C. vellereus* (Sacc. & Speg.) P.M. Kirk 的有效性，同时编制了该属 5 个有效种的检索表。此后不久，Sharma 和 Shouche（2020）在调查印度土壤中的嗜角蛋白真菌时报道了该属的一个新种，即印度栉霉 *C. indicus* Rahul Sharma & Shouche。因此，截至目前，栉霉属共有 6 个有效种。

（三）地丝霉属

1. 地丝霉属的研究历史

地丝霉属隶属于寡囊盘菌目 Thelebolales 假散囊菌科 Pseudeurotiaceae。Traaen（1914）在建立地丝霉属时，描述了该属的 4 个种：金黄色地丝霉 *Geomyces auratus* Traaen、白垩纪地丝霉 *Geomyces cretaceus* Traaen、硫磺地丝霉 *G. sulphureus* Traaen 和普通地丝霉 *G. vulgaris* Traaen，但遗憾的是他并未在上述 4 个种中选定地丝霉属的模式种。Carmichael（1962）注意到毡状金孢的色泽差异极大，且地丝霉属上述 4 个种的主要区别是色泽，遂将其都界定为毡状金孢 *Chrysosporium pannorum* (Link) S. Hughes 的同物异名。由于金黄色地丝霉的菌株可培养，Sigler 和 Carmichael（1976）将其界定为地丝霉属的后选模式种，并指出地丝霉属链状的分生孢子着生于垂直生长的分生孢子梗上呈锐角的分枝菌丝上。在形态与之相近的金孢属中，伴生异形孢子直接着生于菌丝的一端或短梗上，并极少形成由两到三个孢子组成的链状结构，而地丝霉属真菌可以形成短链状结构。

Sigler 和 Carmichael（1976）报道了地丝霉属的两个新种，侧曲地丝霉 *Geomyces asperulatus* Sigler & J.W. Carmich 和毡状地丝霉 *G. pannorum* (Link) Sigler & J.W. Carmich，并进一步概括了地丝霉属的属级特征：营养菌丝窄，无色，分隔。分生孢子梗窄，无色，尖端呈锐角分枝，有时呈轮枝状。繁殖分枝窄，开始时很少，然后有规律地分隔。伴生异形孢子着生于短梗的末端或侧面，中间交替出现节孢子。节孢子比繁殖菌丝宽，呈短或长链，数量多。成熟的孢子由于色素的干预而分散释放，表面光滑或粗糙，无色或黄色，桶状、楔状、近球形或梨形。

van Oorschot（1980）将侧曲地丝霉和酒红地丝霉 *Geomyces vinaceus* Dal Vesco 界定为毡状地丝霉的变种，分别命名为毡状地丝霉侧曲变种 *G. pannorum* var. *asperulatus* (Sigler & J.W. Carmich.) Oorschot 和毡状地丝霉酒红变种 *G. pannorum* var. *vinaceus* (Dal Vesco) Oorschot，同时对地丝霉属的属级特征做了进一步的概括。

之后，Hocking 和 Pitt（1988）、Li 和 Cui（1989）、Gargas 等（2009）、罗韵等（2016）、Chen 等（2017）和 Zhang 等（2020）分别报道了新种粉状地丝霉 *Geomyces pulvereus* A.D. Hocking & Pitt、光滑地丝霉 *G. laevis* Zhong Q. Li & C.Q. Cui、毁害地丝霉 *G. destructans* Blehert & Gargas、贵阳地丝霉 *G. guiyangensis* Y.F. Han, Y. Luo & Z.Q. Liang，福建地丝霉 *G. fujianensis* W.H. Chen, G.P. Zeng, Y. Luo, Z.Q. Liang & Y.F. Han 及倒卵形地丝霉 *G. obovatus* Zhi Y. Zhang, Y.F. Han & Z.Q. Liang。值得注意的是，Minnis 和 Lindner（2013）基于分子系统学研究将毡状地丝霉和毁害地丝霉组合至假裸囊菌属，即毡状假裸囊菌 *Pseudogymnoascus pannorum* (Link) Minnis & D.L. Lindner 和毁害假裸囊菌 *P. destructans* (Blehert & Gargas) Minnis & D.L. Lindner，并表明侧曲地丝霉和酒红地丝霉应隶属于假裸囊菌属。因此，地丝霉属共包括金黄色地丝霉、粉状地丝霉、光滑地丝霉、贵阳地丝霉、福建地丝霉和倒卵形地丝霉等 6 个有效种。

2. 地丝霉属真菌的经济意义

地丝霉属广布全球，且普遍存在于温带和高纬度生态系统的土壤及海洋环境中（Hayes，2012）。该属部分真菌还存在于植物种子中（Chong & Sheridan，1982），或为植物内生真菌（Rosa *et al.*，2010；Ren *et al.*，2016）和内生菌根菌（Vohník *et al.*，2007）。此外，地丝霉属真菌能存在于极地环境的低营养土壤中并休眠，而当新的有机质进入系统时，启动生长和繁殖（Bergero *et al.*，1999）。值得注意的是，地丝霉属真菌还能引起蝙蝠的白鼻综合征（WNS），以及人和其他动物的皮肤感染，是机会病原菌（Gianni *et al.*，2003；Blehert *et al.*，2009；Zelenková，2006；Erne *et al.*，2007）。由于地丝霉属真菌存在于多样性的生态环境中，因而具有丰富的代谢产物。

Krishnan 等（2011）研究南极乔治王岛土壤中产生胞外水解酶的真菌时，共分离出 28 株嗜冷或耐冷真菌。其中，16 株具有淀粉酶活性，23 株具有纤维素酶活性，21 株具有蛋白酶活性。超过 10 株菌同时具有两种以上的酶，而仅有毡状地丝霉菌株同时具有明显的三种酶活性。毡状地丝霉能够在冷、低营养的环境中产生多聚半乳糖醛酸酶、果胶裂解酶、淀粉酶、纤维素酶、几丁质酶、磷酸酯酶、葡萄糖氧化酶、脲酶、蛋白酶及脂肪酶等。其中以淀粉酶、脲酶、脂肪酶 T40 和 T60 含量较高，透明圈直径达 11～25 mm，甚至超过 25 mm（Fenice *et al.*，1997；Duncan *et al.*，2008；梁建东等，2013）。此外，还可产生菊粉内切酶（Gern *et al.*，2001）、木聚糖酶（Del-Cid *et al.*，2014）、β-1,3-葡萄糖酶和β-葡萄糖苷酶（Burtseva *et al.*，2003）及一种新的 α-淀粉酶（Gao *et al.*，2016）。

Gawas-Sakhalkar 和 Singh（2011）对并齿藓 *Tetraplodon mnioides* (Hedw.) Bruch & Schimp.的根际真菌群落及工业生产用酶初步研究时，发现毡状地丝霉在麦芽浸粉琼脂培养基中可产生较高活性的酯酶和磷酸酯酶。

Klimek-Ochab 等（2014）报道了耐冷菌毡状地丝霉 P11 菌株具有氨乙基磷酸代谢酶，可将包含磷碳键的物质 2-氨乙基磷酸矿化，并且该降解过程以磷酸盐的独立方式进行。上述菌株中包含的酶可作为抗菌及杀虫剂的靶标物质。

Li 等（2008）对分离自南极地区的一种地丝霉产生的 asterric acid 衍生物进行研究时，发现该菌株可产生 8 种化合物（图 1），其中 5 种为新的 asterric acid 衍生物，即 ethyl asterrate、*n*-butyl asterrate 和 geomycins A～C。其中 geomycins B 表现出对烟曲霉 *Aspergillus fumigatus* Fresen.有抗真菌活性。geomycins C 对革兰氏阳性菌和革兰氏阴性菌均具有抗细菌活性。

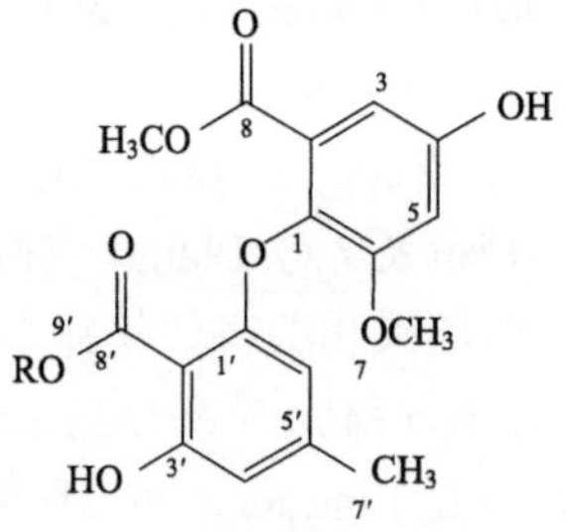

1 R=H
2 R=CH_3
3 R=CH_2CH_3
4 R=$(CH_2)_3CH_3$

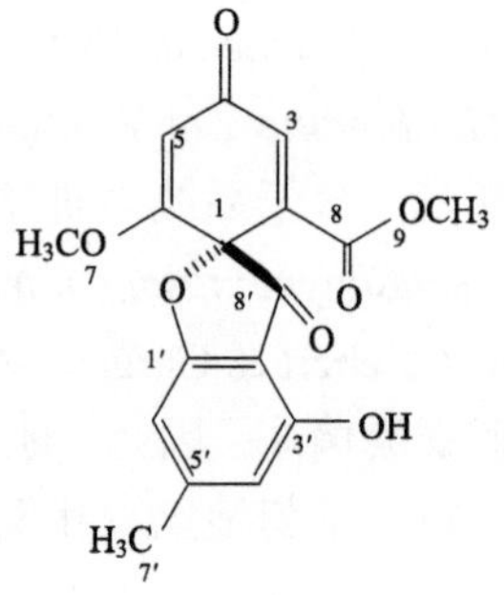

5

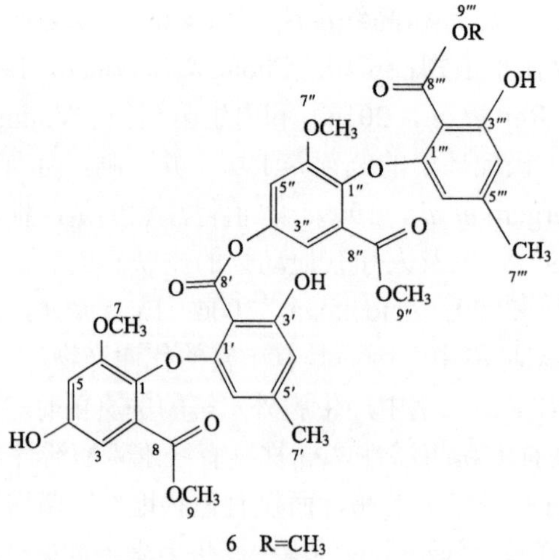

6 R=CH_3
7 R=H

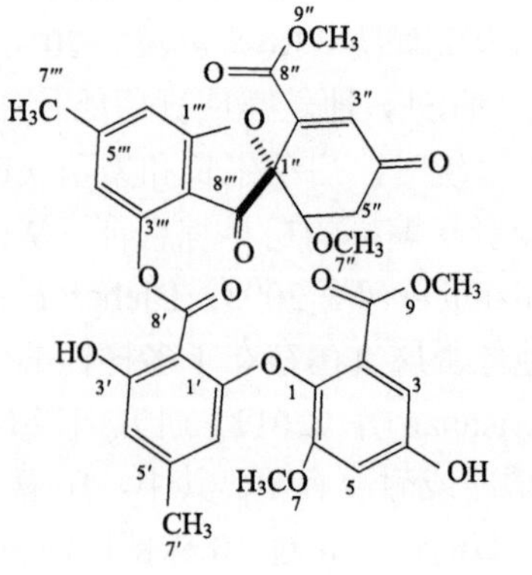

8

图 1 化合物 1～8 的结构式（引自 Li *et al.*，2008）

Na 等（2016）在研究夹竹桃内生真菌中产生吲哚生物碱的菌株时，发现内生真菌 CH1 具有明显的乙酰胆碱酯酶抑制活性，其半抑制浓度（IC_{50}）达到 5.16 μg/mL。经鉴定该菌株为地丝霉。这是关于内生真菌地丝霉产生水甘草碱和乙基-长春胺的首次报道。

李云海等（2017）研究了极地真菌一种地丝霉 3-1 的次级代谢产物，共分离得到 8 个化合物，paulownin、demethylincisterol A3、ergosta-7,22-dienen-3,6-dione、citreoanthrasteroid B、19-norergosta-5,7,9,22-tetraene-3β-ol、(22E)-5α, 8α-epidioxyergosta-6,22-dien-3β-ol、(3β,5α,6β,22*E*)-6-methoxyergosta-7,22-diene-3,5-diol 和 ergosta-7,22-dien-3β-ol（其化学结构分别对应图 2 中的 1～8）。其中化合物 demethylincisterol A3 具有较好的抗菌和细胞毒活性。8 个化合物中第 2～7 个化合物均是在地丝霉属真菌中首次报道（图 2）。

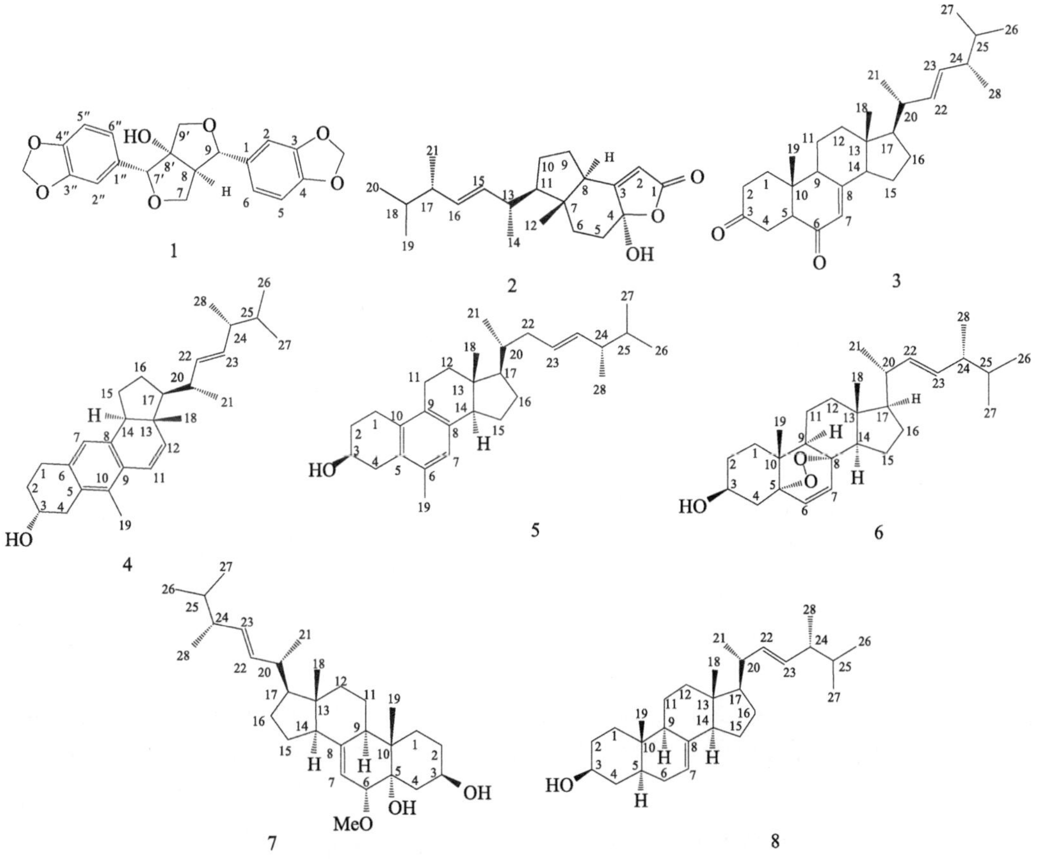

图 2　化合物 1～8 的结构式（引自李云海等，2017）

Hua 等（2017）在对极地海洋菌毡状地丝霉中稀有的化合物及具有纤溶活性的化合物研究中，共分离得到了 6 种化合物（图 3）：tridec-1-ene（1）、methyl (5Z,9Z)-octadeca-5,9-dienoate（2）、(*E*)-icos-4-ene（3）、hexyl 3-(3,5-di-tert-butyl-4-hydroxyphenyl) propanoate（4）、5α,8α-epidioxy ergosta-6,22-diene-3β-ol（5）和 dibutyl phthalate（6）（其化学结构分别对应图 3 中的 1～6）。经鉴定为长链脂肪酸、固醇和芳香烃。纤溶酶活性研究显示化合物 2、3、5 和 6 具有激活纤溶酶的潜力，同时显示出具有较好的纤溶酶活性，

其 EC_{50} 值可分别达到 120 μg/mL、181.3 μg/mL、200 μg/mL 和 179.7 μg/mL。地丝霉可能含有丰富的代谢产物且含有可应用于生物制药的抗血栓药物。

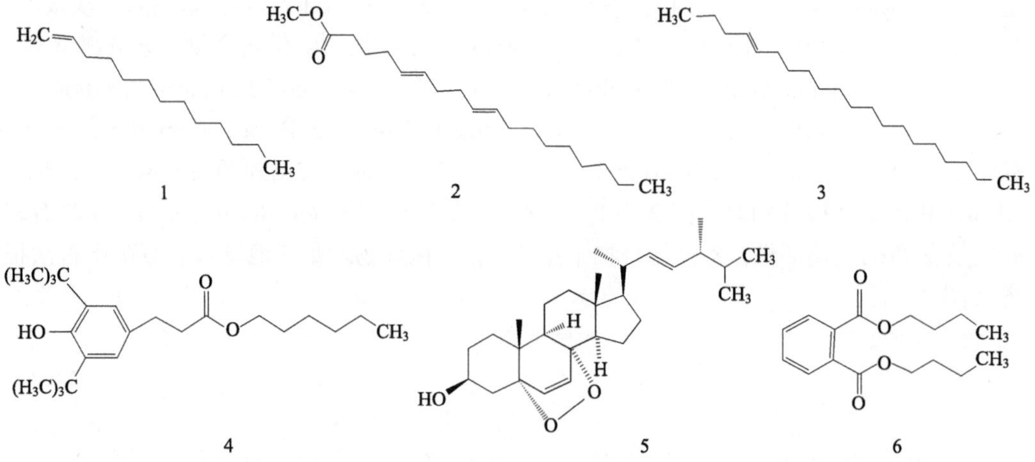

图 3　化合物 1～6 的化学结构（引自 Hua *et al.*，2017）

Parish 等（2009）采用反向筛选法筛选到一株可对 SecA ATP 酶具有抑制作用的毡状地丝霉。通过活性物质为导向的分离方法，从该菌株中分离得到了一个新的 *cis*-十氢萘化合物，并命名为 pannomycin。

Abneuf 等（2016）在对来自于南极土壤中具有抗菌活性真菌的研究中，发现地丝霉菌株 WHB-sp.4 和 YKB-sp.4 具有抑制大肠杆菌 *Escherichia coli* (Migula) Castellani & Chalmers、枯草芽孢杆菌 *Bacillus subtilis* (Ehrenberg) Cohn 和金黄色葡萄球菌 *Staphylococcus aureus* Rosenbach 的活性。

Henríquez 等（2014）在研究南极海洋海绵相关可培养真菌多样性及筛选潜在抗菌、抗肿瘤及抗氧化菌株时，共获得 101 个表型差异的分离株。其中 51%的菌株提取物具有抗菌活性，接近 42%的菌株提取物显示出抗肿瘤活性，同时所试验的菌株均有一定程度的抗氧化活性，并且上述菌株提取物大多来自于地丝霉。因而提出地丝霉含有潜在的抗菌、抗肿瘤及抗氧化活性物质。

李奥兰（2015）发现地丝霉属真菌 WZJ 对马铃薯晚疫病菌 *Phytophthora infestans* (Mont.) de Bary 有抑制作用，抑制率达到 64.4%～78.5%。检测其抑菌谱发现，该菌株对辣椒疫霉 *P. capsici* Leonian、大豆疫霉 *P. sojae* Kaufm. & Gerd.和烟草黑胫病 *P. nicotianae* Breda de Haan 等卵菌病原也具有良好的拮抗效果，抑制率分别达到 68.4%、58.5%、58.7%。该菌的发酵上清液具有良好的抑菌效果，而菌丝体提取物基本无抑菌活性，说明抑菌物质被分泌到胞外起作用。将发酵上清液稀释 176 倍时，该上清液对马铃薯晚疫病菌的抑制效率为 50%。通过追踪抑菌活性，检测发酵上清液中抑菌物质的基本属性，该抑菌物质具有水溶性，且耐高温。

金滨滨（2014）对来自南极土壤的一株地丝霉属真菌所产生的红色素进行了研究。前期通过优化发酵条件，每升发酵液可以得到红色素粗品 1.7 g，其分泌色素的能力远高于同类菌种。从红色素中分离出 7 个单体化合物，鉴定了其中 4 个化合物结构，分别为邻苯二甲酸二丁酯、油酸、2,3-丁二醇和对羟基苯甲酸。还得到一个红色的混合物，通过结构解析，确定其中存在分子量为338、化学结构类似红曲红色素的成分。通过稳定性研究，发现红色素对紫外线稳定；对绝大多数金属离子稳定；在pH为6～10时稳定性好。生物活性方面，当红色素的浓度为1 mg/mL时，超氧阴离子清除率为26.52%；2,2-二苯基-1-苦基肼自由基（DPPH）清除率为73.11%；羟基自由基清除率为67.31%；在低浓度时，红色素的还原性为同浓度下维生素C的50%；红色素不存在抑菌或抗菌作用。研究结果表明，红色素稳定性优于经美国食品和药物监督管理局（FDA）认可的胭脂虫红色素，在生物活性方面要优于其他天然色素，具备开发为食品、纺织、化妆品行业染色用途色素的较大潜力。

董龙龙（2017）对来自南极土壤的高产红色素的地丝霉菌株WNF-15A进行了发酵条件优化和诱变研究，发现以生长4天的WNF-15A菌株最适合制备种子液；摇瓶培养3.5天的种子液以10%的接种量最适合扩大培养。200 L扩大培养10天后，经浓缩、纯化、干燥，得到色价为112.3、产率为1.16 g/L的色素干粉。通过对WNF-15A进行紫外诱变，得到6株产色素能力较强的菌株；对其中的Y-014菌株发酵条件研究表明，pH和糖浓度对发酵影响较大，pH、糖浓度、接种量、转速的最佳水平组合为pH 5.5、糖浓度3%、接种量9%、转速110 r/min。Xing等（2016）在研究提高丹参抗氧化活性和总酚含量时，将土黄地丝霉与丹参进行共培养，可以显著提高丹参中丹酚酸B的含量，且土黄地丝霉可能参与了丹参中丹酚酸的富集过程。Barratt等（2003）研究土壤中降解聚氨基甲酸酯的微生物时，发现真菌是降解该物质的主要类群。在真菌所构成的生物膜中，毡状地丝霉为主要的真菌，并且具有22%～100%的降解聚氨基甲酸酯的能力。综上所示，地丝霉属真菌具有较大的开发利用潜力。

（四）角蛋白癣菌属

角蛋白癣菌属隶属于子囊菌门散囊菌纲爪甲团囊菌目爪甲团囊菌科（Randhawa & Sandhu，1964）。模式种为棕灰角蛋白癣菌 *Keratinophyton terreum* H.S. Randhawa & R.S. Sandhu。迄今，该属包含25个种（Labuda *et al.*，2021；Li *et al.*，2022a），国内已分离报道11个种。

该属传统上以形态特征进行分类鉴定，包括子座、子囊孢子、分生孢子等特征。但为了更准确体现该属的分类地位，采用了形态特征和基于ITS rDNA和LSU rRNA的分子系统发育研究。Labuda等（2021）的研究结果表明，原属于金孢属的部分种与该属成员聚为单系群，因此将其中的10个种移入角蛋白癣菌属。Li等（2022a）同样基于ITS rDNA和LSU rRNA发表了2个新种，其系统发育关系如图4所示。该属成员主要分离自土壤，尤其是富角蛋白基质的土壤环境。由于该属的特征与金孢属类似，其成员分布和应用与金孢属类似，见金孢属章节。

（五）冠癣菌属

冠癣菌属于1899年建立后，仅包含1个种。其形态特征与小孢子菌属相似，但近年的系统发育研究表明两者在系统发育上不在一支，冠癣菌属成为单独一支（de Hoog *et al.*，2017）。

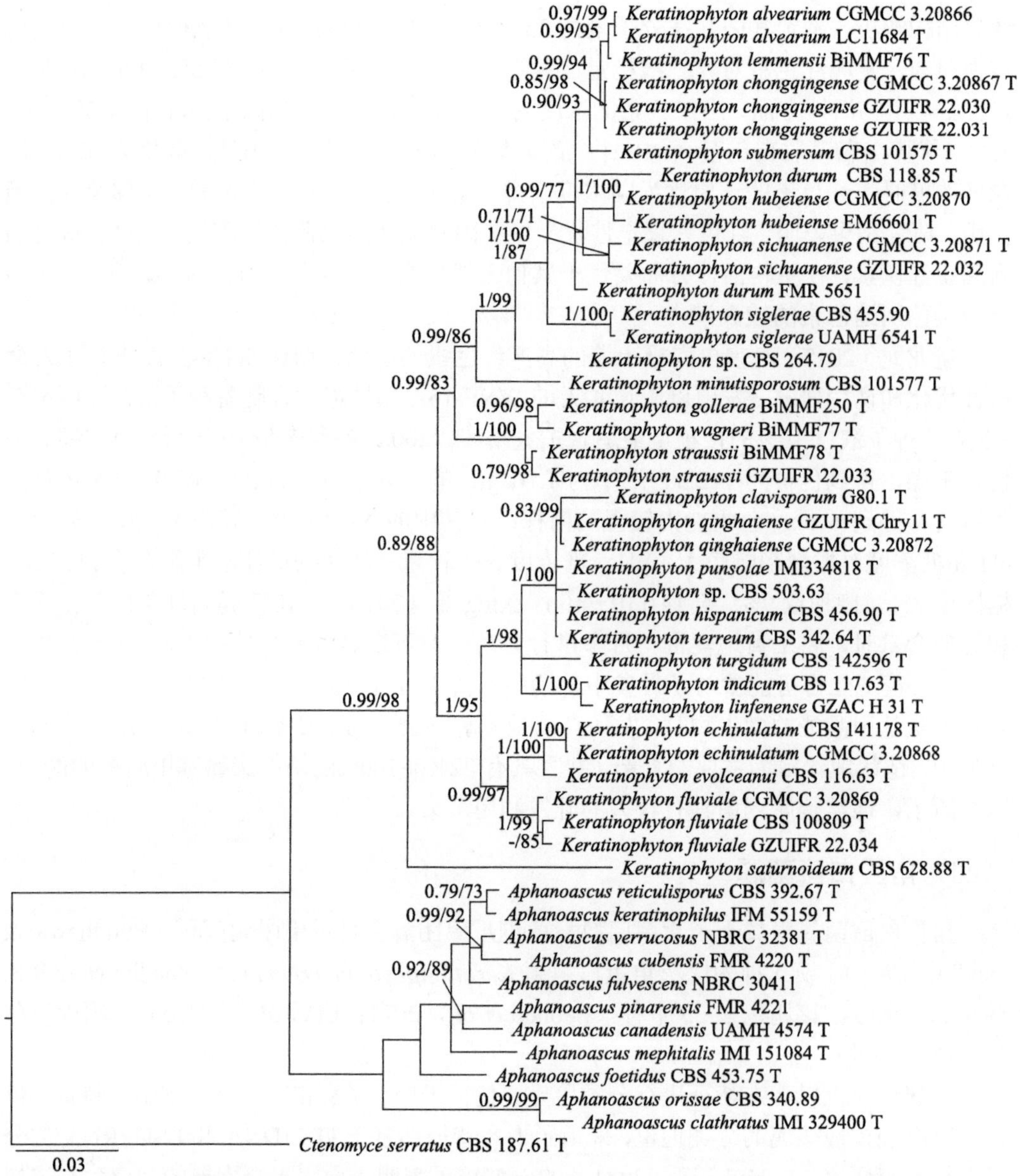

图 4　基于 ITS rDNA 和 LSU rRNA 构建的角蛋白癣菌属的系统发育关系（引自 Li *et al.*，2022a）

（六）小孢子菌属

1. 小孢子菌属的研究历史

小孢子菌属隶属于爪甲团囊菌目裸囊菌科。该属是由 Gruby（1843）以奥杜盎小孢子菌为模式种而建立。Emmons（1934）基于大量文献资料对皮肤癣菌进行整理时，移除了根据该属依据宿主临床反应，而非形态特征鉴定的种，同时还简化了一些鉴定特征，如菌落质地、厚垣孢子、瘤状体及产色素等。最后，仅承认奥杜盎小孢子菌、马小孢子菌 *Microsporum equinum* E. Bodin ex Guég.、猫小孢子菌 *M. felineum* Mewborn、鸡禽小

孢子菌 *M. gallinae* (Mégnin) Grigoraki（现为鸡禽冠癣菌 *Lophophyton gallinae* (Mégnin) Matr. & Dassonv.）、石膏样小孢子菌及昆克纳小孢子菌 *M. quinckeanum* (Zopf) Guiart & Grigoraki 等 6 个种为有效种。此后，基于孢子形态的鉴定被广泛接受，但随后增加的许多生化、生态、流行病学及遗传等鉴别特征，又导致了该类真菌不少种分类地位的混乱。

Ajello（1968）在对皮肤癣菌进行仔细研究后，仅承认 Emmons（1934）接受的 6 个种中的奥杜盎小孢子菌、猫小孢子菌和石膏样小孢子菌等 3 个种为有效种，而将马小孢子菌、鸡禽小孢子菌和昆克纳小孢子菌等 3 种归入发癣菌属。此外，Ajello（1968）另承认亚马逊小孢子菌 *Microsporum amazonicum* Moraes, Borelli & Feo、布拉尔小孢子菌 *M. boullardii* Dominik & Majchr.、库柯小孢子菌 *M. cookei* Ajello（现为库柯帕拉癣菌 *Paraphyton cookei* (Ajello) Y. Gräser, Dukik & de Hoog）、变形小孢子菌 *M. distortum* Di Menna & Marples、铁锈色小孢子菌、粉小孢子菌、猪小孢子菌 *M. nanum* C.A. Fuentes、桃色小孢子菌 *M. persicolor* (Sabour.) Guiart & Grigoraki（现为桃色奈尼兹皮菌 *Nannizzia persicolor* Stockdale）、腊梅黄小孢子菌 *M. praecox* Rivalier、葡萄状小孢子菌 *M. racemosum* Borelli、布鲁盖门小孢子菌 *M. vanbreuseghemii* Georg, Ajello, Friedman & S.A. Brinkm.等 11 个有效种。至此，小孢子菌属已知有效种为 14 个。此后，该属名下新增 7 个种，它们是河岸小孢子菌 *M. ripariae* Hubálek & Rush-Munro、麦哲伦小孢子菌 *M. magellanicum* Caretta & Piont、附属小孢子菌 *M. appendiculatum* Bhat & Miriam、奇妙小孢子菌 *M. mirabile* J.S. Choi, Y. Gräser, G. Walther, Peano, F. Symoens & de Hoog（现为奇妙帕拉癣菌 *Paraphyton mirabile* (J.S. Choi, Y. Gräser, G. Walther, Peano, F. Symoens & de Hoog) Y. Gräser, Dukik & de Hoog）、闪石小孢子菌 *M. aenigmaticum* Hubka, Dobiášová & M. Kolařík、弯小孢子菌 *M. incurvatum* (Stockdale) P.L. Sun & Y.M. Ju（Sun *et al.*，2014）和贵州小孢子菌 *M. guizhouense* Zhi Y. Zhang, X. Zou, Y.F. Han & Z.Q. Liang（张芝元等，2017c；Caretta & Piontelli，1977）。而 Sharma 等（2006）基于 ITS1-5.8S-ITS2 rDNA 序列分析揭示，将附属小孢子菌定为石膏样小孢子菌的同物异名。

该属早期分类研究主要以形态特征为主（de Hoog *et al.*，2000a）。随着分子生物学技术的发展，近年开展了真菌的分子系统学研究，国际上多个实验室的研究评估表明 ITS 是识别大多数真菌成功概率最高的序列，它在真菌鉴定和识别上具有重要意义和实用性（Schoch *et al.*，2012）。de Hoog 等（2017）采用 LSU、ITS、60S 和 TUB 等四个基因片段对小孢子菌进行分子系统学研究时，发现 ITS 的效果是最好的。近年对该属新种的鉴定开展了分子系统学研究，主要以 ITS-5.8S rDNA 为主（Hubka *et al.*，2014；Sun *et al.*，2014），如贵州小孢子菌是基于 ITS 分子系统学和形态学特征发表的一个新种。de Hoog 等（2017）基于 ITS、LSU、TUB 和 60S L10 的多基因数据集，对皮肤癣菌进行了系统发育研究，结果保留了小孢子菌属中的 3 个种，有些种更名或移入奈尼兹皮菌属、帕拉癣菌属和冠癣菌属中。据 Index Fungorum 和 MycoBank 数据库记载，该属共有 108 个菌名，截至目前该属仅承认有效种 20 个，但能进行系统发育研究的不超过 10 个种，该属物种仍有待挖掘。迄今中国报道了该属 5 个种。

2. 小孢子菌属真菌的致病性研究概况

小孢子菌属世界分布，常嗜角蛋白，多为皮肤病原真菌，可引起人类和动物的头癣和体癣（Hubka *et al.*，2014；Ilkit，2010；Sun *et al.*，2014；Krunic *et al.*，2007；罗挺等，1996；虞伟衡和章强强，2008；孙晓红等，2013；朱小红等，2014；张影和郑宝勇，2015；徐宇和杨国玲，2016；张英和金敏，2016；徐宇和杨国玲，2016）。基于其自然生境可分为亲人性、亲动物性和亲土性三类（Ganaie *et al.*，2010）。常见亲人性的有奥杜盎小孢子菌；亲动物性的有亚马逊小孢子菌、犬小孢子菌、猪小孢子菌、粉小孢子菌；亲土性的有库柯小孢子菌、杜伯氏小孢子菌 *Microsporum duboisii* Cif.、石膏样小孢子菌、桃色小孢子菌、腊梅黄小孢子菌、葡萄状小孢子菌和铁锈色小孢子菌（Brasch，2015；王端礼，2005）。

临床医学和动物医学上不断报道小孢子菌属引起的各种流行性病害给人类健康造成了巨大的威胁。例如，上海市第十人民医院皮肤科在 2012 年 6～7 月就发现 30 例由犬小孢子菌引起的体癣病例（马越娥等，2013）。2012 年，江西省九江学院附属医院检验科报道了 20 例由石膏样小孢子菌致儿童脓癣病例（孙晓红等，2013）。Raina 等（2016）报道了 1 例由铁锈色小孢子菌引起的严重体癣病例。宁夏医科大学总医院张小鸣等在银川市及郊区对 1409 名 5～10 岁儿童进行问卷调查，结果显示，1409 名儿童中头癣患病人数 203 例，患病率为 14.4%，其中犬小孢子菌阳性率为 45.0%，铁锈色小孢子菌阳性率为 28.7%，病原菌主要为犬小孢子菌和铁锈色小孢子菌（张小鸣等，2012）。此外，李彩霞和刘维达（2011）对我国 2000 年 1 月至 2010 年 11 月儿童头癣流行情况进行了汇总分析，结果显示，除新疆南部外，犬小孢子菌是头癣的主要致病菌。由此可见，小孢子菌属的真菌严重影响了人类的身体健康。

目前，针对小孢子菌属真菌感染的治疗主要采用生物和化学药物，但随之带来的是抗药性和化学药物对患者产生副作用等问题。研究新型抗小孢子真菌的药物对人类健康有重要的意义。国内外不断地报道由小孢子菌属感染引起的疾病病例，以下为部分常见的由小孢子菌引起的真菌感染特征及临床治疗方法：犬小孢子菌是临床上常见的皮肤癣致病菌，它具有亲动物性的特点，易感染动物的毛发，以及人类头皮、头发、光滑皮肤及指（趾）甲，引起头癣、手足癣、体股癣及甲癣，亦可致头皮深层性感染，从而引起脓癣。一般患者主要发病特征为：发病部位为头皮、面部、颈部、胸部、手臂处和腿部；皮疹为圆形或椭圆形，脱屑性红斑，部分炎症反应明显，皮损表现为肿胀性红斑。发生于头皮者除了脱屑性红斑，亦有断发，伴有不同程度的瘙痒。通常情况下，犬小孢子菌在沙氏葡萄糖琼脂（SDA，简称沙氏）培养基上生长速度快，培养 4～6 天均有菌落长出，菌落中心稍隆起，白色、棉絮状，中央趋向粉末状；背面呈淡棕黄色（马越娥等，2013；曹艳云和徐顺明，2017；杨虹，2013；Daniels，1953）。临床上治疗小孢子真菌感染常使用外用激素类药膏，佟盼琢等（2011）研究发现，盐酸小檗碱与灰黄霉素、特比萘芬、伊曲康唑联合体外使用时，均表现出不同程度的协同抗犬小孢子菌活性的作用。Silva 等（2017）研究发现芳樟醇在临床上对犬小孢子菌引起的皮肤癣菌病具有抑制作用。

石膏样小孢子菌于 1907 年由 Bodin 首次描述，分离自一个 40 岁左右妇女的右脸颊的黄癣病斑中（Bodin，1907）。起初研究发现该菌是一类亲土性真菌，甚少感染人和动物。之后报道的感染案例越来越多，涉及马、猴子、狗、猫、老虎和鸡（Ajello，1953）。不同来源的石膏样小孢子菌株呈现为不同类型的复合体（Stockdale，1963）。石膏样小孢子菌和弯小孢子菌在形态上非常相似，区分困难。菌株交配性测验是区分它们最可靠的方法。然而，这个过程非常耗时，不适合临床应用。因此，这两个种常被作为石膏样小孢子菌。石膏样小孢子菌是人类临床感染不常见的类群（Sun & Ho，2006），黄癣或类似黄癣病仅报道了 13 个病例（Schmitt *et al.*，1996；Bakos *et al.*，1996；Miranda *et al.*，1998；Zhang *et al.*，2002；Prochnau *et al.*，2005；Lu *et al.*，2009；Chen *et al.*，2013）。其中，11 个病例是从阴囊病灶获得的（Zhang *et al.*，2002），原因还不得而知，有可能是高温高湿环境更容易引起该菌的感染（Miranda *et al.*，1998），人类个体免疫差异性也是一个假设原因。这 13 例的病原到底是石膏样小孢子菌，还是弯小孢子菌，还未真正被确定。此外，Fike 等（2018）还报道了一起由石膏样小孢子菌感染引起的灰指甲病例。临床上一般采用口服抗真菌药物，同时外用抗真菌药剂，口服的抗真菌药物有伊曲康唑或特比萘芬等，外用药物有硝酸舍他康唑乳膏、盐酸特比萘芬乳膏或复方酮康唑乳膏等。

铁锈色小孢子菌是 1921 年 Ota 从临床上发现的一种皮肤病原真菌。通常情况下，铁锈色小孢子菌能侵染人类毛发，引起一类不常见的结节肉芽肿性毛囊周围炎皮肤病，常见的症状为感染部位可见多处大小不等的皮下结节，触之活动度好，无红肿，有轻度压痛，质较软，有波动感，部分感染部位可见不同程度的毛发缺损，表皮糜烂，局部隆起高出体表，多个窦道口有淡黄色脓液渗出，可见肉芽组织增生等。在感染初期，临床上给予抗真菌药物及小剂量糖皮质激素治疗，感染严重时一般内服伊曲康唑胶囊和泼尼松片，外用一些如复方酮康唑乳膏等（居哈尔•米吉提等，2009；Qoraan *et al.*，2017）。它也能感染人体皮肤，引起体癣，常见的症状有丘疹和脓疱，临床一般采用灰黄霉素治疗这种皮肤病（Raina *et al.*，2016）。

小孢子菌属真菌的应用价值：随着生物技术、分子生物学技术和微观生态学的不断兴起，越来越多的研究者开始关注真菌的开发与利用。现已研究发现，小孢子菌属真菌可以产生角蛋白酶、纤维素酶、脂肪酶、菊粉酶和 α-半乳糖苷酶等至少 19 种酶类物质（Kushwaha，2000；吕珊珊等，2016；Takiuchi *et al.*，1982；Brouta *et al.*，2001）和一些次生代谢产物，如抗癌、抗氧化、抗真菌、抗寄生虫等抗生素类物质（刘鑫等，2018；郑飞等，2017；李增智，2015）。此外，真菌在食品业也得到应用，特别是在我国酒和茶叶领域得到广泛的开发与利用，如新疆葡萄酒、云南普洱茶和红茶等（张文州和许嵘，2014；冯涛等，2018；林娟等，2015；谢美华等，2013）。

微生物来源的角蛋白酶是目前酶领域研究的热点之一，这类酶在饲料行业、制革工业和环境废弃物处理等多个方面具有广泛的应用前景（杨连等，2015；于晓朋和赵新强，2015；齐志国等，2012）。角蛋白酶在有关行业无可替代的重要作用已成为共识，部分小孢子菌属真菌作为喜角蛋白的真菌类群，在角蛋白酶生产中的潜在价值备受人们关注。张芝元等（2017c）从几所医院采集的土样中分离获得角蛋白降解真菌 774 株，通过形态学特征、生物条码数据系统（Barcode of Life Data Systems）、UNITE 数据库、CBS 数据库和分子系统学分析的多途径方法分析后，发现其中就包含小孢子菌属真菌。还有很多学者对小孢子菌属真菌做了研究，例如，Kunert（1989）研究比较了放线菌弗氏链霉菌（*Streptomyces fradiae*）和石膏样小孢子菌降解角蛋白的生化机制；Brouta 等（2001）报道犬小孢子菌能产生金属蛋白酶。

（七）毁丝霉属

1. 毁丝霉属的研究历史

毁丝霉属是 Costantin 于 1892 年以一种蘑菇病原体——土黄毁丝霉 *Myceliophthora lutea* Costantin 为模式种建立的（Costantin，1892）。其有性型隶属于两个纲中的多个目多个科多个属：散囊菌纲散囊菌目、爪甲团囊菌目裸囊菌科中的节皮菌属和栉霉属；粪壳菌纲 Sordariomycetes 粪壳菌目 Sordariales 粪壳菌科 Sordariaceae 中的梭孢壳属 *Thielavia* Zopf 和棒囊壳属 *Corynascus* Arx（van Oorschot，1980）。

毁丝霉属自建立以来，该属及其种级阶元的分类地位先后被不同的学者进行了多次修正。Carmichael（1962）将毁丝霉属的模式种土黄毁丝霉归入金孢属，定名为土黄金孢 *Chrysosporium luteum* (Costantin) J.W. Carmich.，并将毁丝霉属降为金孢属的异名属。von Arx（1973）认同毁丝霉属的分类地位，将土黄毁丝霉重新订为毁丝霉属的模式种。van Oorschot（1977）基于前人研究，在 *The genus Myceliophthora* 一文中，重新评价了毁丝霉属，并对该属中的 3 个种进行了重新描述。van Oorschot（1980）主要根据菌落和分生孢子的着生方式与形态特征，以及对温度的耐受性和降解角蛋白的能力，对毁丝霉属及其近缘属进行了重新界定，并编制了毁丝霉属与几个近缘属的分属检索表，同时总结了毁丝霉属的有性型和属征，确定了 4 个有效种，即嗜热毁丝霉、棉毛毁丝霉 *M. vellerea* (Sacc. & Speg.) Oorschot、土黄毁丝霉和弗格斯毁丝霉 *M. fergusii* (Klopotek) Oorschot。

20 世纪，除了 Carmichael、Arx、Oorschot 等的工作，Goddard 等（1913）、Doyer（1927）、Burnside（1928）、Awao 和 Udagawa（1983）、Basu（1985）先后发表了 5 个新种：硫磺色毁丝霉 *Myceliophthora sulphurea* Goddard、暗棕毁丝霉 *M. fusca* Doyer、膨大毁丝霉 *M. inflata* Burnside、黄褐毁丝霉 *M. hinnulea* Awao & Udagawa 和印度毁丝霉 *M. indica* M. Basu。但是 Carmichael（1962）认为膨大毁丝霉与膨大拟青霉 *Paecilomyces inflatus* (Burnside) J.W. Carmich.是同物异名，按照《国际藻类、菌物和植物命名法规》，将其定名为膨大拟青霉。2009 年，Liang 等又将其订为膨大戴氏霉 *Taifanglania inflata* (Burnside) Z.Q. Liang, Y.F. Han & H.L. Chu。2013 年，Haybrig 等根据分子系统发育研究，将该种移入了单胞霉属中，命名为膨大单胞霉 *Phialemonium inflatum* (Burnside) Dania García, Perdomo, Gené, Cano & Guarro。Mouchacca（2000）将印度毁丝霉界定为嗜热毁丝霉的同物异名。

Sigler 等（1998）认为根据形态学描述而将棉毛毁丝霉作为毁丝霉属的一个物种很有可能是错误的。van den Brink 等（2012）根据棉毛毁丝霉 *Myceliophthora vellerea* (Sacc. & Speg.) Oorschot 与锯齿栉霉形态特征类似的描述（Guarro *et al.*，1985；Chabasse，1988）而将棉毛毁丝霉归为锯齿栉霉。同时发表了毁丝霉属的 6 个新种，它们是：异宗毁丝霉 *M. heterothallica* (Klopotek) van den Brink & Samson、新几内亚毁丝霉 *M. novoguineensis* (Udagawa & Y. Horie) van den Brink & Samson、瘤孢毁丝霉 *M. sepedonium* (C.W. Emmons) van den Brink & Samson、疣状毁丝霉 *M. verrucosa* (Stchigel, Cano & Guarro) van den Brink & Samson、相似毁丝霉 *M. similis* (Stchigel, Cano & Guarro) van den Brink & Samson、有性毁丝霉 *M. sexualis* (Stchigel, Cano & Guarro) van den Brink & Samson。

Zhang 等（2014）发表了 1 个新种，即水滴状毁丝霉 *Myceliophthora guttulata* Yu Zhang & L. Cai，并且在前人研究的基础上，编制了毁丝霉属 12 个种新的检索表，具体如下：

1. 有性型不存在 ·· 2
1. 有性型存在 ·· 5
  2. 分生孢子光滑或粗糙 ·· 3
  2. 分生孢子光滑 ··· 4
3. 分生孢子 8～9 × 6～7.5 μm ······································ **黄褐毁丝霉 *M. hinnulea***
3. 分生孢子 5.5～6.6 × 3.3～3.6 μm ································ **印度毁丝霉 *M. indica***
3. 分生孢子 4.5～11 × 3～4.5 μm ·································· **嗜热毁丝霉 *M. thermophila***
  4. 分生孢子有水滴状斑点，倒卵形到梨形，（3.8～）4.8～7.2 × 3～5 μm ······························
  ··· **水滴状毁丝霉 *M. guttulata***
  4. 分生孢子无水滴状斑点，梨形到球形，3.8～9 × 3～6 μm ······················ **土黄毁丝霉 *M. lutea***
5. 分生孢子棍棒状到倒卵形或缺失；异宗配合 ····································· 6
5. 分生孢子或多或少球形，始终存在；同宗配合 ·································· 7
  6. 子囊孢子主要有 1 个生殖孔，8～13 × 5～7 μm ······················ **弗格斯毁丝霉 *M. fergusii***
  6. 子囊孢子有 2 个生殖孔，22～32 × 17～23 μm ·················· **异宗毁丝霉 *M. heterothallica***
7. 子囊果多刚毛或具疣状隆起的包被 ·· 8
7. 子囊果无毛 ·· 9
  8. 子囊孢子两端各有 1 个生殖孔，17～23 × 11～13 μm ························ **多毛毁丝霉 *M. setosus***
  8. 子囊孢子有 2 个斜向近顶端的生殖孔，11～18 × 6.5～9 μm ·············· **疣状毁丝霉 *M. verrucosa***
9. 子囊孢子有 2 个侧面接近端点的生殖孔；子囊孢子横向视图为舟状，11～20 × 6.5～9 μm ··········
  ··· **相似毁丝霉 *M. similis***
9. 子囊孢子有 2 个端生生殖孔 ··· 10
  10. 无性型不存在；子囊孢子柠檬形，9～14 × 8～10 μm ······················ **有性毁丝霉 *M. sexualis***
  10. 无性型存在 ··· 11
11. 分生孢子具非常细微的小刺或小疣；子囊孢子 12～19 × 8～11 μm ····· **瘤孢毁丝霉 *M. sepedonium***
11. 分生孢子光滑；子囊孢子 18～24 × 8～10 μm ····················· **新几内亚毁丝霉 *M. novoguineensis***

Yasmina 等（2015）基于 ITS、*EF1A* 和 *RPB2* 对毁丝霉属及其有性型进行了系统发育评价，将该属拆分为 4 个属，建立了 2 个新属，即粗糙壳属 *Crassicarpon* Y. Marín, Stchigel, Guarro & Cano 和嗜热疣霉属，其余两个为毁丝霉属和棒囊壳属。它们的区别是：毁丝霉属成员常常是中温的，通常以无性型形式存在；粗糙壳属成员通常是嗜热的，异宗配合并且产生球形至楔形切壁光滑的分生孢子，闭囊壳壁光滑，子囊孢子两端均具芽孔；棒囊壳属成员通常是同宗配合，中温性并且产生球形且表面大多有纹饰的分生孢子，闭囊壳表面有鳞状纹饰，子囊孢子两端均有 1 个芽孔；嗜热疣霉属常嗜热，异宗配合，与棒囊壳属成员有相似的子座和分生孢子，但它的子囊孢子仅有一个芽孔。

毁丝霉属及其相近属的检索表如下（Yasmina *et al.*，2015）：

1. 50℃条件下不能生长 ·· 2
1. 50℃条件下能生长 ·· 3
  2. 培养基上可见有性型 ··· **棒囊壳属 *Corynascus***
  2. 培养基上有性型缺乏 ·· **毁丝霉属 *Myceliophthora***

3. 分生孢子无色，球形至楔形，壁光滑 ·· **粗糙壳属 *Crassicarpon***

3. 分生孢子褐色，近球形或倒卵圆形至椭圆形，有纹路或少量光滑 ·· **嗜热疣霉属 *Thermothelomyces***

随着现代科学的发展和各学科知识的相互渗透，真菌分类学也从单一的形态特征鉴别发展到以形态特征为主、其他特征并重，将来发展的趋势肯定是向着包括其他类型特征（生理学、生物化学、遗传学、分子生物学等）前进。尤其是近年来，随着基因组研究向各学科的不断渗透，这些学科的进展达到了前所未有的高度。分子生物学的发展对分类学的发展也起了极大的促进作用，形成了新的学科——分子系统学。分子系统学是通过检测生物大分子包含的遗传信息，定量描述、分析这些信息在分类、系统发育和进化上的意义，从而在分子水平上解释生物的多样性、系统发育及进化规律的一门学科。它以分子生物学、系统学、遗传学、分类学和进化论为理论基础，以分子生物学、生物化学和仪器分析技术的最新发展为研究手段，是一门交叉性很强的学科。相对于经典的形态系统分类研究，由于生物大分子本身就是遗传信息的载体，含有庞大的信息量，且趋同效应弱，因而其结论更具可比性和客观性。尤为重要的是，一些缺乏形态性状的生物类群中，它几乎成为探讨其系统演化关系的唯一手段。总的说来，迄今为止分子系统学的研究所获得的生物类群间亲缘关系的结果，大多都和经典的形态系统树相吻合；但是，对一些生物进化谱系不明的类群，它得出的结果却往往和形态系统学的推测不大相同，这需要我们谨慎对待。

毁丝霉属分子系统学方面的研究起步较晚，直到 van den Brink 等（2012）利用 ITS 序列、*EF1A*、*RAPB2* 和扩增片段长度多态性（amplified fragment length polymorphism，AFLP）技术对毁丝霉属和棒囊壳属的 49 个菌株的系统发育关系进行了研究，结果表明这两个属没有明确的分离。因此，他们建议将棒囊壳属所有与无性型毁丝霉属对应的种放入毁丝霉属，遵循优先权和一种真菌一个名称的原则，用毁丝霉属来命名，因此嗜热棒囊壳 *Corynascus thermophilus* (Fergus & Sinden) Klopotek 应改为嗜热毁丝霉（van Oorschot，1977）；将嗜热棒囊壳、新几内亚棒囊壳 *C. novoguineensis* (Udagawa & Y. Horie) Arx、瘤孢棒囊壳 *C. sepedonium* (C.W. Emmons) Arx、有性棒囊壳 *C. sexualis* Stchigel, Cano & Guarro、相似棒囊壳 *C. similis* Stchigel, Cano & Guarro 和疣状棒囊壳 *C. verrucosus* Stchigel, Cano & Guarro 分别相应命名为弗格斯毁丝霉、新几内亚毁丝霉、瘤孢毁丝霉、有性毁丝霉、相似毁丝霉和疣状毁丝霉；同时结合形态描述，证实了棉毛毁丝霉应是锯齿栉霉的同物异名。Zhang 等（2014）用经典方法结合分子手段鉴定发表了水滴状毁丝霉。Yasmina 等（2015）对毁丝霉属及其有性型采用多基因构建了系统发育树并进行了重新评价，将该属分为 4 个属，其中毁丝霉属仅包含有性型为粪壳菌目毛壳科的那些种，共 5 个种：土黄毁丝霉、弗格斯毁丝霉、暗棕毁丝霉、印度毁丝霉和硫磺色毁丝霉。将新几内亚毁丝霉、瘤孢毁丝霉、有性毁丝霉和疣状毁丝霉移入棒囊壳属中，分别命名为新几内亚棒囊壳、瘤孢棒囊壳、有性棒囊壳和疣状棒囊壳 *C. verrucosus* Stchigel, Cano & Guarro。水滴状毁丝霉、异宗毁丝霉、黄褐毁丝霉和嗜热毁丝霉移入新建属嗜热疣霉属中，分别命名为水滴状嗜热疣霉 *Thermothelomyces guttulatus* (Yu Zhang & L. Cai) Y. Marín, Stchigel, Guarro & Cano、异宗嗜热疣霉 *T. heterothallicus* (Klopotek) Y. Marín, Stchigel, Guarro & Cano、黄褐嗜热疣霉 *T. hinnuleus* (Awao & Udagawa) Y. Marín, Stchigel, Guarro & Cano 和嗜热疣霉 *T. thermophilus* (Apinis) Y. Marín, Stchigel, Guarro &

Cano。本卷在介绍开发利用价值和分类历史时按照广义毁丝霉属进行总结和撰写，但在分类描述时按照最新分类进行我国毁丝霉属的撰写，毁丝霉属仅保留了其模式种 1 个种。

国内对毁丝霉属的研究非常有限。目前，据文献资料记载，仅见王冬梅等（2004，2007）、赵春青和李多川（2008）报道了 3 个新记录种，Zhang 等（2014）报道了一个新种，以及韩燕峰等（2016）报道了两个新记录种。但这些种都被移入嗜热疣霉属和栉霉属等属中，仅剩 1 个种，即土黄毁丝霉。

2. 毁丝霉属真菌的开发利用价值

该属真菌分布广泛，从目前研究来看，在多种地质土壤中都有分布；也能从河床污泥、堆肥堆、植物根系、发酵蘑菇基质、鸟巢、动物粪便、动物皮毛和鸟类的体表分离到，还有人从棕榈油污泥中分离到（Domsch *et al.*，1993；Vikineswary，1997；Cooney & Emerson，1964）。而这种强的环境适应能力与其产生不同的、特殊的酶及其他代谢产物有极大的相关性，显示了其独特的潜在应用价值。现已发现毁丝霉属真菌可以产生角蛋白酶、纤维素酶、漆酶、半纤维素酶、β-葡糖苷酶、淀粉水解酶等十来种酶类（Bhat & Maheshwari，1987；Maheshwari *et al.*，2000；Berka *et al.*，1997；Katapodis *et al.*，2003；Subramaniam *et al.*，1999；Canevascini *et al.*，1991；Topakas *et al.*，2004；Le Nours *et al.*，2003；Kaur *et al.*，2004；Johri *et al.*，1990；Adams，1997；Wyss *et al.*，1999；Sheehan & Casey，1993）。微生物来源的角蛋白酶是目前酶领域研究的热点之一。这类酶在医药、饲料、肥料、皮革、食品、纺织、日化等工业生产及环境治理方面都有广阔的应用前景（曾毅等，2004；赵小东等，2005）。角蛋白酶不可替代的重要作用已成为业界的共识，毁丝霉属真菌作为产角蛋白酶的真菌类群，在角蛋白酶生产中的潜在价值备受人们关注，有望从其中找到理想的产角蛋白酶菌株，尤其是毁丝霉对环境的适应能力强，用于工业生产时更具优势。此外，毁丝霉属大多数种是嗜热真菌，能够产生热稳定的酶而吸引人们的关注（Maheshwari *et al.*，2000；Badhan *et al.*，2007；Beeson *et al.*，2011；Bulter *et al.*，2003；Babot *et al.*，2011；Sadhukhan *et al.*，1992；Roy *et al.*，1990），因此其功能酶和细胞外代谢物具有潜在的巨大的开发价值，这也是人们目前关注的重点。目前国内外对毁丝霉属的研究主要致力于对该属新成员的发掘和探索，以及对已有种在医药、纺织、造纸、功能酶的工业化生产等方面的应用进行深入研究。

毁丝霉属真菌在医药研究开发利用等方面的研究现状及利用前景主要体现在以下四方面。

1）致病性

真菌对人或动物的感染因其治疗的顽固性，以及临床易被误诊而忽视的现象越来越受到人们的重视。人们试图通过对感染人畜真菌的种类、特性及感染症状、特点和机制的研究找到应对措施。毁丝霉属的一些种是人或动物的潜在的致病菌或机会致病菌，少数种会对人或动物造成严重的感染，对此类菌的研究将有助于人们对其危害进行防控。目前观察到毁丝霉属真菌感染人类的主要靶组织是肺、心脏、主动脉。研究表明，嗜热毁丝霉已经涉及许多罕见的引起人免疫缺陷的真菌感染。Bourbeau 等（1992）报道了一株嗜热毁丝霉感染了一个白血病患者的左心房、上行主动脉和肺动脉。Farina 等

（1998）报道了嗜热毁丝霉在一个囊性动脉中层坏死患者中引起了致命性的主动脉感染。毁丝霉属的某些菌株是人和动物的机会致病菌，并且是暗色丝孢霉病的病因之一。Tekkok 等（1996）第一次报道了毁丝霉属真菌是能引起脑脓肿患者颅内化脓的病原体。Destino 等（2006）研究显示嗜热毁丝霉可以通过外伤感染而造成严重的骨髓炎。

2）癌症检测和天然药物筛选

生物来源的新药研发，一直是生物技术领域研究的热点之一。毁丝霉在此领域有巨大的开发潜力，尤其是在癌症检测和抗癌活性物质研究方面。由于癌症治疗中手术和放疗的局限性，药物治疗被寄予了厚望。近年来，深部真菌感染的高死亡率和治疗上的困难越来越受到人们的关注，寻找高效、低毒、用于治疗真菌感染的天然药物成了医学科研工作者当前的紧迫任务之一。毁丝霉在新药物的开发中将是一类新的候选对象。目前，在毁丝霉中已发现许多含有抗癌、抗真菌、抗寄生虫等抗生素类物质。Morio 等（2011）研究显示，嗜热毁丝霉的感染模型可用于检测曲霉半乳甘露聚糖交叉反应的血清抗原，进而可用于检测血液恶性肿瘤。Lloret 等（2012）报道了一种从嗜热毁丝霉中分离到的漆酶能共价固定在 Eupergit C 250 L 的 U 形管中，产生 17 个特定的活动位点，每克固定化漆酶 80 U。这样的固定化漆酶在一个填充反应器里能去除内分泌的干扰。

3）在造纸和纺织工业中的应用价值

20 世纪以来生物技术在造纸和纺织工业中的应用日趋广泛，其中生物来源的纤维素酶和半纤维素酶在造纸过程中的生物制浆、生物漂白、废纸生物脱墨、改善浆料性能和树脂生物控制等方面，以及纺织过程中的棉花精炼、棉织物的整理等一系列处理中均有重要应用。毁丝霉属真菌几乎都能产生细胞外木质纤维素酶、漆酶等十多种酶类，而现在纺织的材料大多是选择纯天然的木材原料通过物理化学作用或通过生物发酵得到的。Claus 等（2002）研究表明，来自纤维素降解菌白腐菌、多孔菌和嗜热毁丝霉的漆酶可以对合成染料不同区段脱色。Pereira 等（2005）研究揭示了漆酶能生物氧化莫林类黄酮、毛地黄黄酮、芦丁、槲皮素，可以用来漂白棉花。Kim 等（2011）报道了漆酶介导的对邻二苯酚的氧化聚合在纺织品染色中的运用，能够对亚麻面料更好地染色。他们还报道了漆酶能够催化蛋白质-类黄酮共轭物的形成，这种共轭物能够改善自然蛋白质的性质，能够对亚麻纤维改性，增加亚麻纤维的机械强度，增强抗菌活性和紫外保护能力。

4）在功能酶的工业化生产中的应用

嗜热毁丝霉能产生耐热的酶，可以运用于高温工业生物处理过程中，这种酶的理化性质也可以用于耐热性酶在生物工程范例中的开发研究，β-葡糖苷酶、外切葡聚糖酶和漆酶等都能用于生物发酵分解纤维素。Roy 等（1991）发表了嗜热毁丝霉胞外 β-葡糖苷酶的提纯和鉴定方法，从生长在结晶纤维束上的嗜热毁丝霉上分离出了 β-葡糖苷酶。Sadhukhan 等（1990）报道了嗜热毁丝霉 D14（ATCC 48104）能够产生淀粉水解酶，这种水解酶的活性温度广泛（50～60℃），且在温度是 60℃和 pH 5.6 时，活性最高。Topakas 等（2012）研究了一种来自嗜热真菌中的嗜热毁丝霉的阿魏酸酯酶的表达、特征和结构造型。Zanphorlin 等（2010）描述了嗜热毁丝霉固态发酵生产的蛋白酶和深层发酵生产的蛋白酶特征。Liang 等（2011）对一株嗜热毁丝霉 GZUIFR-H49-1 产角蛋白酶进行了最佳条件优化，进一步提高了其产酶活性，为今后的开发利用提供了科学资料。

（八）奈尼兹皮菌属

奈尼兹皮菌属于 1961 年建立，其形态特征与小孢子菌属很相似，其中部分种曾为小孢子菌属的成员。该属的研究概况可参考本卷小孢子菌属。目前该属包括 17 个种，国内已报道 5 个种，即粉奈尼兹皮菌 *Nannizzia fulva* Stockdale、石膏样奈尼兹皮菌 *N. gypsea* (Nann.) Stockdale、弯奈尼兹皮菌 *N. incurvata* Stockdale、猪奈尼兹皮菌 *N. nana* (C.A. Fuentes) Y. Gräser & de Hoog 和桃色奈尼兹皮菌 *N. persicolor* Stockdale，这几个种属于亲动物性，均有致病性（de Hoog *et al.*，2017）。

（九）帕拉癣菌属

帕拉癣菌属于 2017 年建立，其模式种和部分成员原属于小孢子菌属。de Hoog 等（2017）对能引起皮肤癣菌病的真菌进行了多基因系统发育研究，研究结果将小孢子菌属中 2 个种移出建立该属，并以库柯帕拉癣菌 *Paraphyton cookei* (Ajello) Y. Gräser, Dukik & de Hoog 为模式种，同时发表了一个新种，即皮状帕拉癣菌 *Paraphyton cutaneum* Hubka, Kucerova, Gibas, Kubátová & Hamal。目前帕拉癣菌属包括 4 个种，分别为库柯帕拉癣菌 *P. cookei* (Ajello) Y. Gräser, Dukik & de Hoog、库柯勒文帕拉癣菌 *P. cookiellum* (Clercq) Y. Gräser, Dukik & de Hoog、皮状帕拉癣菌 *P. cutaneum* Hubka, Kucerova, Gibas, Kubátová & Hamal 和奇妙帕拉癣菌 *P. mirabile* (J.S. Choi, Y. Gräser, G. Walther, Peano, F. Symoens & de Hoog) Y. Gräser, Dukik & de Hoog。中国目前报道 2 种，分别为库柯帕拉癣菌和奇妙帕拉癣菌。

（十）瓶霉属

1. 瓶霉属的研究历史

瓶霉属由 Medlar（1915）建立，模式种为疣状瓶霉 *Phialophora verrucosa* Medlar。该模式种是从患病的人的皮肤上分离到的一株真菌。属名来源于产孢细胞是小的浅杯状，产孢瓶体具发育很好的围领。淡领瓶霉属 *Cadophora* Lagerb. & Melin 是 Lagerberg 等（1927）建立的另一个属，强调属的概念为产孢瓶体上产生内生的分生孢子。

Conant（1937）将淡领瓶霉属内的 8 个种与疣状瓶霉做了形态比较，认为是同属，遂将淡领瓶霉属作为瓶霉属的异名属，同时将原淡领瓶霉属内的 7 个种组合为瓶霉属的不同分类单位，将美国淡领瓶霉 *Cadophora americana* Nannf.更名为疣状瓶霉的同物异名。此外 Conant（1937）强调产孢瓶体上产生内生分生孢子，产孢梗顶端着生漏斗状或杯状产孢结构，是该属常见的特点。

Medlar（1915）最初描述模式种时，认为产生两种类型的分生孢子：第一种类型的孢子是典型的瓶梗孢子，半内生孢子；第二种类型的孢子为念珠形增厚的菌丝，它们不分成单个孢子。瓶霉属的概念因为疣状瓶霉的多态性而出现了混乱，Medlar（1915）认真观察了瓶生孢子的形成过程，他发现孢子在组织和水琼脂培养基上长出，同植物的萌芽过程非常近似，这种无性繁殖的特点和机制在该属内木生真菌中是不存在的。

为了使该属与黄油霉属 *Margarinomyces* Laxa、头孢霉属 *Cephalosporium* Corda 和轮枝菌属等产生瓶梗孢子的属区分，Kingma（1943）提出该属的概念应进一步修正，他特别强调了孢子形成特点在分类上的重要性，他不赞成从事病原真菌研究的一些学者将疣状瓶霉和单孢枝霉属 *Hormodendrum* Bonord.混为一谈（Moore & de Almeida，1936；Emmons & Carrión，1937；Carrión，1942）。

根据该属真菌具有性态和无性态 2 个完整阶段的特性，Emmons（1944）修订了该属。Mangenot（1952）描述了灰黑软盘菌 *Mollisia cinerella* Sacc. (Sacc.)和落叶松埋核盘菌 *Pyrenopeziza laricina* Rehm 的无性结构，同时还说明了这些真菌与该属中蜜色瓶霉 *Phialophora melinii* (Nannf.) Conant 和帚状瓶霉 *P. fastigiata* (Lagerb. & Melin) Conant 分离物在形态上的相似性。Ajello 和 Runyon（1953）发现了疣状瓶霉的培养物具有围领状结构，但是没能观察到其完整的发育过程。

随后 Le Gal 和 Mangenot（1956，1961）及 Hütter（1958）论述了该属的近似属——软盘菌属 *Mollisia* (Fr.) P. Karst.和埋核盘菌属 *Pyrenopeziza* Fuckel 的瓶霉状态问题。Hughes（1951）、Le Gal 和 Mangenot（1956）及 Moreau（1963）先后就该属产孢细胞在分类中的作用进行了研究和阐述，其中 Hughes 指出，根据有无产孢瓶体分属是非常困难的，但 Cain（1952）却建议只要具有产孢瓶体都应划归到该属下。Lemaire 和 Ponchet（1963）观察到 *Linocarpon cariceti* (Berk. & Broome) Petr.产生的子囊孢子可以萌发产生与根状瓶霉 *Phialophora radicicola* Cain 相似的分生孢子。

Schol-Schwarz（1970）重新修订了该属的属级概念，认为马加林霉属 *Margarinomyces* Laxa 是该属的异名属，而将那些从合轴产孢细胞上着生芽孢子的种划出该属。这样该属又进一步组合后接受了 12 个种及一个组合种——霍夫曼瓶霉 *Phialophora hoffmannii* (J.F.H. Beyma) Schol-Schwarz。之后，Cole 和 Kendrik（1973）对该属内 6 种木生真菌进行了详细描述并编制了分种检索表。Gams 和 Holubová-Jechová（1976）将属内产生楔形或梨形孢子的种建立为链状组（*Phialophora* sect. *Catenulatae* W. Gams），包括 9 个新种，其他产生椭圆形孢子并聚集在一起形成黏球的种保留在瓶霉组内。

Gams 和 McGinnis（1983）建立了一个新属——单胞霉属 *Phialemonium* W. Gams & McGinnis，包括介于瓶霉属和顶头霉属 *Acremonium* Link 的过渡种类。Iwatsu 等（1985）描述了类似于疣状瓶霉和仙客来瓶霉 *Phialophora cyclaminis* J.F.H. Beyma 的种——日本瓶霉 *P. japonica* Iwatsu & Udagawa。Cabello（1989）报道了与暗色瓶霉 *P. phaeophora* W. Gams 关系紧密的唐克瓶霉 *P. dancoi* Cabello。Millar（1990）从湖泊沉积物中分离到了深海瓶霉 *P. benthica* K.R. Millar。Nirenberg 和 Dalchow 于 1990 年将禾谷膜座霉 *Hymenula cerealis* Ellis & Everth 转到了该属下。

Matsushima（1993）发现了新种钩孢瓶霉 *Phialophora falcatispora* Matsush。Yamamoto（1994）报道了新种——大豆瓶霉 *P. sojae* K. Kobay., Hid. Yamam., Negishi & Ogoshi。de Hoog（1999）报道了新种匍匐瓶霉 *P. reptans* de Hoog。de Hoog 等（2000b）从人的皮肤真菌中分离到了欧洲瓶霉 *P. europaea* de Hoog, Mayser & Haase。Harringto 和 McNew（2003）对淡领瓶霉属的一些种类进行了 ITS 和 28S (LSU) rDNA 序列分析，结果表明其与近似属——瓶霉属在形态上明显不同。之后，又有许多种类被相继描述和报道（Jiang & Wang，2010；Wu & Zhang，2011；Liu *et al.*，2013）。Réblová 等（2013）对刺盾炱目 Chaetothyriales 真菌进行研究时，根据多基因系统发育及二级结构分析，将瓶霉属的迷瓶霉 *P. ambigua* P. Feng & de Hoog、欧洲瓶霉、橄榄瓶霉 *P. olivacea* W.

Gams、匍匐瓶霉和无柄瓶霉 *P. sessilis* de Hoog 移入杯梗孢属 *Cyphellophora* G.A. de Vries。Li 等（2017b）报道了瓶霉属 4 个新种：中华瓶霉 *Phialophora chinensis* Ya L. Li, de Hoog & R.Y. Li、塔尔达瓶霉 *P. tarda* Ya L. Li, de Hoog & R.Y. Li、膨大瓶霉 *P. expanda* Ya L. Li, de Hoog & R.Y. Li 和椭圆瓶霉 *P. ellipsoidea* Ya L. Li, de Hoog & R.Y. Li。目前，该属已报道 97 个名称，一些种名已被摒弃或组合到其他属，共余 49 种。许多种的有性阶段已知为顶囊壳属 *Gaeumannomyces* Arx & D.L. Olivier、锥毛壳属 *Coniochaeta* (Sacc.) Cooke 和紫胶盘菌属 *Ascocoryne* J.W. Groves & D.E. Wilson 及软盘菌属。

2. 瓶霉属真菌的经济意义

瓶霉属广泛分布于世界各地。该属内真菌的一些种常侵染人或动物，也腐生于植物或土壤中，有些为植物寄生菌并引起病害。因其特殊的营养需求，人们一直对其代谢产物进行研究。Aldridge 等（1974）报道了分离自兰格氏瓶霉 *Phialophora lagerbergii* (Melin & Nannf.) Conant 的烯酮类代谢产物，其分子式为 $C_{10}H_{10}O_4$。Lousberg 等（1976）报道了分离自星号瓶霉黄专化形种 *P. asteris* f. sp. *helianthi* Tirilly & C. Moreau 的氯化环戊烯酮类代谢产物(-)-cryptosporiopsin，其分子式为 $C_{10}H_{10}Cl_2O_4$，分子量为 246。该物质对向日葵的重要病原菌——核盘菌 *Sclerotinia sclerotiorum* (Lib.) de Bary 的生长有较强的抑制作用。Lousberg 和 Tirilly（1976）报道了分离自星号瓶霉 *Phialophora asteris* (Dowson) Burge & I. Isaac 的代谢产物 furasterin，其分子式为 $C_{10}H_9ClO_3$，分子量为 212，该物质对华丽腐霉 *Pythium splendens* Hans Braun 具有抑制作用。

Taylor 等（1985）报道了通过高压液相色谱分离出 gregatin A（图 5）。将格瑞瓶霉 *Phialophora gregata* (Allington & D.W. Chamb.) W. Gams NRRL 13198 培养在米饭培养基上，20℃，28 天，最终可从每千克米饭中分离到结晶状态的 gregatin A 247 mg。

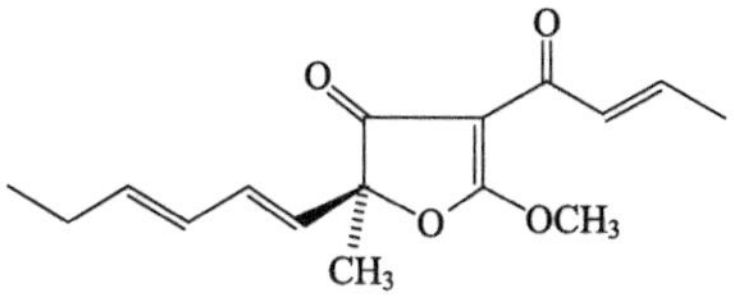

图 5　Gregatin A 的结构图（引自 Taylor *et al.*，1985）

Ayer 和 Trifonov（1994）报道了分离自白色瓶霉 *Phialophora alba* J.F.H. Beyma 的 4 种蒽醌类和蒽酚类化合物（图 6）。其中化合物 1 为 6,8-dihydroxy-1-methoxy-3-methylanthraquinone (1-*O*-methylemodin)，橙色固体，熔点为 257～260℃，之前未见天然存在的此种物质。化合物 2 为 5-chloro-6,8-dihydroxy-1-methoxy-3-methylanthraquinone，橘红色固体，熔点为 250～254℃，分子式为 $C_{16}H_{11}ClO_5$。化合物 3 为 7-chloro-6,8-dihydroxy-1-methoxy-3-methylanthraquinone，橙色固体。化合物 4 为 5-chloro-6,8,1-*O*-trihydroxy-1-methoxy-3-methyl-9(10H)-anthracenone，黄色固体，为二氢蒽酮，分子式为 $C_{16}H_{13}ClO_5$。化合物 1 和 4 均在浓度为 1 mg/mL 时对白杨腐烂病病原菌——窄盖木层孔菌 *Phellinus tremulae* (Bondartsev) Bondartsev & P.N. Borisov 的生长约有 50%的抑制率。

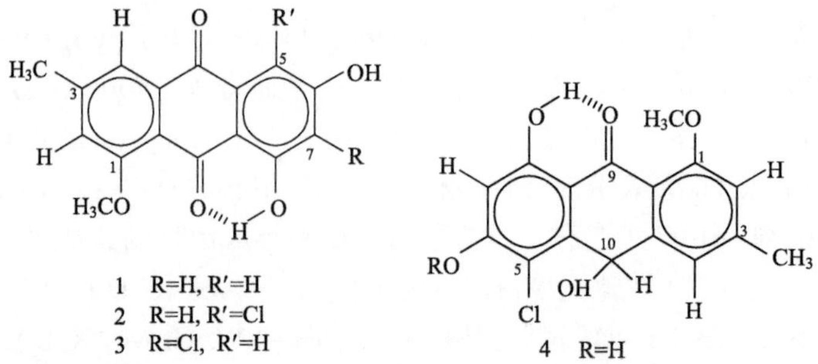

图6　化合物1～4的化学结构（引自 Ayer & Trifonov，1994）

Gray 等（1999）报道了分离自格瑞瓶霉的 2,3-dihydro-5-hydroxy-2-methyl-4H-1-benzopyran-4-one，并对其毒性进行了研究。该物质分离于格瑞瓶霉在 25℃条件下培养 21 天后的米饭培养基中，其可抑制酵母的生长，对大豆细胞未表现出高毒性且具有低的水溶性。

He 等（2012）报道了分离自一株内生真菌瓶霉（96-1-8-1）的 8 种化合物（图 7），其中包含两种新的 heptaketides，即(+)-(2S,3S,4aS)-altenuene（图 7：1a）、(−)-(2S,3S,4aR)-isoaltenuene（图 7：2a），同时还有 6 种已知物质，分别为(−)-(2R,3R,4aR)-altenuene（图 7：1b），(+)-(2R,3R,4aS)-isoaltenuene（图 7：2b），5′-methoxy-6-methyl-biphenyl-3,4,3′-triol（图 7：3），alternariol（图 7：4），alternariol-9-methyl ether（图 7：5）和 4-hydroxyalternariol-9-methyl ether（图 7：6）。其中化合物 1 为无色结晶体，(+)-(2S,3S,4aS)-altenuene（图 7：1a）和(−)-(2R,3R,4aR)-altenuene（图 7：1b），分子式为 $C_{15}H_{16}O_6$，其中包含 8 个不饱和键。化合物 2，(−)-(2S,3S,4aR)-isoaltenuene（图 7：2a）和(+)-(2R,3R,4aS)-isoaltenuene（图 7：2b）均为白色固体，分子式也为 $C_{15}H_{16}O_6$，其中也包含 8 个不饱和键。

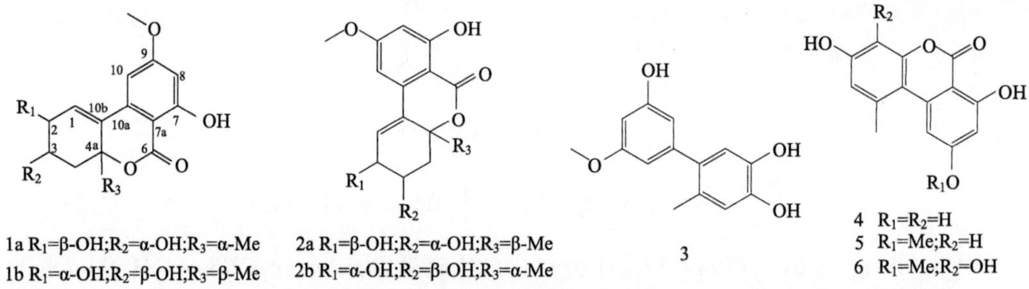

图7　化合物1～6的化学结构（引自 He *et al.*，2012）

Ye 等（2013）报道了分离自瓶霉属成员的代谢产物，共发现 3 种化合物（图 8），化合物 1 为 xinshengin（图 8：1），该化合物为第一个四环交链孢霉素类，并具有交链孢霉烯和四氢呋喃相连的骨架化合物，为白色无定形粉末，分子式为 $C_{20}H_{22}O_9$，其中包含 10 个不饱和键。化合物 2 为 phialophoriol（图 8：2），也为无定形粉末，分子式为 $C_{14}H_{14}O_5$，其中包含 8 个不饱和键。化合物 3 为已知化合物细格菌素（图 8：3），其对人癌细胞株 HL-60 和 A-549 的 IC_{50} 分别为 15.47 μmol/L 和 39.12 μmol/L。

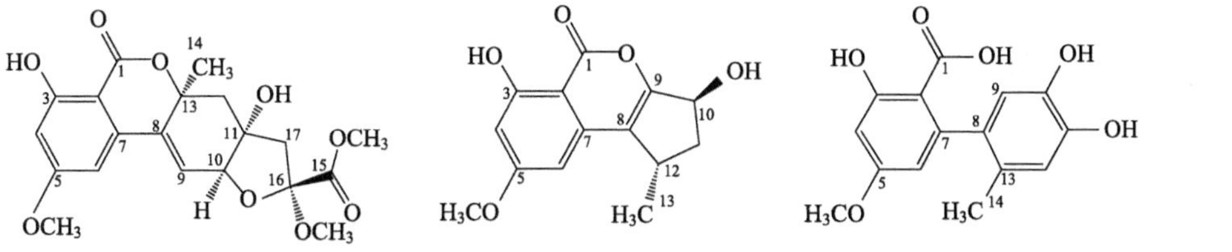

1(6/6/6/5)　2(6/6/5)　3(6-6)

图 8　化合物 1～3 的化学结构（引自 Ye *et al.*，2013）

Nalli 等（2015）报道了分离自慕斯特瓶霉 *Phialophora mustea* Neerg.的 4 种新的生物活性化合物（图 9），phialomustin A～D。其中化合物 1 为黄色黏性油状物，分子式为 $C_{25}H_{30}O_5$，结构组成为(S,2E,4E)-(S)-7-methyl-6,8-dioxo-3-(E)-prop-1-en-1-yl-7,8-dihydro-6H-isochromen-7-yl-4,6-dimethyldeca-2,4-dienoate，命名为 phialomustin A（图 9：1）。化合物 2 为无色油状物，分子式为 $C_{12}H_{20}O_2$，结构组成为(+)(2E,4E,6S)-4,6-dimethyldeca-2,4-dienoic acid，命名为 phialomustin B（图 9：2）。化合物 3 为黄色无定形粉末，分子式为 $C_{25}H_{28}O_7$，结构组成为(E)-3-((S)-7-(((S,2E,4E)-4,6-dimethyldeca-2,4-dienoyl)oxy)-7-methyl-6,8-dioxo-7,8-dihydro-6H-isochromen-3-yl)acrylic acid，命名为 phialomustin C（图 9：3）。化合物 4 也为黄色无定形粉末，分子式为 $C_{25}H_{28}O_8$，结构组成为(E)-3-((7S)-7-(((S,2E,4E)-4,6-dimethyldeca-2,4-dienoyl)oxy)-7-methyl-6,8-dioxo-1a,6,7,8-tetrahydroxireno[2,3-]isochromen-3-yl)acrylic acid，命名为 phialomustin D（图 9：4）。抗菌活性测试显示，化合物 3 和 4 显著抗白色念珠菌，其 IC_{50} 分别为 14.3 μmol/L 和 73.6 μmol/L，而化合物 2 对人类乳腺癌细胞系 T47D 具有显著的细胞毒性，其 IC_{50} 为 1 μmol/L。

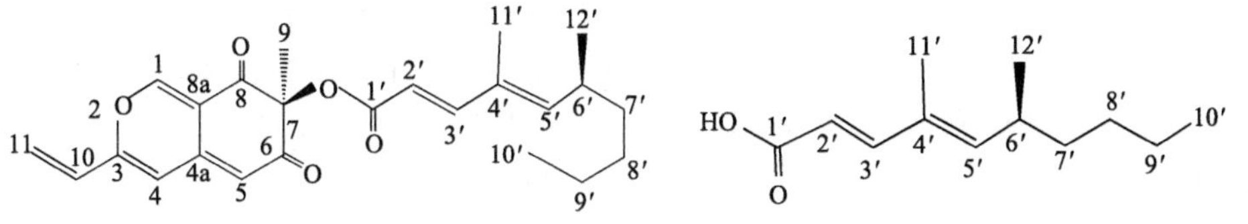

1　2

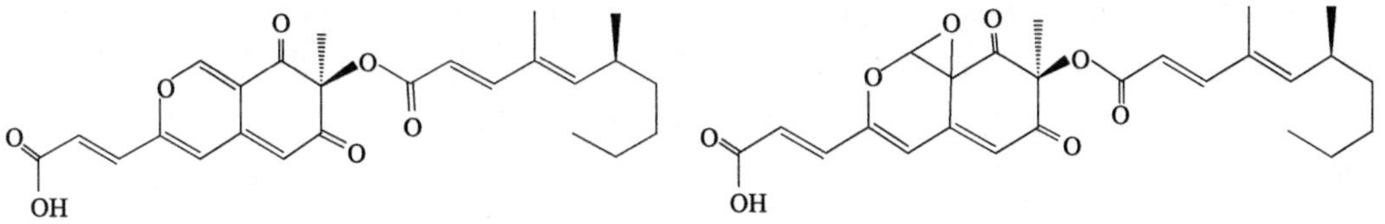

3　4

图 9　化合物 1～4 的化学结构（引自 Nalli *et al.*，2015）

该属的一些真菌常侵染人或动物，也腐生于植物或土壤中。有些为植物寄生菌并引起病害，如玉米瓶霉 *Phialophora zeicola* Deacon & D. B. Scott 作为玉米根部的一种寄生菌，可以引起根组织的腐烂（Deacon & Scott，1983）；星号瓶霉可引起紫菀 *Aster* sp. 的枯萎（Burge & Isaac，1974）；也有的种如寄生瓶霉 *P. parasitica* Ajello, Georg & G.J.K. Wang 既是一种人的皮下组织侵染菌，也可侵染多种树木（Hawksworth，1983）。

（十一）假裸囊菌属

1. 假裸囊菌属的研究历史

假裸囊菌属由 Raillo（1929）基于玫红假裸囊菌 *Pseudogymnoascus roseus* Raillo 及酒红假裸囊菌 *Pseudogymnoascus vinaceus* Raillo 建立，其与裸囊菌属 *Gymnoascus* Baran. 的主要区别特征为：该属物种菌丝外侧未出现刺状突起物。Apinis（1964）对玫红假裸囊菌进行了重新描述，并将假裸囊菌属视为裸囊菌属下的亚属。Cejp 和 Milko（1966）重新恢复了假裸囊菌属，并描述了新种高加索假裸囊菌 *P. caucasicus* Cejp & Milko。Samson（1972）评价了本属，并将酒红假裸囊菌处理为玫红假裸囊菌的同物异名，仅承认假裸囊菌属 2 个种，即玫红假裸囊菌和高加索假裸囊菌，此外还描述了 1 个新种，即巴蒂假裸囊菌 *P. bhattii* Samson。上述 3 个种均具有光滑的子囊孢子和地丝霉无性型，故作为属征与其他属相区别（Samson，1972；Orr，1979；Currah，1985）。Currah（1985）认为上述 3 个种为同一个种且玫红假裸囊菌具有优先权。Locquin-Linard（1982）及 Müller 和 von Arx（1982）分别描述了分离自阿尔及利亚牛粪的树状假裸囊菌 *P. dendroideus* Locq.-Lin.和欧石南根际土壤的高山假裸囊菌 *P. alpinus* E. Müll. & Arx，且这两个种均具有有凸纹的子囊孢子和未发育成熟的有性型结构。Udagawa 等（1993）提出裸星孢属 *Gymnostellatospora* Udagawa, Uchiy. & Kamiya，用于容纳子囊孢子具有纹路和缺乏有性型或有性型结构未能发育成熟的物种。Udagawa（1997）将树状假裸囊菌和高山假裸囊菌组合至裸星孢属，即树状裸星孢 *G. dendroidea* (Locq.-Lin.) Udagawa 和高山裸星孢 *G. alpina* (E. Müll. & Arx) Udagawa。Rice 和 Currah（2006）基于形态学特征和 ITS 序列分析，又报道了两个假裸囊菌属的新种，即附属假裸囊菌 *Pseudogymnoascus appendiculatus* A.V. Rice & Currah 和疣状假裸囊菌 *P. verrucosus* A.V. Rice & Currah。Minnis 和 Lindner（2013）基于多基因系统发育分析，将毁害地丝霉、卡尼侧孢 *Sporotrichum carnis* F.T. Brooks & Hansf.、毡状侧孢 *Sporotrichum pannorum* Link 组合至假裸囊菌属，即毁害假裸囊菌、卡尼假裸囊菌 *Pseudogymnoascus carnis* (F.T. Brooks & Hansf.) Minnis & D.L. Lindner 和毡状假裸囊菌 *P. pannorum* (Link) Minnis & D.L. Lindner。至此，假裸囊菌属包含 6 个种。Crous 等（2019）报道了林氏假裸囊菌 *P. lindneri* Rea, Smyth & Overton 和特纳假裸囊菌 *P. turneri* Rea, Smyth & Overton。Crous 等（2020）报道了分离自洞穴沉积物的新种——帕尔梅假裸囊菌 *P. palmeri* A.E. Rea-Ireland, Smyth, D.L. Lindner & Overton。Zhang 等（2020）报道了分离自土壤的 3 个新种：陕西假裸囊菌 *P. shaanxiensis* Zhi Y. Zhang, Y.F. Han & Z.Q. Liang、贵州假裸囊菌 *P. guizhouensis* Zhi Y. Zhang, Y.F. Han & Z.Q. Liang 和中国假裸囊菌 *P. sinensis* Zhi Y. Zhang, Y.F. Han & Z.Q. Liang，并将巴蒂假裸囊菌组合至裸星孢属，即巴蒂裸星孢 *Gymnostellatospora bhattii* (Samson) Zhi Y. Zhang, Y.F. Han & Z.Q. Liang。Villanueva 等（2021）报道了分离自极地生境的 4 个新种：南极假裸囊菌 *Pseudogymnoascus antarcticus* Vaca & R. Chávez、澳洲假裸囊菌 *P. australis* Vaca & R. Chávez、灰假裸囊菌 *P. griseus* Vaca & R. Chávez 和绒毛假裸囊菌 *P. lanuginosus* Vaca & R. Chávez。Zhang 等（2021b）报道了分离自中国城市土壤的 4 个新种：链状假裸囊菌 *P. catenatus* Zhi Y. Zhang, Y.F. Han & Z.Q. Liang、福建假裸囊菌 *P. fujianensis* Zhi Y. Zhang, Y.F. Han & Z.Q. Liang、云南假裸囊菌 *P. yunnanensis* Zhi Y.

Zhang, Y.F. Han & Z.Q. Liang 和浙江假裸囊菌 *P. zhejiangensis* Zhi Y. Zhang, Y.F. Han & Z.Q. Liang。随后，Zhang 等（2023a，2023b）又报道了分离自中国城市土壤的 5 个新种：葡状假裸囊菌 *P. botryoides* Zhi Y. Zhang, Y.F. Han & Z.Q. Liang、校园假裸囊菌 *P. campensis* Zhi Y. Zhang & Y.F. Han、香樟假裸囊菌 *P. camphorae* Zhi Y. Zhang, Y.F. Han & Z.Q. Liang、构树假裸囊菌 *P. papyriferae* Zhi Y. Zhang, Y.F. Han & Z.Q. Liang 和宗琦假裸囊菌 *P. zongqii* Zhi Y. Zhang, Y.F. Han & Z.Q. Liang。Becker 等（2023）报道了 *P. cavicola* F. Baert, E. D'hooge & P. Becker；同年，*Ciliciopodium hyalinum* Dasz.被移入该属，更名为 *P. hyalinus* (Dasz.) Minnis & D.L. Lindner，*Gymnoascus rhousiogongylinus* Wener & Cain 被移入该属，更名为 *P. rhousiogongylinus* (Wener & Cain) Minnis & D.L. Lindner；随后，Zhang 等（2024）发表了野生动物园假裸囊菌 *P. wildlifiparkensis* Zhi Y. Zhang & Y.F. Han；Childress 等（2025）报道了 3 个新种，即 *P. irelandiae* Childress & C.A. Quandt、*P. ramosus* Childress & C.A. Quandt 和 *P. russus* Childress & C.A. Quandt。Childress 等（2025）基于 ITS + LSU + *MCM7* + *RPB2* + *EF1A* 序列构建了该属的系统发育树（图 10），包含 29 个种。作者根据这些文献序列信息，构建了假裸囊菌属成员的系统发育树，包括 30 个种。至此，假裸囊菌属共有 30 个分类单元。

2. 假裸囊菌属真菌的研究现状

该属真菌目前共有 30 个种，主要分离自牛粪、洞穴沉积物、矿石沉积物、植物根际土壤、附生植物根际土壤和城市绿地土壤。大量类群还有待于系统调查研究。研究表明，该属的某些种是动物的病原菌，如毁害假裸囊菌是蝙蝠白鼻综合征的病原体，这种疾病已导致北美数百万蝙蝠死亡（Donoghue *et al.*，2015）。因此目前人们对该属主要在以下几个方面进行了研究。

1）生物学特性

从冬虫夏草子座上分离得到了菌种 Pseu-F，经形态学及分子鉴定，确定为玫红假裸囊菌。经温度试验研究，发现该菌株的适宜生长温度为 17.5～20.0℃；采用正交试验法对该菌株进行了液体发酵培养基筛选，试验因素包括马铃薯、黄豆、蔗糖+葡萄糖、蛋白胨、酵母膏、矿物盐、维生素等，筛选出优化的液体发酵培养基为（g/L）：蔗糖 20，葡萄糖 10，蛋白胨 10，酵母膏 5，黄豆 50，马铃薯 100（何苏琴等，2011）。

2）可产生不同的酶

（1）蛋白酶：Deepak 等（2017）对从一株假裸囊菌 BPF6 菌株中分离的蛋白酶进行了研究，发现其最适 pH 为 9.0，最适温度为 60℃；不同的金属离子和有机物对酶活影响不同，如 Ca^{2+}对酶的活性有轻微的抑制作用；Co^{2+}对酶的刺激作用最大；Zn^{2+}对酶的抑制作用最大，可使其活性降低 90%；乙二胺四乙酸（ethylenediaminetetra-acetic acid，EDTA）和十二烷基硫酸钠（sodium dodecylsulfate，SDS）分别使酶的活性降低 60%和 50%。白鼻综合征（WNS）是一种蝙蝠皮肤真菌病，是北美洞穴蝙蝠大量死亡的原因。研究表明，毁害假裸囊菌分泌的胞外蛋白酶在 WNS 感染过程中对翅膀坏死起到一定作用。经等电聚焦法和亲和层析法从该菌的胞外分泌液中分离得到一种丝氨酸蛋白酶，其活性被苯甲基磺酰氟（phenylmethylsulfonyl fluoride，PMSF）强烈抑制。因此丝氨酸蛋白酶是进一步研究 WNS 宿主-病原体相互作用的候选者（Pannkuk *et al.*，2015）。

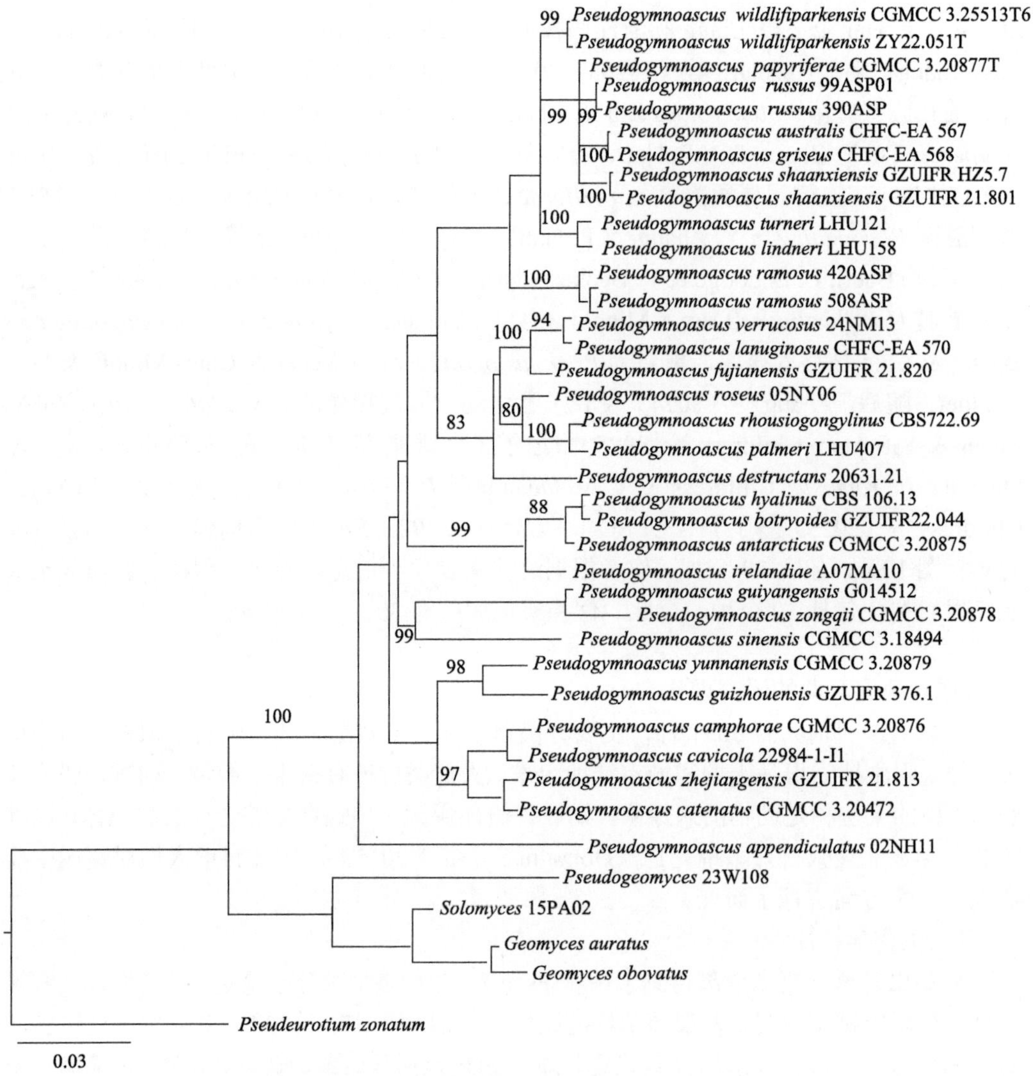

图 10　基于多基因序列构建的假裸囊菌属内各种的系统发育关系

（2）纤维素酶：韩龙等（2013）从四川省阿坝藏族羌族自治州黄龙沟的高山腐殖土中筛选出一株产纤维素酶的耐冷菌 HD1031，鉴定为玫红假裸囊菌。该菌可在 4～25℃条件下生长，最适生长温度为 16～17℃。研究发现，该菌在以微晶纤维素和玉米芯粉为碳源、硫酸铵和胰蛋白胨为氮源的培养基中，17℃、160 r/min 摇瓶发酵 8 天后产生纤维素酶，其中内切葡聚糖酶酶活为 366.67 U/mL，滤纸酶酶活为 87.6 U/mL，β-葡萄糖苷酶酶活为 90.8 U/mL，酶最适反应 pH 为 6.0，最适反应温度为 50℃。结果表明此菌株所产纤维素酶在 20～40℃条件下活性较高，对热敏感，具有低温纤维素酶的特点。

（3）葡聚糖酶和木聚糖酶：玫红假裸囊菌 HD1031 菌株进行低温发酵培养，可产生内切葡聚糖酶和木聚糖酶，其性质研究表明，内切葡聚糖酶的最适作用温度为 50℃、最适反应 pH 为 5.0，木聚糖酶的最适温度为 50℃、最适反应 pH 为 6.0（韩龙，2013）。

（4）氧化还原酶和水解酶：毁害假裸囊菌是造成蝙蝠白鼻综合征的病原体，经研究发现其可分泌氧化还原酶和水解酶；进一步研究后的结果表明，一个分泌的丝氨酸肽酶——destructin-1，是该菌分泌蛋白组的主要成分；该肽酶被纯化并显示可降解哺乳动物结缔组织的主要结构蛋白——胶原蛋白。此外，destructin-1 的化学抑制能阻止胶原蛋白在条件培养基中降解。因此，该研究表明丝氨酸内肽酶有助于真菌毁害假裸囊菌在蝙蝠上的侵袭性生长和组织破坏，该结果有望作为白鼻综合征治疗干预的潜在靶点（Donoghue *et al.*，2015）。

（5）脂解活性和脂肪酸的利用：毁害假裸囊菌作为病原体，导致几种穴居北美蝙蝠数量急剧下降。Gabriel 等（2019）从其脂解活性和脂肪酸的角度，阐述了该菌如何分解和利用蝙蝠翅膀膜中的脂质，进而引发侵袭性感染的过程。采用体外生长实验进行研究，用 3 种最常见的脂肪酸：油酸、软脂酸和硬脂酸组成蝙蝠皮脂。结果表明，在一些情况下观察到了形态上的差异，但没有观察到任何脂肪酸对平均生长有显著的影响。随后，Veselská 等（2020）也进行了此方面的研究，结果表明，脂肪酶是允许真菌菌丝消化和侵入皮肤的重要酶，毁害假裸囊菌偏爱蝙蝠皮肤和脂类中的含氮底物，这可能是引起感染的原因之一。

3）发现新化合物

Shi 等（2021a）从来自南极的一株嗜冷致病真菌假裸囊菌 HSX2#-11 菌株中分离出了一种新型聚酮——伪酚酮 A，以及 6 个已知的类似物。新化合物的结构被广泛地进行了光谱研究和单晶 X 射线衍射研究，发现化合物 1 和 2 表现出对一系列菌株的抗菌活性。Figueroa 等（2014）从膜虫属的南极海绵上分离到一株假裸囊菌，并从其培养物中获得了 4 个新的亚硝基甲酸衍生物，假裸囊菌素 A～C 和 3-nitroasterric acid，以及两个已知的化合物，单甲醚（questin）和吡嗪酰胺（pyrazinamide）。Shi 等（2021b）首次从一株假裸囊菌 HSX2#-11 中发现吡啶、嘧啶和双酮哌嗪类化合物，其中包含一个新的吡啶衍生物——4-(2-甲氧基羰基乙基)-吡啶-2-羧酸甲酯、一个嘧啶——胸腺嘧啶，以及 8 种二酮哌嗪，即环(dehydroAla-L-Val)、环(dehydroAla-L-Ile)、环(dehydroAla-L-Leu)、环(dehydroAla-L-Phe)、环(L-Val-L-Phe)、环(L-Leu-L-Phe)、环(L-Trp-L-Ile)和环(L-Trp-L-Phe)。之后，Shi 等（2021c）又从该菌 HSX2#-11 菌株中分离出 6 个新的倍半萜类（tremulanes A～F）、1 个已知类似物（11,12-环氧-12β-羟基-1-tremulen-5-one）和 5 个已知甾体，部分甾体化合物对人乳腺癌（MDAMB-231）、大肠癌（HCT116）和肝癌（HepG2）细胞具有细胞毒活性，对海洋污垢细菌——沙门氏菌也具有抗菌活性。这是首次在该属中发现萜类和甾类化合物。该属的分类、代谢和应用价值等还有待于继续深入研究和挖掘。

（十二）赛多孢属

1. 赛多孢属的研究历史

Saccardo（1911）从意大利一位足癣患者身上分离得到一个新种，并命名为尖端单孢菌 *Monosporium apiospermum* Sacc.。但该菌株仅出现无性阶段而未见有性阶段，因而归为半知菌。数年之后，Saccardo 建议使用赛多孢属命名上述真菌，但他未描述或正式提出新属用于容纳该种。Castellani 和 Chalmers（1919）确证赛多孢属的有效性，并将尖端单孢菌组合为尖端赛多孢。

Neveu-Lemaire（1921）将菌核单孢菌 *Monosporium sclerotiale* Pepere 组合到赛多孢

属并命名为菌核赛多孢 *Scedosporium sclerotiale* (Pepere) Neveu-Lem.。Conant 等（1954）认为尖端赛多孢、波氏假阿利什菌 *Allescheria boydii* Shear 和菌核单孢菌为同物异名，而尖端赛多孢为波氏假阿利什菌的无性型形式。

Malloch 和 Salkin（1984）报道了赛多孢属未描述的一个新种，该种分离自一位患者骨组织的活检标本中，并命名为膨大赛多孢 *Scedosporium inflatum* Malloch & Salkin。Guého 和 de Hoog（1991）研究发现膨大赛多孢与 Hennebert（1974）所报道的来自温室土壤的多育节荚孢霉 *Lomentospora prolificans* Hennebert & B.G. Desai 相同，而根据《国际藻类、菌物及植物命名法规》，多育节荚孢霉优先于膨大赛多孢，故而新组合为多育赛多孢 *Scedosporium prolificans* (Hennebert & B.G. Desai) E. Guého & de Hoog。

Gilgado 等（2005）采用分子系统学方法结合形态特征研究波氏假阿利什菌 *Pseudallescheria boydii* (Shear) McGinnis, A.A. Padhye & Ajello（无性型尖端赛多孢）复合种时，橘黄赛多孢 *Scedosporium aurantiacum* Gilgado, Cano, Gené & Guarro 独立为一亚分支，明显与该属其他种相区别，确证其为赛多孢属一新种。

Gilgado 等（2008）采用形态、生理及分子（微管蛋白）研究 141 株波氏假阿利什菌（无性型尖端赛多孢）复合种和其相近种时，命名了一新种——分生赛多孢 *Scedosporium dehoogii* Gilgado, Cano, Gené & Guarro，同时认为尖端赛多孢和波氏假阿利什菌为不同的种，因而命名为波氏赛多孢 *S. boydii* (Shear) Gilgado, Gené, Cano & Guarro 并作为有性型种波氏假阿利什菌的无性型。

Rainer 和 Kaltseis（2010）在研究分生赛多孢的多样性时，发现该种的一些临床分离株的 ITS 序列与模式菌株存在差异，而此前有大量的分生赛多孢分离自土壤，上述临床和土壤中的菌株 ITS 序列与 BT2（微管蛋白）的序列都聚成不同的两个分支。同时，在聚乙烯醇琼脂培养基中加入柴油、菜籽油和无添加时生长速率均不同，因而将原以为是分生赛多孢的临床分离株命名为新种——缺陷赛多孢 *Scedosporium deficiens* J. Rainer & Kaltseis。

Lackner 等（2012）在研究假阿利什菌属 *Pseudallescheria* Negr. & I. Fisch.、赛多孢属及其相关分类单元时，认为赛多孢属在假阿利什菌属前发表，因而新种小孢赛多孢 *Scedosporium minutisporum* (Gilgado, Gené, Cano & Guarro) Lackner & de Hoog 和沙生赛多孢 *S. desertorum* (Arx & Moustafa) Lackner & de Hoog 用于组合假阿利什菌属中的小孢假阿利什菌 *Pseudallescheria minutispora* Gilgado, Gené, Cano & Guarro 和皮特里霉属 *Petriellidium* Malloch 中的沙生皮特里霉 *Petriellidium desertorum* Arx & Moustafa。多育赛多孢在系统发育上与赛多孢属其他成员差异较大，因而重新使用其原命名——多育节荚孢霉。

Crous 等（2016）报道了赛多孢属一新种——蜡状孢赛多孢 *Scedosporium cereisporum* Rougeron, Stchigel, S. Giraud, Bouchara & Cano。韩燕峰等（2017b）报道了来自中国广西和海南的赛多孢属新种——少孢赛多孢 *S. rarisporum* Y.F. Han, H. Zheng, Y. Luo, Y.R. Wang & Z.Q. Liang 和三亚赛多孢 *S. sanyaense* Y.F. Han, H. Zheng, Y. Luo, Y.R. Wang & Z.Q. Liang。

Abrantes 等（2021）在对送往阿根廷国家实验室的 49 株赛多孢菌株进行表型验证、系统发育分析及体外抗菌药敏实验研究时，发现 92%的菌株都属于尖端赛多孢复合种，包括波氏赛多孢和尖端赛多孢，但有一株产生多胞分生孢子的菌株，而在已知种中仅有尖端赛多孢具有多胞分生孢子。基于系统发育分析，该菌株又与尖端赛多孢有区别，因而认为该菌株为一新种，随后命名为美国赛多孢 *Scedosporium americanum* Abrantes, Refojo, Hevia, J. Fernández & Stchigel。Zhang 等（2021b）在研究来自中国城市土壤中嗜角蛋白真菌多样性时，报道了 3 个赛多孢属新种：海口赛多孢 *S. haikouense* Zhi Y. Zhang, Y.F. Han & Z.Q. Liang，海南赛多孢 *S. hainanense* Zhi Y. Zhang, Y.F. Han & Z.Q. Liang 和多孢赛多孢 *S. multisporum* Zhi Y. Zhang, Y.F. Han & Z.Q. Liang。迄今赛多孢属有 18 个有效种。

2. 赛多孢属真菌的经济意义

该属真菌多发现于土壤中，营腐生生活，在热带地区较为少见。它们通常耐热，可在低氧条件下生存，并能耐受高盐和高渗透压的环境（Cortez *et al.*，2008；Guarro *et al.*，2006；de Hoog *et al.*，1994）。赛多孢通常出现在营养丰富的基质处，如土壤（Ajello，1953，1956；Niu *et al.*，2017）和家畜、家禽的粪便（Rippon & Carmichael，1976；Bell，1978）。同时也发现存在于蝙蝠、乌鸫等的粪便中（Brandsberg *et al.*，1969；Ajello *et al.*，1977；Vissiennon，1999），竹园和竹鼠的洞穴中（Gugnani *et al.*，2007），富营养的厚污泥中（Cooke & Kabler，1955；Fisher *et al.*，1982），污水处理厂的污水中（Janda-Ulfig *et al.*，2008），海岸、海水及咸淡水滩涂的污泥中（Dabrowa *et al.*，1964；Kirk，1967）。

由于赛多孢属真菌在富营养的土壤及水体中经常被发现，因而其作为潜在的生物修复剂被广泛研究。

1）脂肪族化合物的降解

波氏赛多孢 ATCC 58400 可在气态烷（又称气态饱和烃）如丙烷、正丁烷中正常生长（Zajic *et al.*，1969；Volesky & Zajic，1971；Davies *et al.*，1973）。该种在原油中生长良好，并可将中高分子量的烷烃（n-C_{12} 到 n-C_{26}）作为唯一的碳源和能量来源进行降解。波氏赛多孢将 n-烷烃转化为毒性较低的中间化合物。当烃类是唯一的碳源和能源时，波氏赛多孢菌株会优先降解脂肪族化合物（April *et al.*，1998，1999）。分离自委内瑞拉原油污染的土壤的两株赛多孢，尽管产生的降解木质素的胞外酶的能力较低，但可利用多芳香环烃作为唯一的碳源和能源，并能在芘、菲、二苯噻吩、萘和环己烷中生长（Naranjo *et al.*，2007）。

2）芳香族污染物的降解

赛多孢降解芳香污染物的能力最早从德国汉堡农业土壤中分离的菌株中得到证实，该菌株能在矿物盐培养基中将苯酚或对苯酚作为唯一的碳源和能源（Claußen & Schmidt，1998；Santos *et al.*，2003）。赛多孢也能生长在含有苯酚、1,2,4-三羟基苯、邻苯二酚、1,4-羟基苯醇、1,4-羟基苯甲醛和 4-羟基苯甲酸的培养基中。此外，赛多孢还能利用对甲基苯甲酸和 4-氯苯基苯甲酸（Claußen & Schmidt，1999）。脱色实验显示，赛多孢在培养 20 天后，脱色率达到 10%～25%（Tigini *et al.*，2014）。

3. 产酶及其他代谢物方面

Kuroda 等（1993）从一株赛多孢 SPC-15549 中发现了一种新的化合物——AS-183

（图 11），它可抑制酰基辅酶 A（胆固醇转移酶）的活性，其对胆固醇转移的半抑制浓度为 0.94 μmol/L。AS-183 对于 HepG2、CaCo2 和 THP-1 细胞的胆固醇酯形成的半抑制浓度分别为 18.1 μmol/L、25.5 μmol/L 和 34.5 μmol/L。

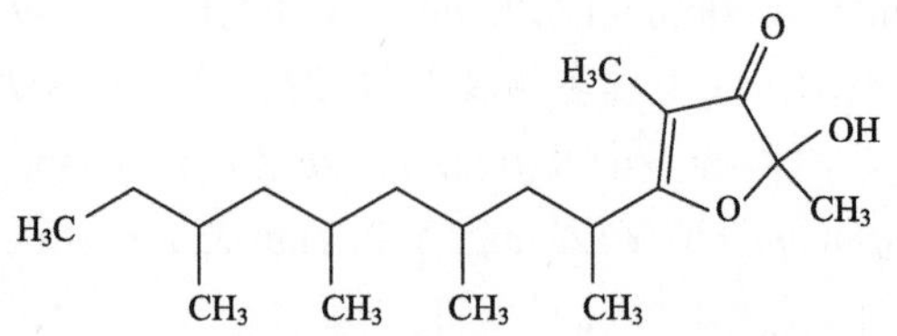

图 11　化合物 AS-183 的化学结构式（引自 Kuroda *et al.*，1993）

Larcher 等（1996）报道了分离自尖端赛多孢的一种新的丝氨酸蛋白酶，为一个分子质量约为 33 kDa 的单肽链。该酶的物理及生化性质与分离自烟曲霉的丝氨酸蛋白酶有许多相似之处。该酶未糖基化，等电点 pI 为 9.3，酶活的最适温度为 37～50℃，最适 pH 为 9.0，但在 pH 处于 7～11 时仍然保持高酶活。通过抑制部位和 N 端测序表明其属于枯草杆菌蛋白酶。其含有独特的合成物 *N*-琥珀酰-Ala-Ala-Pro-Phe-*p*-nitroanilide，其米氏常数为 0.35 mmol/L。

Pereira 等（2009）研究了尖端赛多孢菌株 HLPB 的分生孢子和菌丝所产生的蛋白质和肽酶，菌丝所合成的蛋白质的分子质量在 62～48 kDa 和 22～18 kDa 范围内，而分生孢子提取物中未检测到类似蛋白质。分生孢子产生一个分子质量为 28 kDa 的肽酶，能消化共聚白蛋白，而菌丝则可合成 6 个各不相同且分子质量为 90～28 kDa 的肽酶。上述所有的蛋白水解酶活性均在酸性 pH 及 1,10-菲咯啉下完全被抑制，这些特性表示其为金属类肽酶。

Han 等（2017）研究了橘黄赛多孢临床分离株和普通环境分离株所分泌的蛋白酶的差异，发现枯草杆菌蛋白酶和胰蛋白酶均存在。临床菌株包含肽酶 S8，其属于枯草杆菌蛋白酶家族，理论等电点 pI 为 8.2，理论分子质量为 41.2 kDa；假定亮氨酸肽 M28-Zn-肽酶，理论等电点 pI 为 5.22，理论分子质量为 53.7 kDa；PA SaNapH 类蛋白酶，理论等电点 pI 为 4.84，理论分子量为 52.7；肽酶 C48，理论等电点 pI 为 10.46，理论分子量为 100.3；天门冬氨酸类胃蛋白酶，理论等电点 pI 为 5.66，理论分子量为 51.8；类胰蛋白酶丝氨酸蛋白酶，理论等电点 pI 为 4.46，理论分子量为 32.7。环境分离株除了与临床株相同的肽酶 S8、假定亮氨酸肽 M28-Zn-肽酶、PA SaNapH 类蛋白酶，其所含的天门冬氨酸类胃蛋白酶理论等电点 pI 为 5.75，理论分子量为 103.4；还有一个未命名的蛋白酶 KLLA0E06711p，理论等电点 pI 为 6.94，理论分子量为 83.1。

Krasny 等（2011）报道了通过数据库搜索获得的赛多孢和阿利什菌的低分子量的代谢物，研究显示多育赛多孢 CBS 116904 所产生的重要脂肪酸，包括 4 种单己糖神经酰胺，元素组成分别为 $C_{41}H_{76}NO_8^+$、$C_{43}H_{80}NO_8^+$、$C_{41}H_{77}NO_9Na^+$、$C_{43}H_{81}NO_9Na^+$；11 种甘油磷脂类，元素组成分别为 $C_{40}H_{81}NO_8P^+$、$C_{42}H_{79}NO_8P^+$、$C_{42}H_{81}NO_8P^+$、$C_{42}H_{83}NO_8P^+$、$C_{44}H_{79}NO_8P^+$、$C_{44}H_{81}NO_8P^+$、$C_{44}H_{83}NO_8P^+$、$C_{44}H_{85}NO_8P^+$、$C_{46}H_{79}NO_8P^+$、$C_{46}H_{81}NO_8P^+$、$C_{44}H_{84}NO_8PNa^+$。

Hu 等（2016）研究了来自于海洋的分生赛多孢 F41-4 菌株产生的香木兰烷类似物，发现了 9 种化合物（图 12），其中 6 种为新的香木兰烷类倍半萜化合物 scedogiine A～

F（图 12：1～3，6～8）。scedogiine A（图 12：1）的元素组成为 $C_{15}H_{22}O_2$；scedogiine B（图 12：2）的元素组成为 $C_{15}H_{22}O_2$，但与 scedogiine A 的立体构型不一致；scedogiine C（图 12：3）的元素组成为 $C_{15}H_{22}O_3$；scedogiine D（图 12：6）的元素组成为 $C_{17}H_{24}O_3$；scedogiine E（图 12：7）与化合物 1 的元素组成相同，为 $C_{15}H_{22}O_2$，但立体构型存在差异；scedogiine F（图 12：8）的元素组成为 $C_{15}H_{22}O_2$；化合物 9 的元素组成为 $C_{20}H_{34}O_3$，为聚缩酮类的 scedogiine G（图 12：9）；pseuboydones A（图 12：4）和 pseuboydones B（图 12：5）为已知的倍半萜类化合物。

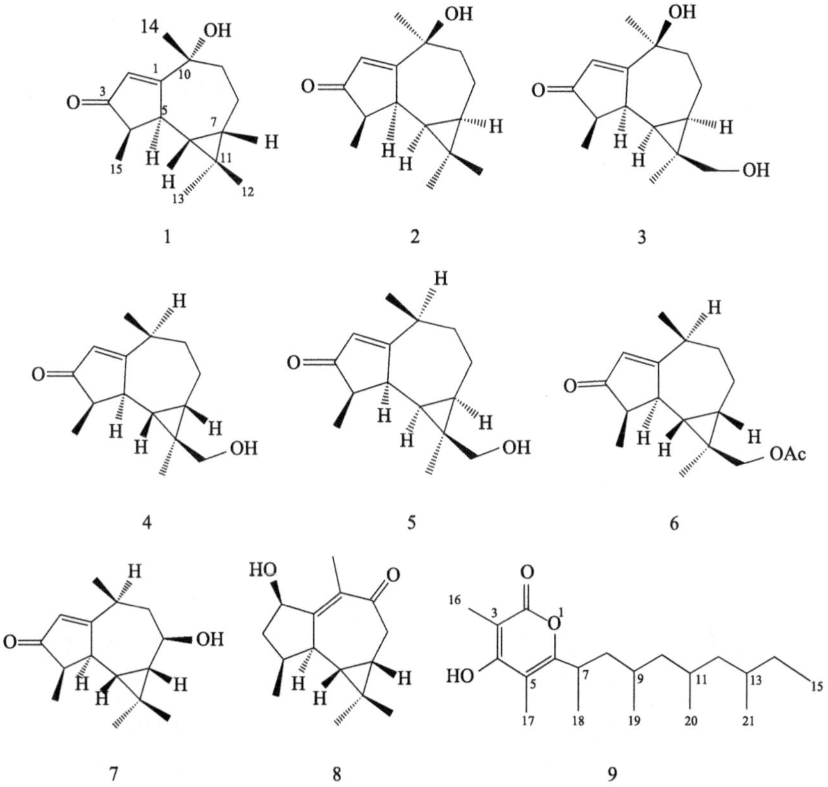

图 12　化合物 1～9 的化学结构式（引自 Hu *et al.*，2016）

Yu（2016）研究了赛多孢在体外产生的致病性物质，发现多育赛多孢可产生金合欢烯（farnesene）、异香树烯（alloaromadendrene）、没药烯（bisabolene）和牻牛儿烯（germacrene）；尖端赛多孢可产生柏木烯（cedrene）、香橙烯（aromadendrene）、芹子烯（selinene）和异牻牛儿烯（isogermacrene）；波氏赛多孢可产生柏木烯、愈创木烯（guaiene）、古芸烯（gurjunene）、倍半香桧烯水合物（sesquisabinene hydrate）。

Huang 等(2017)研究了通过氨基酸定向策略促进源于海洋的菌株尖端赛多孢 F41-1 菌株产生生物碱，共发现 21 个化合物，其中 14 个新化合物，分别为 scedapin A～G 和 scequinadoline A～G，其中 scedapin A 的元素组成为 $C_{26}H_{24}N_4O_5$；scedapin B 的元素组成与 scedapin A 相同，但立体构型存在差异；scedapin C 的元素组成为 $C_{28}H_{28}O_7N_4S$；scedapin D 的结构与 neosartoryadin A 一致；scedapin E 与从费希新萨托菌 *Neosartorya fischeri* (Wehmer) Malloch & Cain 分离得到的 tryptoquivaline T 的结构一致；

scequinadoline A 和 scequinadoline B 的立体构型存在差异；scequinadoline C 的元素组成为 $C_{28}H_{31}N_5O_4$；scequinadoline D 和 scequinadoline E 的元素组成分别为 $C_{27}H_{27}N_5O_4$ 和 $C_{28}H_{29}N_5O_4$；norquinadoline A 和 quinadoline A 与 scequinadoline D 相似，仅在 C-22 上有差异；scequinadoline F 与 quinadoline A 相似；scequinadoline G 的元素组成为 $C_{23}H_{20}N_4O_4$；scedapin F 和 scedapin G 为甲酰胺类，均含两个苯基、一个脂肪族的次甲基、一个甲酰胺和一个羰基酮。在上述化合物中，化合物 scedapin A～E 为稀有的由四氢呋喃环连接的 pyrazinoquinazolinedione 和 imidazoindolone 或吲哚酮；化合物 scedapin C 为第一例包含磺酰胺的 fumiquinazoline；化合物 scedapin C 和 scequinadoline D 对丙型肝炎表现出明显的抗病毒活性。

Staerck 等（2017）报道了分离自波氏赛多孢的聚酮化合物 boydone A，该化合物表现出对金黄色葡萄球菌的抑制作用。通过结构分析知道，该化合物含有 5 个不饱和键；17 个碳，包括 5 个甲基、4 个亚甲基、2 个次甲基和 6 个季碳。

（十三）帚霉属

1. 帚霉属的研究历史

帚霉属是 Bainier（1907）以短帚霉 *Scopulariopsis brevicaulis* (Sacc.) Bainier 为模式种建立的，同时还包含其他两个种，即朱红帚霉 *Scopulariopsis rubellus* Bainier 和淡红帚霉 *S. rufulus* Bainier。此外，该属还包含 3 个种，分别为：短帚霉、普通帚霉 *Scopulariopsis communis* Bainier、匍匐帚霉 *S. repens* Bainier。Westling（1911）强调了青霉属的种都产生绿色的分生孢子，因而确证短帚霉的正确性。

Thom 和 Raper（1932）等对该属进行了详细的描述，但他们都是把该属和青霉属紧密结合起来，称其产孢细胞为产孢小梗。Hughes（1953）明确指出帚霉属内所谓的产孢瓶体或产孢小梗在结构上与青霉属及曲霉属的产孢瓶体非常不同。该属的产孢结构为环痕式产孢梗，其明确含义描述为“产孢细胞随着分生孢子的产生而延长，每次都留下一个疤痕，这种产孢细胞称为环痕”。至此，该属的概念才明确。

在帚霉属中，分生孢子梗一般很短或分化不明显，形成不规则分枝或轮生，每个分枝的顶部产生环痕梗。通常情况下，环痕梗直接着生于表生菌丝或菌索上，虽然环痕梗在形态和大小上有很大变化，但在种内该特征是相对稳定的，因而环痕梗作为帚霉属属内区分种的主要依据。Morton 和 Smith（1963）将帚霉属内环痕梗大致分为 4 种类型。①短帚霉 *Scopulariopsis brevicaulis* 组：环痕梗近圆柱形，仅产孢的环痕区域稍窄些。②球孢帚霉 *S. sphaerospora* 组：环痕梗的基部宽，至顶部产孢的环痕带逐渐变窄。③布鲁姆帚霉组：瓶形的环痕梗，即基部或近基部明显膨大，顶部具较窄的环痕带。④绳生帚霉 *S. fimicola* 组：环痕梗非常长，而且较窄，类似粘鞭霉属 *Gliomastix* Guég.内的产孢瓶体。与多拉霉属 *Doratomyces* Corda 的环痕梗相比，多拉霉属的环痕梗很少圆柱形，一般基部膨大，然后骤然变窄形成环痕带，常呈锥形。

Morton 和 Smith（1963）发表的著作是研究帚霉属分类最重要的依据，尤其是对于本卷不产生有性阶段的种。在该著作中，共承认帚霉属 24 个种，分别为短帚霉、黄帚霉 *Scopulariopsis flava* (Sopp) F.J. Morton & G. Sm.、车叶草帚霉 *S. asperula* (Sacc.) S. Hughes、康宁帚霉 *S. koningii* (Oudem.) Vuill.、白帚霉 *S. candida* Vuill.、暗色帚霉 *S. fusca*

Zach、顶帚霉 *S. acremonium* (Delacr.) Vuill.、球孢帚霉 *S. sphaerospora* Zach、雪白帚霉 *S. nivea* Demelius、布鲁姆帚霉 *S. brumptii* Salv.-Duval、番红花帚霉 *S. croci* J.F.H. Beyma、加拿大帚霉 *S. canadensis* F.J. Morton & G. Sm.、黑帚霉 *S. carbonaria* F.J. Morton & G. Sm.、纸帚霉 *S. chartarum* (G. Sm.) F.J. Morton & G. Sm.、细小帚霉 *S. parvula* F.J. Morton & G. Sm.、绳生帚霉 *S. fimicola* (Costantin & Matr.) Vuill.、柏伦帚霉 *S. baarnensis* F.J. Morton & G. Sm.、嗜朊帚霉 *S. proteophila* Y. Horie & Udagawa、缓生帚霉 *S. tardifaciens* Y. Horie & Udagawa、林生帚霉 *S. silvatica* (Oudem.) Apinis、草生帚霉 *S. hibernica* A. Mangan、小麦帚霉 *S. tritici* K.B. Deshp. & K.S. Deshp.、玫瑰疹帚霉 *S. roseola* N. Inagaki 和辣椒帚霉 *S. capsici* J.F.H. Beyma。

许多研究已经通过培养、配接型及分子方法证明帚霉属的有性阶段为微囊菌属 *Microascus* Zukal（Curzi，1930，1931；Abbott *et al.*，1998；Abbott & Sigler，2001；Issakainen *et al.*，2003）。该属在建立之初，只有长喙微囊菌 *Microascus longirostris* Zukal 一个种，随后又有许多种被描述，它们的无性阶段都被认为是帚霉属。但微囊菌属与青霉属的有性阶段中发现的子囊壳明显不同。

Sandoval-Denis 等（2016）统计帚霉属容纳的种共 77 个，微囊菌属为 32 个，但是其中许多已描述的种都存在疑问，因为这些种的模式标本丢失或者原始文献描述得不够清楚。此外，新的《国际藻类、菌物及植物命名法规》不再允许同时出现有性型和无性型的双重命名。而在解决命名优先问题时，无论是在属级还是种级水平，都需要了解物种之间的关系，以及一个稳定的通用规则。因而在这种情况下，如果帚霉属和微囊菌属之间的命名是一致的，则优先考虑前面的帚霉名称。通过对帚霉属、微囊菌属及其相关属的研究，作者对帚霉属进行了总结：将匍匐帚霉、阿诺德帚霉 *Scopulariopsis arnoldii* (L. Mangin & Pat.) Vuill.、*S. ivorensis* H. Boucher、暗棕帚霉 *S. fusca* Zach 和玫瑰疹帚霉，视为车叶草帚霉的同物异名；将青霉状帚霉 *S. penicillioides* H.Q. Pan, Y.L. Jiang, H.F. Wang & T.Y. Zhang、康宁帚霉、朱红帚霉、人类帚霉 *S. hominis* (Brumpt & Langeron) Sartory、食虫帚霉 *S. insectivora* (Sopp) Thom、黄粪帚霉 *S. stercoraria* (Ehrenb.) S. Hughes 和黄白帚霉 *S. alboflavescens* Zach 视为短帚霉的同物异名；将短帚霉光滑变种 *S. brevicaulis* var. *glabra* (Thom) Thom 视为白帚霉的同物异名；将短帚霉白色变种 *S. brevicaulis* var. *alba* (Thom) Thom、金色帚霉 *S. aurea* Sartory、卡塞伊短帚霉 *S. casei* Loubière 和蝗帚霉 *S. grylli* Sartory 视为黄帚霉的同物异名；将普通帚霉视为顶帚霉的同物异名；将绳生帚霉视为粪生帚霉 *S. coprophila* (Cooke & Massee) W. Gams 的同物异名。

此外，作者还列举出帚霉属中应排除或是存在疑问的种，包括顶帚霉、银纹帚霉 *Scopulariopsis argentea* Svilv.、贝塔西尼帚霉 *S. bertaccini* Redaelli、加拿大帚霉、黑帚霉、卡氏帚霉 *S. castellanii* M. Ota & Komaya、粪生帚霉、芬氏帚霉 *S. finkii* Sartory & R. Sartory ex Vuill.、嗜盐帚霉 *S. halophilica* Tubaki、韩氏帚霉 *S. hanii* Moustafa & Abdul-Wahid、棉毛帚霉 *S. lanosa* J.F.H. Beyma、淡紫帚霉 *S. lilacea* Svilv.、舌色帚霉 *S. lingualis* Neto bis & C. Martins、长梗帚霉 *S. longipes* H.Q. Pan & T.Y. Zhang、马杜拉帚霉 *S. maduramycosis* Q.T. Chen、梅西瑞帚霉 *S. mencieri* C.W. Dodge、极小帚霉 *S. minima* Sartory, Hufschm. & J. Mey.、莫特帚霉 *S. mottai* Vuill.、芭蕉帚霉 *S. musae* Matsush、烟草帚霉 *S. nicotianae* J.F.H. Beyma、雪白帚霉、橄榄帚霉 *S. olivacea* Svilv.、小帚霉 *S. parva*

(A.H.S. Br. & G. Sm.) Samson、青霉状帚霉、多色帚霉 *S. polychromica* Svilv.、红斑帚霉 *S. rosacea* Svilv.、朱红帚霉、思念帚霉 *S. sehnsuchta* Mello、林生帚霉、多刺帚霉 *S. spinosa* E. Müll. & Pacha-Aue、针状帚霉 *S. sputicola* (Galippe) C.W. Dodge、小麦帚霉、韦内尔帚霉 *S. venerei* Greco、轮生帚霉 *S. verticilloides* Kamyschko、具瘤帚霉 *S. verrucifera* H.F. Wang & T.Y. Zhang、维尼鲁塔帚霉 *S. vignolo-lutatii* (Matr.) C.W. Dodge 和云南帚霉 *S. yunnanensis* Q.T. Chen & C.L. Jiang。

Woudenberg 等（2017）在研究室内环境出现的帚霉及其类似种时，报道了 4 个新种：非洲帚霉 *Scopulariopsis africana* Woudenb. & Samson、带白帚霉 *S. albida* Woudenb. & Samson、奶酪生帚霉 *S. caseicola* Woudenb. & Samson 和有性帚霉 *S. sexualis* Woudenb. & Samson。Zhang 等（2017b）研究喀斯特山洞中可培养真菌时，发现了帚霉属一新种：粗糙帚霉 *S. crassa* Z.F. Zhang, F. Liu & L. Cai。

2. 帚霉属真菌的经济意义

该属真菌分布广泛，植物种子、纤维物质、粪、土壤和腐殖质、空气等都可带该属真菌，还能参与纤维素、粪素的分解和腐殖质的形成（姜成林等，1985）。此外，帚霉属中的部分种被视为空气、储藏谷物、药品和食品中的污染物，其中短帚霉和白帚霉作为人类的致病菌，可引起人类指甲上的一些疾病，尤其是甲真菌病（Čapek *et al.*，1975）。在帚霉属的应用研究中，其产生的胞外酶较为多见，此外，帚霉属还可参与堆肥（Bosso *et al.*，2015）及作为植物病原菌的拮抗剂（de Stefano *et al.*，2002；Bosso *et al.*，2016）。

Anbu（2005）报道了分离自短帚霉的角蛋白酶的酶学性质。该酶通过硫酸铵沉淀法、DEAE-纤维素和葡聚糖凝胶 G-100 色谱技术进行分离纯化。纯化得到的酶经 SDS 聚丙烯酰胺凝胶电泳（SDS-polyacrylamide gel electrophoresis，SDS-PAGE）和凝胶过滤分离，得到分子质量分别为 39 kDa 和 36 kDa 的单体蛋白形成的角蛋白酶。最适产酶条件为：40℃、pH 为 8.0。同时，利用 Box-Behnken 设计实验，优化了角蛋白酶在 pH、温度和盐浓度的最优条件，结果表明，利用二阶式回归模型检测到最大酶活，其中 R^2 值为 0.9957，F 值为 178.32。预测得到 100%角蛋白酶产量的最佳条件为：5 mmol/L $CaCl_2$，pH 为 8.0，温度为 40℃。该酶活性可被苯甲基磺酰氟强烈抑制，表明该酶具有丝氨酸残基或附近有活性位点。

Anbu 等（2007）对印度纳马卡尔倾倒羽毛的土壤中的嗜角蛋白真菌进行调查，发现了 34 种真菌。在以鸡羽毛为唯一碳源和氮源的平板上生长时，大多数真菌在产胞外角蛋白酶时表现出了不同的效率。其中，包括短帚霉在内的 8 种真菌都可降解羽毛。在所有的这些种中，短帚霉在各种方法下均产生大量的角蛋白酶。对其产酶条件进行优化发现，其产酶活性最高可达 3.2 KU/mL，同时在鸡羽毛为培养基的条件下，降解率可达 79%。

Malviya 等（1992）对短帚霉的两种胞外角蛋白酶进行了纯化及部分酶学性质的研究，角蛋白酶的分离纯化采用凝胶过滤层析和 SDS-PAGE 技术。结果表明，角蛋白酶（KⅠ和 KⅡ）分别被纯化了 33 倍和 29 倍。凝胶过滤色谱产物（KⅠ和 KⅡ）的 DS-PAGE 上只产生一条带，表明它们具有同质性。角蛋白酶的最适 pH 为 7.8，最适温度分别 40℃（KⅠ）和 35℃（KⅡ）。通过估算得到它们（KⅠ和 KⅡ）的分子质量分别为 40～45

kDa 和 24～29 kDa。两种角蛋白酶都被磺氟化物所抑制，表明在活性部位或附近有丝氨酸残基。

Singh 和 Vézina（1971）报道了分离自短帚霉的胞外蛋白酶。该蛋白酶存在于短帚霉 *Scopulariopsis brevicaulis* 的液体培养液中，通过硫酸铵沉淀、DEAE-纤维素、DEAE-葡聚糖凝胶和羟磷灰石进行分馏。纯化得到的蛋白酶通过离心仪显示一个沉降成分，其分子量约为 24 000。该蛋白酶可水解酪蛋白、尿素变性血红蛋白、明胶、纤维蛋白、纤维蛋白原、胰岛素链 A 和 B，但不能水解人血清白蛋白或蛋白及牛奶。该酶对于各种多肽均无活性，同时表现出较低的酯酶活性。该酶降解酪蛋白、血红蛋白和明胶的最适 pH 分别为 10.5～11、7.0～9.5 和 6.0～8.0。该酶活性不受大多数金属离子、巯基（SH 反应）和一些天然胰蛋白酶抑制剂的影响，但受二异丙基氟磷酸盐（diisopropylfluorophosphate）和苯甲基磺酰氟的强烈抑制。虽然该酶在某些方面类似于从米曲霉中分离出的 CA-7 及其他碱性蛋白酶，但短帚霉与其中的任何一种都是不完全相同的。

Cai 等（2006）报道了分离自短帚霉的几丁质脱乙酰酶，并对其酶学性质进行了研究。研究发现，该酶是在以几丁质作为培养基中的唯一碳源时产生的。当该菌在 29℃条件下 200 r/min 培养 96 h 后，酶活可达每毫升培养液 10～11 单位。通过硫酸铵沉淀，交联葡聚糖凝胶 G-25 和 G-100 柱层析后，该酶纯化了 74 倍，回收效率提高了 38%。经 SDS-PAGE 和凝胶过滤色谱法测定，该酶的分子质量约为 55 kDa，且为单组分存在。该酶的活性糖基至少具有两个 *N*-乙酰氨基葡萄糖残基，但酶活性随着 *N*-乙酰氨基葡萄糖残基数量的增加而增大。当以 6-*N*-acetylchitohexaose 作为基质测定酶活时，该酶的最适 pH 为 7.5，最适温度为 55℃。在上述条件下，研究了该酶对水溶性壳聚糖、来自于黑曲霉的几丁质和虾晶甲壳素的酶活，并采用红外光谱、X 射线衍射（XRD）和电位滴定等对产物的结构进行了研究。结果表明，基质的结晶度对酶活具有重要影响。该酶对非晶甲壳素（37%脱乙酰作用）和水溶性壳聚糖（33%）有较高的乙酰活性，但对虾晶甲壳素具有较低的活性（3.7%）。

Afzal 等（2005）对分离自一株帚霉的木聚糖酶进行了研究。研究发现，两种吲哚-1,4-β-木聚糖酶通过硫酸铵沉淀、疏水作用、阴离子交换和凝胶过滤色谱的结合而分离。低酸性木聚糖酶（LAX）和高酸性木聚糖酶（HAX）的分子质量分别为 25 kDa 和 144 kDa，亚基的分子质量分别为 25 kDa 和 36 kDa。在 40℃，pH 为 6.5 的条件下，燕麦木聚糖的 LAX 和 HAX 的 Kcat 分别为 95 000 min^{-1} 和 9900 min^{-1}，LAX 和 HAX 的 K_m 值分别为 30 mg/mL 和 3.3 mg/mL。木聚糖水解的热力学活化参数表明，与 HAX 相比，LAX 的高活性不是由于 $\Delta H\#$的减少，而是熵驱动的。对降解产物的高效液相色谱分析表明，LAX 形成了 xylotrioses 和 xylobioses，而 HAX 主要形成 xylotrioses。在 50℃、pH 为 6.5、50 mmol/L 的 2-*N*-吗啉乙磺酸（MES）中，LAX 和 HAX 的半衰期分别为 267 min 和 69 min。热力学分析表明，在较低的温度下，LAX（$\Delta H\# = 306$ kJ/mol）比 HAX（$\Delta H\# = 264$ kJ/mol）的热稳定性高是由于更多非共价键的相互作用。在更高温度时，LAX[$\Delta S^* = -232$ J/(mol·K)]比 HAX[$\Delta S^* = 490$ J/(mol·K)]更耐热是由于其有更有序的过渡态构象。因而在涉及木聚糖水解的生物技术过程中，可同时使用耐热性和高度活性的 LAX。

Mudau 和 Setati（2008）报道了白帚霉两个菌株产生的吲哚-β-1,4-甘露聚糖酶。该种的两个菌株 LMK004 和 LMK008 分离自盐场，通过伏地尔培养基，以氯化钠和半乳甘露聚糖为碳源，可诱导产生甘露聚糖酶。LMK004 和 LMK008 所产生的甘露聚糖酶活性分别为 180 nkat/mL 和 116 nkat/mL。当采用纤维素作为碳源时，产酶活性明显下降。通过硫酸铵沉淀和阴离子交换色谱对两种酶进行了部分纯化。LMK004 和 LMK008 所产生的 β-甘露聚糖酶分子质量分别约为 41 kDa 和 28 kDa。LMK004 所产生的 β-甘露聚糖酶的最适 pH 和温度分别为 5 和 50℃，且在 pH 为 5～6.5，4℃时孵育 24 h 后，仍具有 80%的酶活。相比之下，LMK008 产生的 β-甘露聚糖酶在 pH 为 6～7 时，仅保留 60%的酶活性。两种酶在 30～40℃可保持酶活达 3 h，而在较高温度下则会使活性丧失。LMK008 产生的 β-甘露聚糖酶可耐受高浓度的氯化钠，其在 20%的氯化钠中孵育 2 h 后，酶活性保留 70%，而 LMK004 所产生的 β-甘露聚糖酶仅能在 0～10%氯化钠中保持活性。

Takeda 等（1994）报道了短帚霉（菌株为 N-01）会产生 1-蔗果三糖，并对该菌产生 1-蔗果三糖的最适条件进行了优化，优化后的条件为：500 mL 的圆锥烧瓶中，含有 25 mL 培养基，（每升培养基包含）150 g 蔗糖，15 g 酵母提取物，0.6 g 尿素，1 g K_2HPO_4，0.3 g $MgSO_4·7H_2O$（pH 7.0），30℃培养 72 h。当菌株在 2.5 L 发酵罐中发酵时，理论产生量为 85%，共产生 95 g 1-蔗果三糖。从上述培养液中，1-蔗果三糖通过结晶，再结晶纯度为 98%和 99.8%的收益率分别达到 71%和 78%。特别是再结晶时的第一个晶体纯度超过了 99.9%。

Sarao 等（2010）报道了顶帚霉深层液体发酵产纤维素酶和木聚糖酶，研究了温度、pH 和接种量等工艺参数对其的影响，并用异丙醇纯化这些酶，用于酶活性测定。结果表明，在 30℃孵育下，纤维素酶和木聚糖酶的活性最高分别可达 694.45 IU 和 931.25 IU。达到上述酶活的最适 pH 为 5.5，接种量为 $1×10^5$ 孢子/mL。纤维素酶和木聚糖酶的失活温度分别为 80℃和 70℃。

Hatakeyama 等（1996）研究了来自于短帚霉的 β-呋喃果糖苷酶的动力学参数。该酶经过 Toyopearl 650M、DEAE-葡聚糖凝胶 A-50、Toyopearl HW-55F 和羟基磷灰石的一系列柱层析色谱纯化后，从短帚霉的细胞提取物中的回收率为 4.6%，纯化了 1100 倍。该酶的最适温度和 pH 分别为 40℃和 6～9。该酶为同源二聚体蛋白，亚单位的分子质量为 110 000 kDa。动力学研究采用纯化后的 β-呋喃果糖苷酶，分别测定了 1-蔗果三糖、蔗果四糖、聚木糖，以及来自蔗糖、1-蔗果三糖、果糖、木糖和聚木糖的 difructosylxyloside 的 V_{max} 值。蔗糖可抑制蔗果四糖和 1-蔗果三糖的形成，葡萄糖可抑制 1-蔗果三糖、聚木糖和 difructosylxyloside 形成，且两者之间具有竞争性。所有上述寡糖的水解动力学常数均是在较低浓度下测定的，因为在较高浓度下，水解被抑制。

Takeda 和 Kinoshita（1995）报道了短帚霉产生的两种寡糖：fructosylxyloside（FX）和 fructosylfructosylxyloside（FFX）。其中，FX 为众所周知可抗龋齿的糖，而 FFX 则为一种新发现的糖。FX 可抑制不溶性葡聚糖的产生，其存在于口腔，可抑制由变形链球菌产生蔗糖而导致的龋齿发展，而 FFX 则不具有。通过培养条件优化，得到了寡糖的最适培养条件：80 g/L 蔗糖，80 g/L D-木糖，初始 pH 为 7，培养温度为 30℃。上述最适培养条件下可产生 50.3 g/L FX 及 9.8 g/L FFX。

Niyonzima 和 More（2014）报道了来自于帚霉产生的碱性蛋白酶。将该菌的碱性蛋白酶从凝集素-琼脂糖柱上回收，回收率为32.2%，含量为138.1 U/mg。通过SDS-PAGE测得该碱性蛋白酶的分子质量约为（15±1）kDa。该酶由单体糖蛋白构成，并经聚丙烯凝胶电泳为单一条带。该酶可在pH为8.0～12.0时保持酶活稳定，最适pH为9.0。具有酶活的最高温度为50℃，且保持24 h后，仍具有活性。Co^{2+}和Mn^{2+}浓度在10 mmol/L时可促进酶活力，但Ba^{2+}、Mg^{2+}、Cu^{2+}、Na^{+}、K^{+}和Fe^{2+}不影响酶活。Ca^{2+}和Fe^{3+}能适量降低酶活（约18%）。而Zn^{2+}和Hg^{2+}则可降低近40%的酶活。该碱性蛋白酶可被5 mmol/L的苯甲基磺酰氟（PMSF）完全抑制，以及部分被*N*-bromosuccinimide（NBS）和tocylchloride methylketone（TLCK）所抑制。因此，该酶的活性部位可能包含丝氨酸、色氨酸和组氨酸。与酪蛋白、纤维蛋白原、鸡蛋白蛋白、牛血清白蛋白相比，蛋白酶对明胶更有效。以酪蛋白为基质，该酶的K_m和V_{max}分别为4.3 mg/mL和15.9 U/mL。浓度为2%时，十二烷基硫酸钠（SDS）、吐温80和Triton X-100对该酶有活化作用。而H_2O_2和次氯酸则不影响酶活。在不同的存储温度（20℃、4℃和 28℃±2℃）测试下均具有较好的稳定性，且存储40天后可保留超过85%的酶活性。该酶作为商业酵素和酶洗涤剂（0.7%，*w*/*v*）时，处于28℃、60℃和90℃时，仍保留超过50%的酶活性。

Choi等（1998）报道了短帚霉产生的α-半乳糖苷酶。该酶的最适培养条件为：胰蛋白胨1.5%，硝酸铵0.2%，棉子糖2.5%，磷酸二氢钾0.5%，酵母提取物0.5%，pH 7.0，温度为27℃。上述菌株产α-半乳糖苷酶的最适pH和温度分别为7.0和27℃，其酶活在pH为6.0～8.0和温度低于40℃时，保持相对稳定。酶活受到金属离子Ag^{2+}、Hg^{2+}、Cu^{2+}及对氯汞苯甲酸和碘的抑制。这些结果表明在酶的催化部位存在SH基团。对于硝基苯基-α-D-吡喃半乳糖，其K_m值为1.9 mmol/L，V_{max}值为9.66×10^2 μmol/(L·min)。用高效液相色谱法测定了糖和糖中蔗糖、葡萄糖、果糖的酶解，该酶显著降低了发酵液的糖分成分。

Sharaf和Khalil（2011）报道了分离自短帚霉的碱性蛋白酶。通过SDS-PAGE对该碱性蛋白酶的纯度及分子量进行评价，得到该酶具有同质性，具有单一的蛋白质条带，且分子质量约为28.5 kDa。纯化蛋白酶的氨基酸显示，它是由14种氨基酸组成，谷氨酸（20.86%）、丙氨酸（14.52%）、甘氨酸（14.21%）、亮氨酸（8.59%）和丝氨酸（7.81%）的比例较高。该酶含有适量的缬氨酸（6.01%）、苏氨酸（5.58%）和苯基丙氨酸（5.22%）。纯化后的碱性蛋白酶具有很强的角蛋白酶活性，能够水解不同的角蛋白材料，在鸡羽毛上的活性最高，其次是人的指甲和人的毛发。

帚霉属的应用研究除了产酶，其活性物质也研究较多，主要包含环状胜肽物质、生物碱、萜类化合物及萘的衍生物等。Yu等（2008）报道了分离自短帚霉的两种新的scopularide A和scopularide B，同时，也从该菌中检测到已知的代谢物——蕈青霉素。通过核磁、质谱和化学衍生作用得到scopularide A和scopularide B的结构分别为cyclo-(3-hydroxy-4-methyldecanoyl-Gly-L-Val-D-Leu-L-Ala-L-Phe)和cyclo-(3-hydroxy-4-methyloctanoyl-Gly-L-Val-D-Leu-L-Ala-L-Phe)（图13）。根据计算得到元素组成为$C_{36}H_{58}N_5O_7$。抗菌实验显示对革兰氏阴性菌无抑制作用，而对革兰氏阳性菌的抑制作用弱，但是在浓度为10 μg/mL时对肿瘤细胞具有明显的抑制作用。

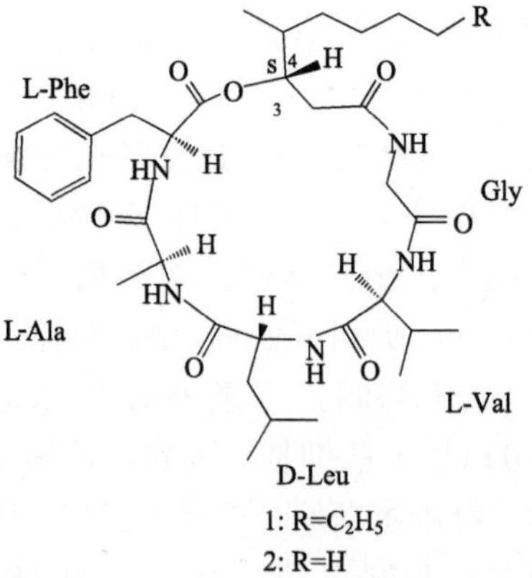

图 13　scopularide A（1）和 scopularide B（2）的结构式（引自 Yu *et al.*，2008）

Yang 等（2012）报道了分离自一株帚霉的两种新萘的衍生物：1-(4′-hydroxy-3′,5′-dimethoxy-phenyl)-1,8-dimethoxynaphthalen-2(1H)-one（图 14：1）和 1,8-dimethoxynaphthalen-2-ol（图 14：2）。前者为黄色的针状晶体，其元素组成为 $C_{20}H_{20}O_6$（其中包含 7 个不饱和键），其紫外光谱的最大吸收峰为 250 nm 和 336 nm；后者为黄色油状物，其元素组成为 $C_{12}H_{12}O_3$（其中包含 7 个不饱和键），其紫外光谱的最大吸收峰为 227 nm 和 285 nm。

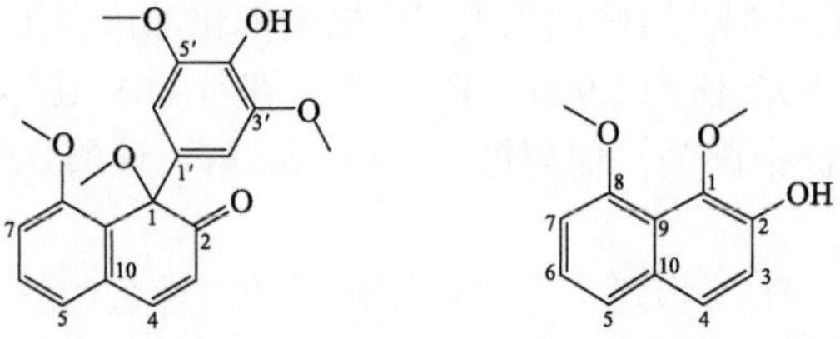

1　　2

图 14　分离自一株帚霉的两种新萘的衍生物的化学结构（引自 Yang *et al.*，2012）

Shao 等（2013）报道了分离自一株帚霉的生物碱 fumiquinazoline L（图 15：1）的结构和构型。该物质与柄曲霉素（sterigmatocystin）一起被发现，其元素组成为 $C_{26}H_{25}N_5O_4$(其中包含 17 个不饱和键),其与文献中已报道的两种生物碱 fumiquinazoline C（图 15：2）和 fumiquinazoline H（图 15：3）相关。最明显的区别在于存在不同的氨基酸残基，分别为缬氨酸、丙氨酸和亮氨酸。化合物 1 对于人肺癌细胞 A549 活性不显著（IC_{50}> 10 μmol/L），且在浓度为 100 μmol/L 时，对人肺腺癌细胞 SPCA1、人肝肿瘤细胞 BEL7402、人胃癌细胞 SGC7901 和人 erythromyeloblastoid 白血病细胞 K562 未表现出任何活性。同时化合物 1 在浓度为 50 μmol/L 时，表现出对致病菌枯草芽孢杆菌、金黄色葡萄球菌和溶血弧菌 *Vibrio parahaemolyticus* (Fujino *et al.*) Sakazaki, Iwanami & Fukumi 较弱的抑制性；而对蜡状芽孢杆菌 *Bacillus cereus* Frankland & Frankland、藤黄微球菌 *Micrococcus luteus* Cohn、四联微球菌 *Micrococcus tetragenus* Koch & Gaffky、大肠杆菌和鳗弧菌 *Vibrio anguillarum* Bergeman 未表现出任何抑制活性。

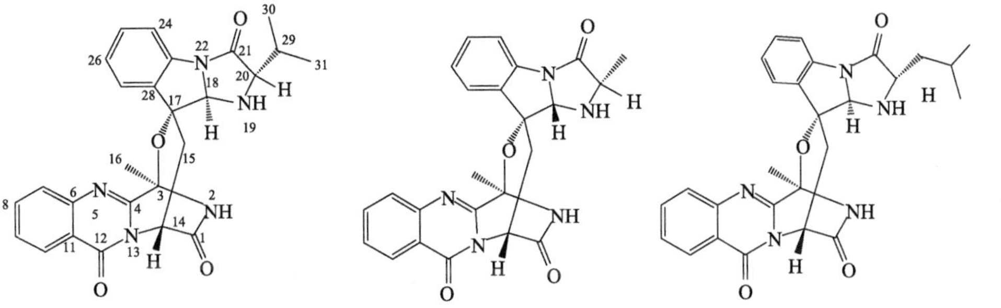

1　　2　　3

图 15　生物碱 fumiquinazoline L（1）、fumiquinazoline C（2）和 fumiquinazoline H（3）的化学结构（引自 Shao *et al.*，2013）

Tamminen 等（2014）对短帚霉深层培养产 scopularide A 的条件进行了优化。在生物反应器搅拌罐中，培养 90 h 可产生（29±3）mg/L scopularide A，无论是体积还是 scopularide A 的浓度相对于在摇瓶中的产量均有提高。当该菌生长在添加了 20 g/L 的葡萄糖而未添加硫酸铵的 YMP 培养基时，scopularide A 的产量减至 8.1 mg/L。当在添加了 20 g/L 葡萄糖的 YMP 培养基中加入 1 g/L 硫酸铵时，可明显提高 scopularide A 的产量（58 mg/L）。当该菌由非搅拌环境转移至搅拌环境时，产生 scopularide A 的菌丝量增加了近 6 倍，体积增加了近 29 倍。

Lukassen 等（2015）报道了从短帚霉基因组中合成 scopularide 的合成基因簇。scopularide A 是缩短的碳链（3-hydroxy-methyldecanoyl）连接了 5 个氨基酸（甘氨酸、L-缬氨酸、D-亮氨酸、L-丙氨酸和 L-苯丙氨酸）。采用最新测序得到的短帚霉的基因组序列，鉴定潜在的合成基因簇发现，scopularide A 的结构与来源于构巢曲霉 *Aspergillus nidulans* (Eidam) G. Winter 和小麦冠腐病菌 *Fusarium pseudograminearum* O'Donnell & T. Aoki 的 emericellamide A 类似。scopularide A 的合成基因簇包括一个核糖体肽合成酶（NRPS1）、一个聚合酶（PKS2）、一个 CoA 连接酶（ligase）、一个脂肪酰基转移酶（acyltransferase）和一个转录因子。短帚霉中的同源重组低，局部转录因子随机分布，从而导致 scopularide 的产量增加了 3～4 倍。

Elnaggar 等（2015）报道了分离自一株帚霉的新呫吨酮和倍半萜类衍生物。将该菌株培养于固体米饭培养基中，其乙酸乙酯提取物对 L5178Y 和 A2870 细胞具有细胞毒性，且表现出抗结核杆菌活性。乙酸乙酯提取物中发现了 7 种新的次级代谢产物，包括 3 种呫吨酮衍生物、2 种酚类、1 种药烷型倍半萜、1 种生物碱。此外还发现了 14 种已知化合物，包括 6 种呫吨酮衍生物、3 种药烷型倍半萜、4 种 2-苯基醚衍生物和 1 种麦角甾醇。

Elnaggar 等（2016）进一步研究了一株帚霉所产生的呫吨酮和倍半萜类衍生物。其中发现了两种新的呫吨酮衍生物，即 12-dimethoxypinselin 和 12-*O*-acetyl-AGI-B4；两种药烷型倍半萜，即 11,12-dihydroxysydonic acid 和 1-hydroxyboivinianic acid；一种新的生物碱，即 scopulamide；一种新的 α-吡喃酮衍生物，即 scopupyrone。此外还有 23 种已知化合物。化合物 1 为淡黄色固体，其元素组成 $C_{18}H_{16}O_8$，紫外最大吸收光谱为 203 nm、237 nm、265 nm 和 380 nm，与 pinselin 相似。化合物 2 为黄色固体，紫外光谱显示与

AGI-B4 相近，其元素组成为 $C_{18}H_{16}O_8$，分子量要高于化合物 3。化合物 15 为白色无定形粉末，紫外最大吸收光谱为 220 nm、246 nm 和 302 nm，与 sydonic acid 和 11-hydroxysydonic acid 类似，元素组成为 $C_{15}H_{22}O_6$，其比化合物 14 多一个氧原子。化合物 16 的元素组成为 $C_{12}H_{12}O_5$，其中包含 7 个不饱和键，其结构与 1-hydroxyboivinianin A 相似，不同在于其用羧基取代了芳香甲基。化合物 scopulamide 为黄色无定形粉末，元素组成为 $C_{16}H_{21}NO_6$，紫外最大吸收光谱为 205 nm、251 nm 和 299 nm，与 pyrenochaetic acid 衍生物相似。化合物 26 的紫外最大吸收光谱为 205 nm 和 300nm，显示具有 α-呋喃基团，其元素组成为 $C_{12}H_{18}O_4$，同时包含 4 个不饱和键。23 种已知化合物分别为 AGI-B4、huperxanthone C、pinselin、sydowinin B、13-*O*-acetylsydowinin B、2,11-dihydroxy-1-methoxycarbonyl-9-carboxylxanthone、sydowinin A、8-(methoxycarbonyl)-1-hydroxy-9-oxo-9H-xanthene-3-carboxylic acid、methyl-3,8-dihydroxy-6-methyl-9-oxo-9H-xanthene-1-carboxylate、sydowic acid、sydonic acid、11-hydroxysydonic acid、violaceol Ⅰ、violaceol Ⅱ、diorcinol、rikuzenol、lumichrome、WIN 64821、scopularide A、scopularide B、pyrenochaetic acid A、7-hydroxy-2,5-dimethylchromone 和麦角甾醇。与阳性对照 kahalalide F（IC_{50} 为 4.3 μmol/L）相比，化合物 AGI-B4、scopularide、violaceol Ⅰ和 violaceol Ⅱ表现出十分显著的细胞毒性，IC_{50} 分别为 1.5 μmol/L、1.2 μmol/L、9.5 μmol/L 和 9.2 μmol/L。

Mou 等（2018）报道了分离自一株帚霉中的一种新的单萜类化合物，即 dihydroquinolin-2(1H)-one（图 16）。该化合物为淡黄色固体，元素组成为 $C_{26}H_{31}NO_5$，其中包含有 12 个不饱和键，其与 aflaquinolone B 相似。该物质的抗污能力（EC_{50} = 0.103 μmol/L，LC_{50}/EC_{50} = 222）要高于阳性对照 SeaNine 211（EC_{50} = 4.36 μmol/L，LC_{50}/EC_{50} = 20）。因而该化合物具有强的抗污活性及高的治理效率。

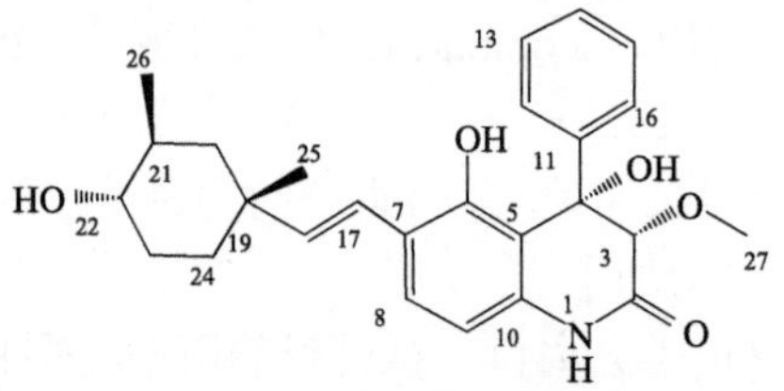

图 16　dihydroquinolin-2(1H)-one 的化学结构式（引自 Mou *et al.*，2018）

Elnaggar 等（2017）报道了一株帚霉产生 17 种不同的化合物（图 17），包括新的三萜化合物（图 17：1、2）及一种新的萘醌类似物（图 17：8）此外，还有 14 种已知化合物，其中包含 3 种三萜衍生物（图 17：3～5），两种倍半萜类化合物（图 17：6、7），两种聚酮化合物（图 17：9、10）和 7 种含氮化合物（图 17：11～17）。化合物 1 为白色无定形粉末，元素组成为 $C_{30}H_{46}O_6$，紫外最大吸收光谱为 249 nm，表明其包含一个共轭烯酮基团，并命名为 3β,7β,15α,24-tetrahydroxyolean-12-ene-11,22-dione。化合物 2 的元素组成为 $C_{30}H_{46}O_4$，其中包含 8 个不饱和键，命名为 15α,22β,24-trihydroxyolean-11,13-diene-3-one。化合物 8 为黄色无定形固体，紫外最大吸收光谱为 203 nm、269 nm 和 351 nm，元素组成为 $C_{12}H_{10}O_3$，其中包含 8 个不饱和键，命名为 6-hydroxy-2,7-dimethyl-

1,4-naphthoquinone。14 种已报道的化合物为 7β,15α,24-trihydroxyolean-12-ene-3,11,22-trione（图 17：3）、15α,24-dihydroxyolean-12-ene-3,11,22- trione（图 17：4）、soyasapogenol B（图 17：5）、(2E,4E)-4′-dihydrophaseic acid（图 17：6）、(2Z,4E)-4′-dihydrophaseic acid（图 17：7）、6-hydroxy-2,2-dimethyl-2H chromene（图 17：9）、scoparone（图 17：10）、5-methyluracil（图 17：11）、4-hydroxy-3-methoxy-2(1H)-quinolinone（图 17：12）、4-hydroxyphenylgly oxylic acid amide（图 17：13）、indole-3-carboxaldehyde（图 17：14）、indole-3-carboxylic acid（图 17：15）、(1H-indol-3-yl) oxoacetamide（图 17：16）和 *N*-acetyl-β-oxotryptamine（图 17：17）。

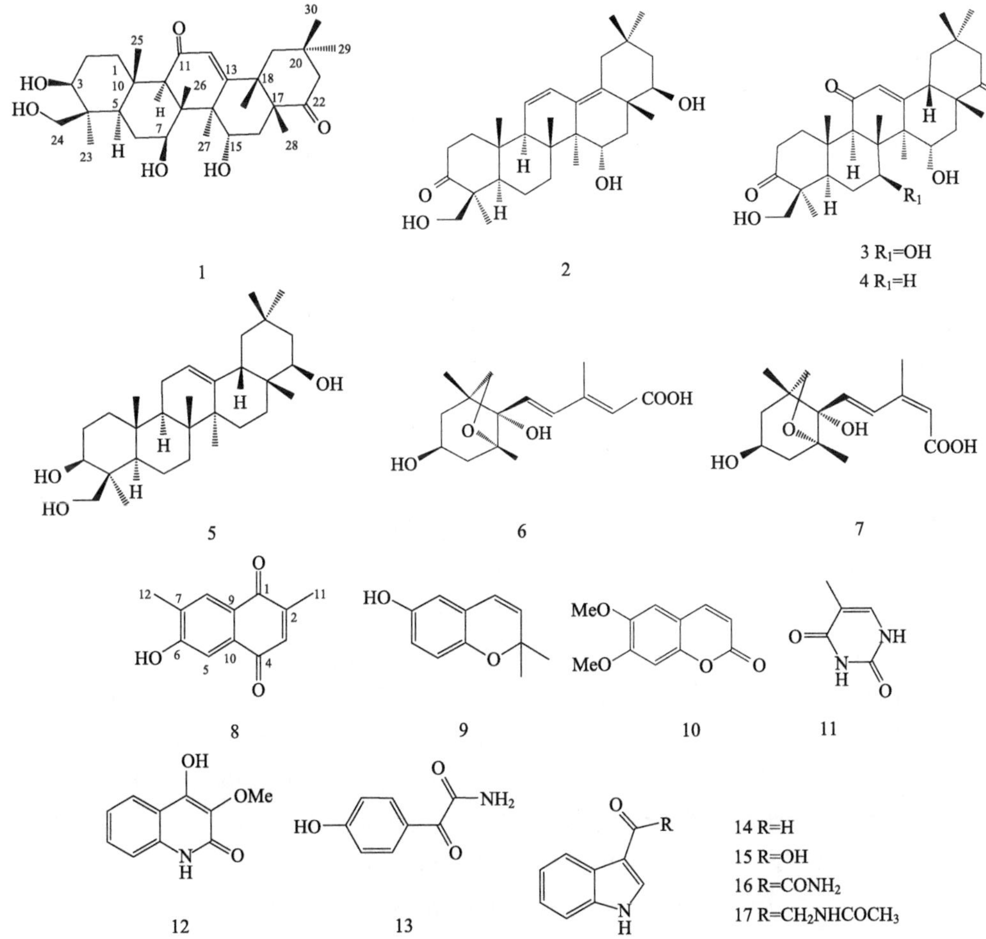

图 17　化合物 1～17 的化学结构式（引自 Elnaggar *et al.*，2017）

（十四）嗜热疣霉属

嗜热疣霉属于 2015 年建立，其大多成员原属于毁丝霉属，隶属于子囊菌门粪壳菌纲粪壳菌目毛壳科。Yasmina 等（2015）根据系统学研究，将毁丝霉属的 4 个种移入该属，后 Wang 等（2022）发表 1 个新种，并重新描述弗格斯毁丝霉为弗格斯嗜热疣霉。因此目前该属有 6 个种，分别为弗格斯嗜热疣霉 *Thermothelomyces fergusii* X. Wei Wang & Houbraken、水滴状嗜热疣霉 *T. guttulatus* (Yu Zhang & L. Cai) Y. Marín, Stchigel,

Guarro & Cano、异宗嗜热疣霉 *T. heterothallicus* (Klopotek) Y. Marín, Stchigel, Guarro & Cano、黄褐嗜热疣霉 *T. hinnuleus* (Awao & Udagawa) Y. Marín, Stchigel, Guarro & Cano、球状嗜热疣霉 *T. myriococcoides* X. Wei Wang & Houbraken 和嗜热疣霉 *T. thermophilus* (Apinis) Y. Marín, Stchigel, Guarro & Cano。该属形态与毁丝霉属类似，同时其研究应用现状参见毁丝霉属。

（十五）发癣菌属

1. 发癣菌属的研究历史

发癣菌属是 Malmsten 在 1848 年以断发发癣菌 *Trichophyton tonsurans* Malmsten 为模式种建立的。之后，发癣菌属新的分类单元不断被报道。Emmons（1934）对皮肤真菌基于纯形态学的方法进行了整理，为癣菌的分类学基础研究奠定了基础。Dodge（1935）进一步对之前报道的大量皮肤真菌进行了整理，仅承认 3 个属的 19 个种。其中包括发癣菌属的 11 个种，分别为白色发癣菌 *Trichophyton album* Sabour.、同心发癣菌 *T. concentricum* R. Blanch.、脱毛发癣菌 *T. epilans* Mégnin、铁锈色发癣菌 *T. ferrugineum* (M. Ota) Langeron & Miloš. ex Talice、马尼氏发癣菌 *T. megninii* R. Blanch.、须癣发癣菌、红色发癣菌、萨氏发癣菌 *T. sabouraudii* R. Blanch.、许兰氏发癣菌、硫磺色发癣菌 *T. sulfureum* Sabour. & Fox 和紫色发癣菌。

Vanbreuseghem（1952）以阿耶罗角蛋白霉 *Keratinomyces ajelloi* C.O. Dawson & Gentles 为模式种建立了新属角蛋白霉属 *Keratinomyces* Vanbreus.。Ajello（1968）认为角蛋白霉属的建立是不必要的，因为该属的属征已被包括在发癣菌属内，应作为发癣菌属的同物异名，因而组合了发癣菌属的 3 个种：阿耶罗发癣菌 *Trichophyton ajelloi* (Vanbreus.) Ajello、长梭形发癣菌 *T. longifusum* (Flórián & Galgoczy) Ajello 和阿耶罗发癣菌猪变种 *T. ajelloi* var. *nanum* (J. Kunert & Hejtm.) Ajello；将萨氏发癣菌、脱毛发癣菌和硫磺色发癣菌视为断发发癣菌的同物异名；将白色发癣菌视为疣状发癣菌的一个变种（Georg，1950；Ainsworth & Georg，1954）；将铁锈色发癣菌重新组合到小孢子菌属；将鸡禽发癣菌 *T. gallinae* (Mégnin) Marg. Silva & Benham 错误置于小孢子菌属；将马发癣菌视为须癣发癣菌的同物异名；将猴发癣菌从皮诺耶拉属 *Pinoyella* Castell. & Chalm. 组合到发癣菌属（Castellani & Chalmers，1919）。Vanbreuseghem（1950a，1950b，1951）和 Vanbreusegheme 等（1966）对 Langeron 和 Milochevitch（1930b）所提出的分类系统进行了修正，为容纳苏丹发癣菌 *Trichophyton soudanense* Joyeux 而建立了新属兰氏属 *Langeronia* Vanbreus.。Ajello（1968）认为兰氏属是没有必要的，其模式种苏丹兰氏 *L. soudanensis* (Joyeux) Vanbreus.根据 Joyeux（1912）的原始描述，应属于发癣菌属，同时由于缺少拉丁文描述，因而该种不算有效发表。此外，发癣菌属的一些分类单元也应该排除，包括柠檬发癣菌 *Trichophyton citreum* Szathmáry、依沃生发癣菌 *T. evolceanui* H.S. Randhawa & R.S. Sandhu、河泥发癣菌 *T. fluviale* Szathmáry、奔马发癣菌 *T. gallopavum* Metianu, A. Lucas & Drouhet、印度发癣菌 *T. indicum* H.S. Randhawa & R.S. Sandhu、库瑞安发癣菌 *T. kuryangei* Vanbreus. & S.A. Rosenthal、产脂发癣菌 *T. lipoferum* Kominami、奥塔发癣菌 *T. otae* Dowding、伯维斯发癣菌 *T. pervesi* Catanei、牧草发癣菌 *T. pratense* Szathmary、长梗发癣菌 *T. radicosum* Catanei、罗氏发癣菌 *T. rodhainii*

Vanbreuseghem 和雅温德发癣菌 *T. yaoundei* G. Cochet & Doby-Dub.。

Gräser 等（1999）采用 ITS 序列、PCR 指纹图谱和 AFLP 对须癣发癣菌和断发发癣菌进行分子分类。通过上述方法确证了 Georg 的判断，将网发癣菌 *Trichophyton areolatum* Negroni 视为断发发癣菌的同物异名，巴吞鲁日发癣菌 *T. batonrougei* Castellani（Castellani，1939）和枝状发癣菌 *T. candelabrum* Listemann（Listemann，1973）视为指间发癣菌 *T. interdigitale* Priestley 的同物异名；将扁发癣菌 *T. depressum* L. MacCarthy（MacCarthy，1925）视为须癣发癣菌的同物异名，马发癣菌自养变种 *T. equinum* var. *autotrophicum* J.M.B. Sm., Jolly, Georg & Connole（Smith *et al.*，1968）视为马发癣菌的同物异名；将刺猬发癣菌 *T. erinacei* (J.M.B. Sm. & Marples) Quaife 视为须癣发癣菌的一个变种（Smith & Marples，1964），结论和其他作者一样，不将其视为一个独立的种（Padhye & Carmichael，1971）；花状发癣菌 *T. floriforme* Beintema 在形态特征上与断发发癣菌一致，分子印证了 Georg 的结果（Georg，1956）；将瑞氏发癣菌 *T. krajdenii* J. Kane, J.A. Scott & Summerb（Kane *et al.*，1992）视为指间发癣菌的同物异名；将兰格发癣菌 *T. langeronii* (E.A. Baudet) Nann.（Baudet，1930；Milochevitch，1931）视为须癣发癣菌的同物异名；将须癣发癣菌戈茨变种 *T. mentagrophytes* var. *goetzii* Hantschke（Hantschke，1969）重新鉴定为指间发癣菌；将须癣发癣菌结节变种 *T. mentagrophytes* var. *nodulare* Georg & Meachling（Georg & Maechling，1949）视为指间发癣菌的同物异名；将多疣发癣菌 *T. papillosum* Lebasque（Lebasque，1934）视为须癣发癣菌的同物异名；将多育发癣菌 *T. proliferans* M.P. English & Stockdale（English & Stockdale，1968）视为刺猬发癣菌的同物异名；将圆形发癣菌 *T. rotundum* L. MacCarthy（MacCarthy，1925）视为指间发癣菌的同物异名；将沙凯斯维发癣菌 *T. sarkisovii* L.G. Ivanova & I.D. Poljakov（Ivanova，1983）视为须癣发癣菌的同物异名；将枣红发癣菌 *T. spadiceum* (Katoh) Nannizzi（Pollacci & Nannizzi，1930）视为红色发癣菌的同物异名；将硫磺色发癣菌（Sabouraud，1910）视为断发发癣菌的同物异名，且结果与 Ajello（1974）一致；将疣状发癣菌自养变种 *T. verrucosum* var. *autotrophicum* D.B. Scott 视为指间发癣菌的同物异名。因此，发癣菌属的 24 个种和变种就组合成了 6 个，即断发发癣菌、指间发癣菌、须癣发癣菌、猴发癣菌、马发癣菌和刺猬发癣菌。

Nenoff 等（2007）将须癣发癣菌作为猴发癣菌的同物异名，并作为猴节皮菌 *Arthroderma simii* Stockdale, D.W.R. Mack. & Austwick 的无性型。Heidemann 等（2010）认为猴发癣菌不应作为须癣发癣菌的同物异名，而应该作为单独的一个种，命名为须癣发癣菌昆克纳变种 *Trichophyton mentagrophytes* var. *quinckeanum* (Zopf) J.M.B. Sm. & Austwick。此外，其认为同心发癣菌、须癣发癣菌刺猬变种 *T. mentagrophytes* var. *erinacei* J.M.B. Sm. & Marples 和疣状发癣菌应为系统发育种而单独存在。

de Hoog 等（2017）对皮肤真菌采用新的多位点系统发育进行了分类研究，重新对发癣菌属进行了梳理，将本哈米节皮菌 *Arthroderma benhamiae* Ajello & S.L. Cheng 组合到发癣菌属，并命名为本哈米发癣菌 *Trichophyton benhamiae* (Ajello & S.L. Cheng) Y. Gräser & de Hoog；将网发癣菌和马发癣菌自养变种视为马发癣菌的同物异名；将巴吞鲁日发癣菌、枝状发癣菌、瑞氏发癣菌、长梗发癣菌视为指间发癣菌的同物异名；将阿比西尼亚发癣菌 *T. abissinicum* (A. Agostini) Nann.视为须癣发癣菌的同物异名；认为昆

克纳发癣菌 *T. quinckeanum* (Zopf ex Guég.) D.M. MacLeod & Muende、石膏样发癣菌昆克纳变种 *T. gypseum* var. *quinckeanum* (Zopf ex Guég.) Frágner 和须癣发癣菌昆克纳变种为同一个种，即昆克纳发癣菌；将巴克勒发癣菌 *T. balcaneum* Castellani、罗氏发癣菌、河海发癣菌 *T. fluviomuniense* Pereiro、费氏发癣菌 *T. fischeri* J. Kane、鲁比切克发癣菌 *T. raubitschekii* J. Kane, Salkin, Weitzman & Smitka、红色发癣菌鲁比切克变种 *T. rubrum* var. *raubitschekii* (J. Kane, Salkin, Weitzman & Smitka) Brasch、凯恩发癣菌 *T. kanei* Summerbell 视为红色发癣菌的同物异名；将花状发癣菌、浸水发癣菌 *T. immergens* Milochevitch 视为断发发癣菌的同物异名；将雅温德发癣菌和库瑞安发癣菌视为紫色发癣菌。因而，发癣菌属现在承认 16 种，分别为本哈米发癣菌、大疱性发癣菌 *T. bullosum* Lebasque、同心发癣菌、马发癣菌、羊毛发癣菌、刺猬发癣菌、指间发癣菌、须癣发癣菌、昆克纳发癣菌、红色发癣菌、许兰氏发癣菌、猴发癣菌、苏丹发癣菌、断发发癣菌、疣状发癣菌和紫色发癣菌。

2. 发癣菌属真菌的经济意义

皮肤癣菌病主要是由发癣菌属、小孢子菌属和表皮癣菌属三类丝状真菌引起的皮肤感染（Achterman & White，2012）。皮肤癣菌侵入已死亡的皮肤组织或其附属器（角质层、甲、毛发）并保留在角蛋白的表面进而引发疾病（吴润标等，2017）。体外培养皮肤癣菌，其培养液具有角蛋白水解活性，它在真菌的自身营养、组织入侵及控制宿主的防御机制上均起着十分重要的作用（张颖和罗庆录，2007）。而目前发癣菌是皮肤癣菌病主要的致病菌，占浅部感染的大部分（Cheng & Chong，2002；Lupa *et al.*，1999），而其外分泌的蛋白酶是最为重要的毒力因子（陈剑等，2009）

Apodaca 和 McKerrow（1989）报道了来自红色发癣菌蛋白酶的活性。将该菌株培养于包含容易代谢的碳源、氮源、硫源和磷源的培养基中，当碳源、氮源或硫在培养基中被消耗尽，则蛋白酶的活力将被抑制。在所有情况下，会产生一个高分子量（ >200 000）的蛋白酶；当培养基中氮源耗尽时，产生分子量为 71 000 的蛋白酶；当硫源消耗尽后，产生分子量为 124 000 和 27 000 的蛋白酶；当添加无论是无机硫源（$MgSO_4$、Na_2SO_3、$Na_2S_2O_3$）还是有机硫源（甲硫氨酸、半胱氨酸）至硫源耗尽时，蛋白酶的表达均受到抑制。对比可知，角蛋白分解活性不被碳源、氮源或硫源所抑制，但当最简培养基包含蛋白质时，蛋白酶活性反而被诱导。当红色发癣菌培养进入稳定期时，都会产生蛋白降解酶。与处于对数生长期不同，处于稳定期时，该菌会产生具有蛋白质降解、促弹性组织离解及角蛋白降解的酶。这些酶的活性在最简培养基中的碳源、氮源和磷源耗尽时下降，但在硫源耗尽时仍保持较高的活性。

Apodaca 和 McKerrow（1990）研究了培养基和葡萄糖对红色发癣菌产生蛋白酶的影响，并探讨了这些酶在培养中的作用。红色发癣菌在所有所试的培养基中均表达较高的蛋白降解酶、促弹性组织离解酶及角蛋白酶活性。葡萄糖会部分抑制蛋白酶的水解活性，且其作用依赖于培养时间和条件。一般而言，蛋白降解酶的酶活最高点出现在培养后的第 3 周，促弹性组织离解酶的酶活最高点出现在培养后的第 1 或 2 周，角蛋白酶酶活最高点的出现时间随培养基成分的变化而变化。

Kacinová 等（2014）报道了阿耶罗发癣菌产生胞外角蛋白酶。该菌株由毛发钓饵法获得。Kadhim 等（2015）研究了红色发癣菌的临床分离株的生长条件及蛋白酶的活性。在 SDA 培养基上，红色发癣菌的菌落呈絮状或粉状，白色或浅姜黄色，平坦或隆起；背面产生或未见色素。其最适培养温度为 30℃，pH 6，培养基为 SDA。当以酪蛋白为底物时，红色发癣菌的蛋白酶产量为 10.5～80.1 U/mL。此外，蛋白酶的活性随着 pH、温度、培养时间和基质浓度的变化而变化。

赵小东等（2004）研究了角蛋白与毛发诱导须癣发癣菌产生的角蛋白酶，发现纯化人角蛋白诱导须癣发癣菌产生角蛋白酶的能力高于毛发，以角蛋白为底物得到人角蛋白诱导角蛋白酶和毛发诱导角蛋白酶的比活分别为 1020 U/mg、910 U/mg，以毛发为底物得到人角蛋白诱导角蛋白酶和毛发诱导角蛋白酶的比活分别为 850 U/mg、790 U/mg。

冷文川等（2005）对红色发癣菌分泌的蛋白酶进行了分析，通过构建红色发癣菌的 cDNA 文库及生物信息学分析得到 18 个可能的分泌性蛋白酶，包括 4 个分泌性肽酶、1 个分泌性金属蛋白酶、2 个细胞外丝氨酸蛋白酶、1 个分泌性天冬氨酸蛋白酶、9 个分泌性枯草杆菌蛋白酶样丝氨酸蛋白酶、1 个空泡丝氨酸蛋白酶。

肖媛媛等（2008）研究了皮肤癣菌体外蛋白水解酶活性。来自于红色发癣菌的 20 株菌的蛋白水解酶 OD 值为 5.56 ± 2.39，须癣发癣菌的 3 株菌的蛋白水解酶 OD 值为 40.80 ± 8.01。而分离自不同部位的红色发癣菌中，5 株分离自足部的蛋白水解酶 OD 值为 5.56 ± 3.66，10 株分离自股部的蛋白水解酶 OD 值为 5.92 ± 2.24，5 株分离自指甲等甲部的蛋白水解酶 OD 值为 4.82 ± 1.22，而 2 株发癣菌肉芽肿类型的蛋白水解酶 OD 值为 3.00 ± 0.20。因而，须癣发癣菌的体外蛋白水解酶活性显著高于红色发癣菌。红色发癣菌的细胞外蛋白水解酶活性在分离自浅部不同感染部位的菌株之间无差异，但明显高于引起发癣菌肉芽肿的菌株。

Wirth 等（1965）报道了来自红色发癣菌的黄色麦格尼发癣菌素，它分离自该菌菌丝体提取物的结晶。Blank 等（1966）报道了分离自紫色发癣菌两种新的色素：vioxanthin 和 viopurpurin 及可能的结构。vioxanthin 的分子式为$(C_{15}H_{13}O_5)_2$。viopurpurin 是一种非常难溶的深红色的色素，分子式为 $C_{35}H_{26}O_{14}$。Ng 和 Just（1969）报道了分离自紫色发癣菌的 xanthomegnin（图 18：1），vioxanthin（图 18：2）和 viopurpurin（图 18：3）的分子结构。

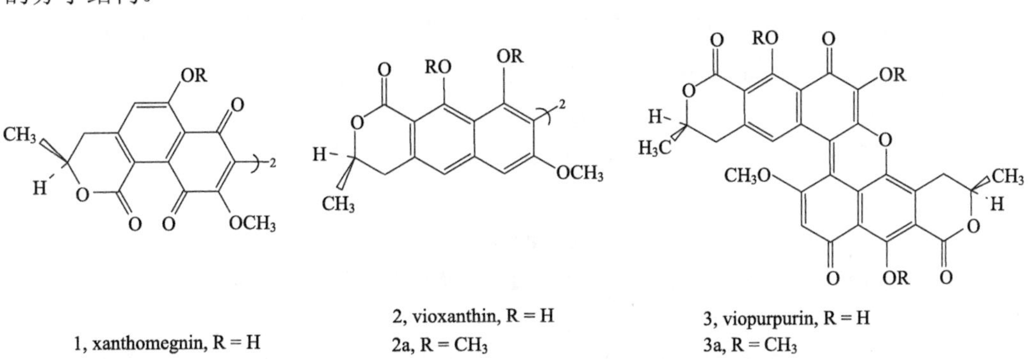

1, xanthomegnin, R = H
1a, R = Ac

2, vioxanthin, R = H
2a, R = CH_3
2b, R = Ac

3, viopurpurin, R = H
3a, R = CH_3
3b, R = Ac

图 18　xanthomegnin（1）、vioxanthin（2）和 viopurpurin（3）的化学结构（引自 Ng & Just，1969）

Floersheim 等（1982）报道了疣状发癣菌菌株 632/77 的冻干提取物对于尤氏肉瘤具

有明显的抑制作用，与对照相比，体积仅为 5.6%。此外，该菌株的提取物与其他真菌提取物相比，在培养 31 天后，抑制率可达 11 倍。疣状发癣菌的其他菌株 93/78 和 3636/77 的提取物对人的结肠癌细胞具有明显的抑制作用，但对于尤氏肉瘤未出现明显抑制作用。上述菌株的提取物未表现出明显的细胞毒性。

Jung 等（2002）报道了分离自木质素降解菌红色发癣菌 LKY-7 的漆酶的酶学性质。该菌在葡萄糖蛋白胨液体培养基中分泌大量的漆酶。2,5-二甲苯胺对漆酶的产生几乎没有诱导作用。漆酶通过三步色谱法纯化至具有同质性，其总回收率为 40%。通过 SDS-PAGE 得到纯化后漆酶的分子质量约为 65 kDa。纯化后的漆酶明显呈蓝色及典型的基本光谱特征，即两个最大的吸收值为 278 nm 和 610 nm，峰肩值为 338 nm。对漆酶的 N 端测序得到，其与木腐真菌如虫拟蜡菌 *Ceriporiopsis subvermispora* (Pilát) Gilb. & Ryvarden 的漆酶具有高度同源性，而与非木腐真菌如双孢蘑菇 *Agaricus bisporus* (J.E. Lange) Imbach 和新型隐球菌 *Cryptococcus neoformans* (San Felice) Vuill.相似性较低。该酶具有较低的氧化还原点位，约为 0.5 V，但它是氧化一系列基质/介质中最活跃的漆酶。与其他漆酶相比，当 3-ethylbenzothiazoline-6-sulfonic acid 作为基质时，具有非常低的 K_m 值；但当紫尿酸作为基质时，则具有非常高的 K_m 值。漆酶的等电点为 4.0，该漆酶的最适 pH 处于酸性（3～4），即使其在中性环境中稳定性优于酸性环境，该酶对于对苯二酚的氧化速度也要快于邻苯二酚和邻苯三酚。叠氮化钠和氟化钠对漆酶有强烈的抑制作用。

Mohanty 和 Prakash（2004）报道了分离自钓饵法得到的土壤嗜角蛋白菌——阿耶罗发癣菌的代谢物对斯氏按蚊和五带淡色库蚊幼虫的作用。阿耶罗发癣菌的胞外代谢物来自于沙氏培养基培养 21 天的发酵液，并经 Whatman-Ⅰ滤纸过滤后得到。五带淡色库蚊的幼虫感染率是斯氏按蚊的 0.25 倍。两种蚊子的 1 龄期要比 2、3 龄更易感染。因而阿耶罗发癣菌的胞外代谢物可以作为合成农药的潜在替代品。

Murugesan 等（2009）报道了须癣发癣菌的胞外代谢物对埃及伊蚊的作用。该菌株分离自钓饵法的土壤，培养 2 天后的滤出液（10～100 μL/mL）对埃及伊蚊 3 龄幼虫表现出毒性，其 LC_{50} 和 LC_{90} 值分别为（110±11.5）μL/mL 和（200±20.7）μL/mL。该菌的胞外代谢物具有蛋白酶的性质，且对蚊子幼虫的几丁质具有专一性。

Bertrand 等（2013）研究了红色发癣菌和淡色生赤壳菌 *Bionectria ochroleuca* (Schwein.) Schroers & Samuels 共培养时产生的代谢物，共发现了 5 种化合物，但仅对化合物 4 进行了鉴定，其分子式为 $C_{32}H_{35}O_{13}S$，其中包含 17 个不饱和键。上述结构与前人研究报道的来自于顶头霉属 *Acremonium* Link 一个种的化合物较类似，PS-990［4-(4′-(4′-Hydroxy-2′-methoxy-3′,5′,6′-trimethylbenzyloxy)-2′-hydroxy-3′,5′,6′-trimethylbenzyloxy)-2-methoxy-3,5,6-trimethylbenzoic acid］（图 19）。化合物的结构与报道自构巢曲霉 *Aspergillus nidulans* (Eidam) G. Winter 与吸水链霉菌 *Streptomyces hygroscopicus* (Jensen) Waksman & Henrici 共培养时的产物 4,6-二羟-2-甲苯甲酸相关。

Xu 等（2015）对红色发癣菌的全蛋白质组研究发现，红色发癣菌的全蛋白质组共包含 35 874 种多肽，此外还发现了 323 种新的多肽。其中，104 种新多肽的编码基因被确认，127 种基因结构编码的新多肽与现有的已注释的多肽相冲突，需要通过手动注释。通过 RNA-Seq 确证了 95%的总肽。

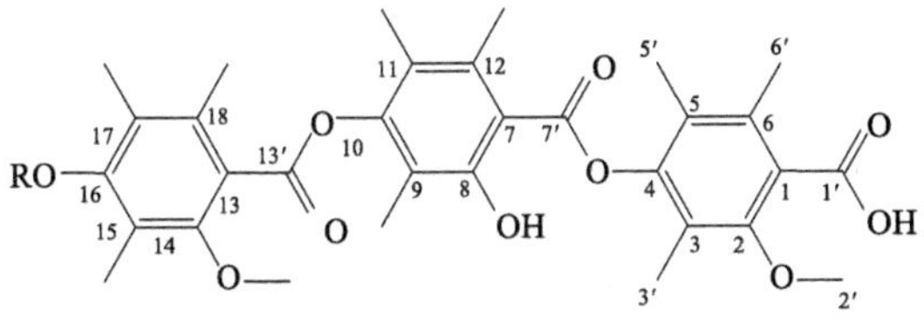

图 19　共培养 *T. rubrum* 和 *B. ochroleuca* 分离获得的化合物（引自 Bertrand *et al.*，2013）

李治建等（2009）采用高效液相色谱分析了红色发癣菌麦角甾醇的含量。他们建立了一种测定红色发癣菌中麦角甾醇含量的方法，具体条件为：使用 Waters symmetry shield TM RP C18 色谱柱（4.6 mm × 150 mm，5 μm），流动相为100%甲醇，流速为1.0 mL/min，检测波长为282 nm，通过保留时间进行定性，并采用外标法定量。实验结果表明，麦角甾醇的保留时间为8.9 min；麦角甾醇浓度为0.93～29.75 mg/L 时，峰面积与浓度呈线性关系，相关系数为 0.9998；麦角甾醇的平均添加回收率为 94.8%～96.3%，相对标准偏差为1.8%～3.3%，麦角甾醇的最低检出浓度为0.008 mg/L。

陈先进等（2009）对红色发癣菌降解动物毛发进行了研究，发现培养2周后，红色发癣菌和须癣发癣菌在豚鼠和小鼠毛发上均能生长，菌落生长后，光镜下可见动物毛发正常结构被破坏，横纹变模糊或消失。

刘艳等（2009）对导致脓癣病的须癣发癣菌体外酶活性进行了研究，发现2株须癣发癣菌产生的胞外分泌酶主要为碱性磷酸酶、白氨酸芳胺酶、β-葡萄糖苷酶、*N*-乙酰-葡萄糖胺酶、α-甘露糖苷酶。

赵明月等（2012）对红色发癣菌中羊毛甾醇含量采用高效液相色谱法进行了分析，得到羊毛甾醇的保留时间为12.1 min；在0.1～10.0 mg/L 浓度范围内，峰面积与浓度呈良好的线性关系，其相关系数为0.9981；羊毛甾醇的平均添加回收率为93.7%，相对标准偏差为1.9%，其最低检出浓度为0.05 mg/L。

刘丽和赖维（2016）研究了采用不同底物体外诱导不同基因型红色发癣菌产生角蛋白酶的活性，结果表明，不同底物诱导红色发癣菌产生的角蛋白酶活性中，（指、趾）甲显著高于皮屑，皮屑显著高于头发，指甲和趾甲间无明显差异。

张芳芳等（2016）研究了须癣发癣菌中的烯醇化酶基因，通过克隆该菌的 cDNA 全长获得了须癣发癣菌烯醇化酶（enolase）基因全长序列（1491 bp），其拥有一个1317 bp的开放阅读框，编码438个氨基酸，5′非编码区为106 bp，3′非编码区为68 bp，同源比对发现与断发发癣菌烯醇化酶同源性达100%，与红色发癣菌同源性达98%。

专　论

金孢属 **Chrysosporium** Corda

Deutschl. Fl., 3 Abt. (Pilze Deutschl.) 3(13): 85

菌落常扩展，可呈多种颜色，白色、奶油色、淡褐色或黄色，绒毛状、毡状和/或粉状。菌丝大多无色，壁光滑，具不规则的、或多或少的垂直分枝。可育菌丝没有或很少有差别。有时具球拍状菌丝。顶生和侧生的分生孢子产生于整个菌丝，直接着生于菌丝或短突起上或侧生分枝上，单生或成短链，无色或半透明或淡黄色，薄或厚壁，光滑或粗糙，近球形、棒状、梨形或倒卵形，单胞，偶有双胞，基部平截。间生孢子有时存在，单生，偶见串生，半透明或淡黄色，比其菌丝宽，常单胞，两端平截。厚垣孢子偶见。

模式种：粪生金孢 *Chrysosporium merdarium* (Ehrenb.) J.W. Carmich.。

中国金孢属分种检索表

1. 间生孢子缺或少 ······ 2
1. 间生孢子丰富 ······ 11
　2. 间生孢子少 ······ 3
　2. 间生孢子无 ······ 4
3. 菌落白色 ······ **热带金孢 *C. tropicum***
3. 菌落亮黄色或粉色或绿色 ······ **粪生金孢 *C. merdarium***
　4. 具球拍状菌丝 ······ **雷公山金孢 *C. leigongshanense***
　4. 无球拍状菌丝 ······ 5
5. 菌落为棕红色 ······ **乔治金孢 *C. georgiae***
5. 菌落为浅色或白色 ······ 6
　6. 分生孢子粗糙 ······ 7
　6. 分生孢子光滑 ······ 8
7. 分生孢子较大，3.5～6.0 × 1.5～2.5 μm ······ **江苏金孢 *C. jiangsuense***
7. 分生孢子较小，2.7～3.8 × 1.2～2.2 μm ······ **轮带金孢 *C. zonatum***
　8. 分生孢子有色，呈金淡黄色至淡褐色 ······ **法斯特金孢 *C. fastidium***
　8. 分生孢子无色 ······ 9
9. PDA 培养基上菌落背面淡黄色 ······ 10
9. PDA 培养基上菌落背面沙棕色 ······ **卵孢金孢 *C. ovalisporum***
　10. 分生孢子倒卵形或柱状至棒状 ······ **开阳金孢 *C. kaiyangense***
　10. 分生孢子倒梨形、椭圆形或圆形 ······ **侧孢金孢 *C. laterisporum***
11. 具球拍状菌丝 ······ 12
11. 无球拍状菌丝 ······ 19
　12. 培养基上有性结构大量存在 ······ **苍白金孢 *C. pallidum***
　12. 培养基上未见有性结构 ······ 13

13. 产孢细胞与分生孢子连接处具有独特的“领状”结构…………………… **山西金孢 *C. shanxiense***
13. 无“领状”结构……………………………………………………………………………… 14
  14. 分生孢子近球形………………………………………………………………………… 15
  14. 分生孢子非近球形……………………………………………………………………… 16
15. 菌落有明显的轮带，白色……………………………………………… **贵州金孢 *C. guizhouense***
15. 菌落无轮带，白色至黄绿色……………………………………………… **三亚金孢 *C. sanyaense***
  16. 分生孢子长度大多数超过 10 μm ………………………………… **荆州金孢 *C. jingzhouense***
  16. 分生孢子长度不超过 10 μm ……………………………………………………………… 17
17. 分生孢子大多数长卵圆形………………………………………… **广西金孢 *C. guangxiense***
17. 分生孢子非卵圆形………………………………………………………………………… 18
  18. 间生孢子椭圆形………………………………………………… **四川金孢 *C. sichuanense***
  18. 间生孢子梨形………………………………………………………… **甘肃金孢 *C. gansuense***
19. 分离自冬虫夏草内菌核…………………………………………………… **中国金孢 *C. sinense***
19. 腐生……………………………………………………………………………………… 20
  20. 分生孢子无色或淡褐色至中等褐色………………………… **嗜毛金孢 *C. keratinophilum***
  20. 分生孢子无色……………………………………………………………………………… 21
21. 顶侧生孢子呈梭形…………………………………………………… **梭形金孢 *C. fusiforme***
21. 顶侧生孢子非梭形………………………………………………………………………… 22
  22. 顶侧生孢子为长卵形……………………………………………… **节状金孢 *C. articulatum***
  22. 顶侧生孢子非长卵形……………………………………………………………………… 23
23. 顶侧生孢子形状多样，卵形、梨形和棒状………………………… **绒毛金孢 *C. villiforme***
23. 顶侧生孢子为梨形………………………………………………………………………… 24
  24. 间生孢子椭圆形……………………………………………… **不规则金孢 *C. irregularum***
  24. 间生孢子非椭圆形………………………………………………………………………… 25
25. 间生孢子桶状…………………………………………………………… **裂叶金孢 *C. lobatum***
25. 间生孢子长卵形至梭形……………………………………………… **多形金孢 *C. multiforme***

节状金孢 图 20

Chrysosporium articulatum Scharapov, Nov. Sist. Niz. Rast. 15: 146, 1978. Han *et al*., Microbiology China 46(9): 2210, 2019.

在 PDA 培养基上，28℃培养 7 天，菌落直径约 70 mm，毯状，边缘规则，隆起；背面半透明。菌丝有隔，光滑，无色至半透明，宽 1.5～2.3 μm。分生孢子单生或串生，表面光滑，无色，着生于菌丝上，单胞或双胞，长卵形、5.8～15 × 2.5～5.5 μm，或棒状、10.5～17 × 3～4.3 μm；有间生孢子，单生或链状，光滑，长卵形，6.8～7.5 × 2.5～3.5 μm；基痕宽 0.5～1 μm。

研究菌株：甘肃：玉门关市，雅丹地质公园，骆驼刺根际土样，2017 年 7 月 20 日，韩燕峰，HMAS 255374，GZAC H5.11（菌株）；贵州：遵义市，遵义医学院附属医院，绿化地土，2018 年 7 月 15 日，韩燕峰，GZAC H4.12。

世界分布：中国、泰国、俄罗斯。

讨论：该种与中国金孢和梭形金孢在形态上类似，但中国金孢分离自冬虫夏草上；梭形金孢的孢子呈梭形；本种分离自土壤，未见厚垣孢子，分生孢子呈长卵形或棒状，能与相似种区分（韩燕峰等，2019）。

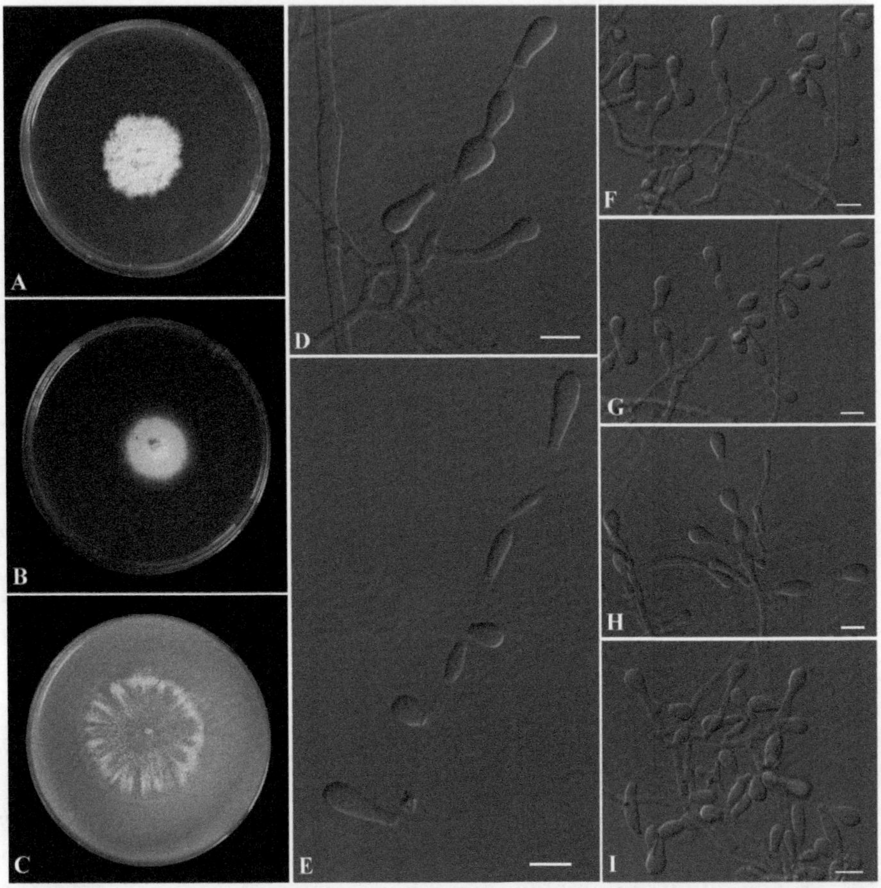

图 20　节状金孢 *Chrysosporium articulatum* Scharapov（标本号 HMAS 255374）

A～C. PDA、MEA 和 OA 培养基上的菌落；D、E. 球拍状菌丝；F～I. 产孢结构。标尺：D～I = 10 μm

梭形金孢　图 21

Chrysosporium fusiforme Y.F. Han, W.H. Chen, J.D. Liang & Z.Q. Liang, Phytotaxa 539(1): 3, 2022.

在 PDA 培养基上，26℃培养 7 天，菌落直径 35 mm，白色，毡状，中部一圈白色毡状轮纹，边缘半透明，规则；背面奶酪色至白色。菌丝分隔，光滑，无色，宽 1.0～2.5 μm。球拍状菌丝未见。分生孢子大量产生，单生，光滑，无色，顶生或侧生于菌丝或短轴上，梭形、5.5～10.5 × 2～4 μm，或卵形、3.5～5 × 1～3 μm，两端钝圆；基痕宽 1～1.5 μm；间生孢子梭形，6.5～9 × 2.5～3 μm。

研究菌株：山西：晋中市祁县，土样，2017 年 8 月 5 日，韩燕峰，GZAC I9（主模式），GZU I9（模式菌株）；山西：临汾市吉县，树下土样，2017 年 8 月 3 日，韩燕峰，GZAC I8 和 GZAC L17.2。

世界分布：中国。

讨论：基于 ITS rDNA 分子序列构建的系统发育，梭形金孢与江苏金孢在亲缘关系上相近，但在形态上，江苏金孢不产生间生孢子。梭形金孢的鉴别特征是分生孢子梭形或卵形，具间生孢子（Han *et al.*，2022）。

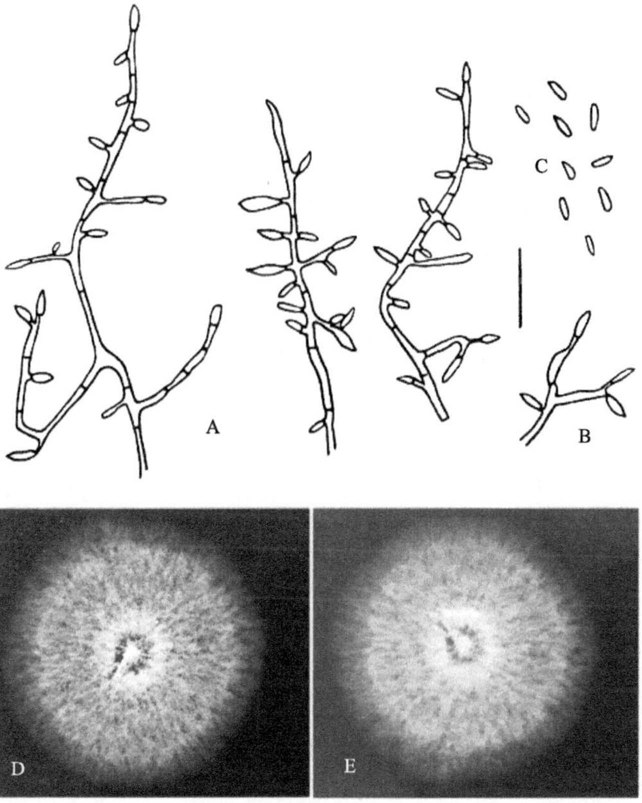

图 21　梭形金孢 *Chrysosporium fusiforme* Y.F. Han, W.H. Chen, J.D. Liang & Z.Q. Liang（标本号：GZAC I9）

A、B. 产孢结构；C. 分生孢子；D、E. PDA 培养基上的菌落。标尺：A～C = 20 μm

甘肃金孢　图 22

Chrysosporium gansuense Y.F. Han, W.H. Chen, J.D. Liang & Z.Q. Liang, Phytotaxa 539(1): 5, 2022.

在 PDA 培养基上，26℃培养 14 天，菌落直径约 40 mm，中心微隆，白色，粉状，平展，薄，边缘绒毛状，近圆形；背面奶酪色至白色。菌丝分隔，无色至半透明，宽 0.5～2 μm，无球拍状菌丝。分生孢子大量产生，单生，表面粗糙，无色，顶生或侧生于菌丝或短突起上，梨形或棒状、3.5～6 × 1.5～3.5 μm，或椭圆形、3.5～4.5 × 2.5～4.0 μm，两端钝圆；基痕宽 0.5～1.0 μm；间生孢子多见于侧分枝上，椭圆形，3.0～5 × 1.5～2.5 μm。

研究菌株：甘肃：嘉峪关市，松树根际土土样，2017 年 7 月 10 日，王健健，GZAC C4.1（主模式），GZU C4.1（模式菌株），GZU C4.2。

世界分布：中国。

讨论：基于 ITS rDNA 分子序列构建的系统发育，甘肃金孢与山西金孢和热带金孢相近，但山西金孢在产孢细胞和分生孢子之间有明显的膨大的“领状”结构（张延威等，2016），热带金孢有柱状的间生孢子和球形的厚垣孢子（van Oorschot，1980）。甘肃金孢的主要特征是分生孢子表面粗糙，梨形或棒状，具间生孢子。

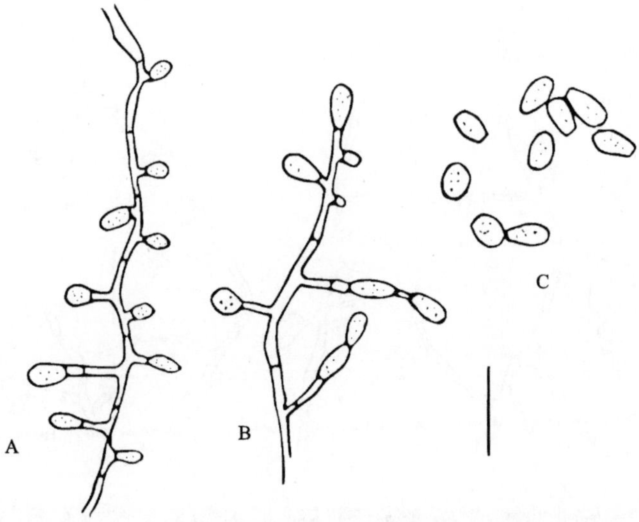

图 22　甘肃金孢 *Chrysosporium gansuense* Y.F. Han, W.H. Chen, J.D. Liang & Z.Q. Liang（标本号：GZAC C4.1）

A、B. 产孢结构；C. 分生孢子。标尺：A～C = 10 μm

乔治金孢　图 23，图版 I：1

Chrysosporium georgiae (Varsavsky & Ajello) Oorschot [as '*georgii*'], Stud. Mycol. 20: 31, 1980. Han *et al*., Microbiology China 46(9): 2210, 2019.

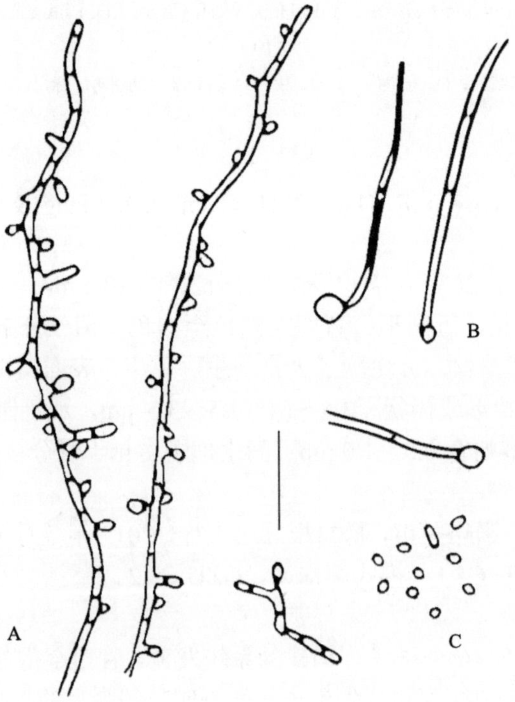

图 23　乔治金孢 *Chrysosporium georgiae* (Varsavsky & Ajello) Oorschot（标本号：HMAS 255377）

A. 产孢结构；B. 厚垣孢子；C. 分生孢子。标尺：A～C = 20 μm

在 PDA 培养基上，26℃培养 14 天，菌落直径 40～43 mm，绒毛状，中心微隆，边缘不规则，近圆形；背面棕红色。菌丝分隔，光滑，无色，宽 0.8～2.6 μm；无球拍状菌丝。分生孢子单生或 2 个串生，表面光滑，无色，直接着生于菌丝上或短柄上，棒状、2.6～5.1 × 1.2～1.6 μm，卵形、2.2～3.5 × 0.9～2.5 μm，不规则椭圆形、1.8～3.8 × 1.5～2.6 μm；无间生孢子；厚垣孢子粗糙，4.3～4.9 × 4.6～4.7 μm；基痕宽＜ 1 μm。

有性型为西费氏节皮菌 *Arthroderma ciferrii* Varsavsky & Ajello。

研究菌株：甘肃：玉门关市，雅丹地质公园，骆驼刺根际土样，2017 年 7 月 15 日，王健健，GZAC H10.10，HMAS 255377。

世界分布：中国、埃及、美国。

讨论：乔治金孢的近似种为卡氏金孢，但前者不产生间生孢子能与后者相区别（韩燕峰等，2019）。

广西金孢　图 24

Chrysosporium guangxiense Y.F. Han, W.H. Chen, J.D. Liang & Z.Q. Liang, Phytotaxa 539(1): 7, 2022.

在 PDA 培养基上，26℃培养 14 天，菌落直径 45～50 mm，白色，平展，毡状，致密，中心处有一圈沟纹，边缘规则；背面奶酪色至淡黄色。菌丝有隔，光滑，无色至半透明，宽 1.0～3.0 μm，菌丝上有膨大突起，有球拍状菌丝，17～50 × 2.5～5.5 μm。分生孢子大量产生，单生或两个簇生，偶见两个串生，光滑，无色，顶生或侧生于菌丝或短突起上，单胞，大多数长卵圆形、5.0～8.5 × 3.5～7.0 μm，或棒状、7～13 × 2.5～3.0 μm，两端平截；基痕宽 0.5～1 μm；间生孢子椭圆形，5.5～10 × 2.0～2.5 μm。

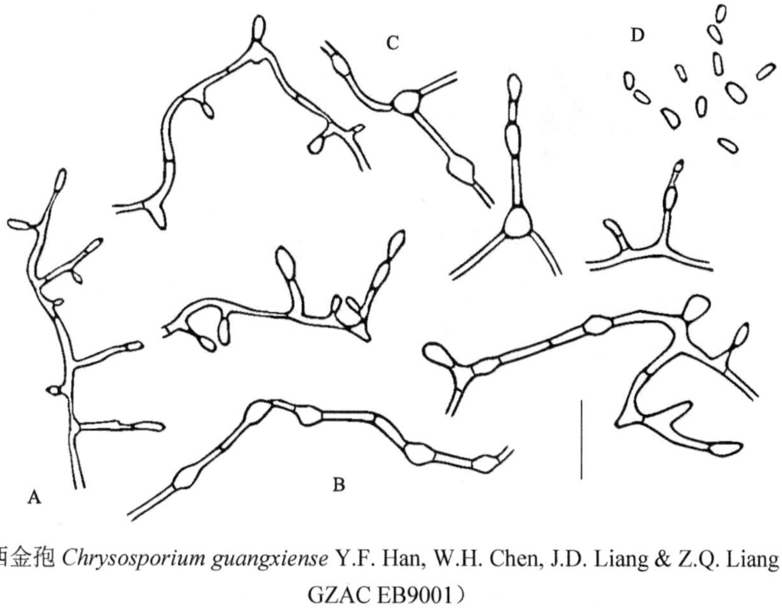

图 24　广西金孢 *Chrysosporium guangxiense* Y.F. Han, W.H. Chen, J.D. Liang & Z.Q. Liang（标本号：GZAC EB9001）

A. 产孢结构；B. 膨大突起结构；C. 间生孢子；D. 分生孢子。标尺：A～D = 20 μm

研究菌株：广西：农田土样，2015 年 7 月 8 日，王玉荣，GZAC EB9001（主模式），

GZU EB9001M。

世界分布：中国。

讨论：广西金孢的形态近似种有贵州金孢、水生金孢和荆州金孢。但贵州金孢在 PDA 上的菌落具有明显的轮带；水生金孢具有明显的厚垣孢子，荆州金孢的分生孢子较长，一般超过 10 μm；广西金孢菌落白色、无轮带，未见厚垣孢子，且分生孢子的长度一般小于 10 μm。

贵州金孢 图 25

Chrysosporium guizhouense Y.W. Zhang, Y.F. Han & Z.Q. Liang, Phytotaxa 270(3): 212, 2016.

在查氏培养基上，25℃培养 14 天，菌落直径达 18～20 mm，白色，绒毛状，圆形。在 PDA 培养基上，菌落直径达 44～54 mm，白色，绒毛状，中部稍致密，有一个明显的轮带，边缘稀疏；背面浅黄色。菌丝分隔，无色至半透明，光滑，1.2～4.3 μm；有球拍状菌丝，6.5～19.4 × 4.3～7.6 μm。分生孢子侧生或顶生，着生于菌丝上或短的突起或短柄上，单生，无色，光滑，大多数单胞，偶见双胞，近球形、2.2～4.3 μm，倒卵圆形至椭圆形、5.4～6.5 × 3.2～4.3 μm；间生孢子丰富，产生于长侧枝上，桶状、不规则柱状或椭圆形，2.2～24.9 × 1.3～4.3 μm；基痕宽 0.8～2.5 μm。

研究菌株：贵州：贵阳市，开阳县，蛇蜕下土样，2015 年 5 月 1 日，王玉荣，GZAC EM14.2002（主模式），贵州大学真菌资源研究所标本馆（GZUIFR）GZUIFR EM14.2002。

世界分布：中国。

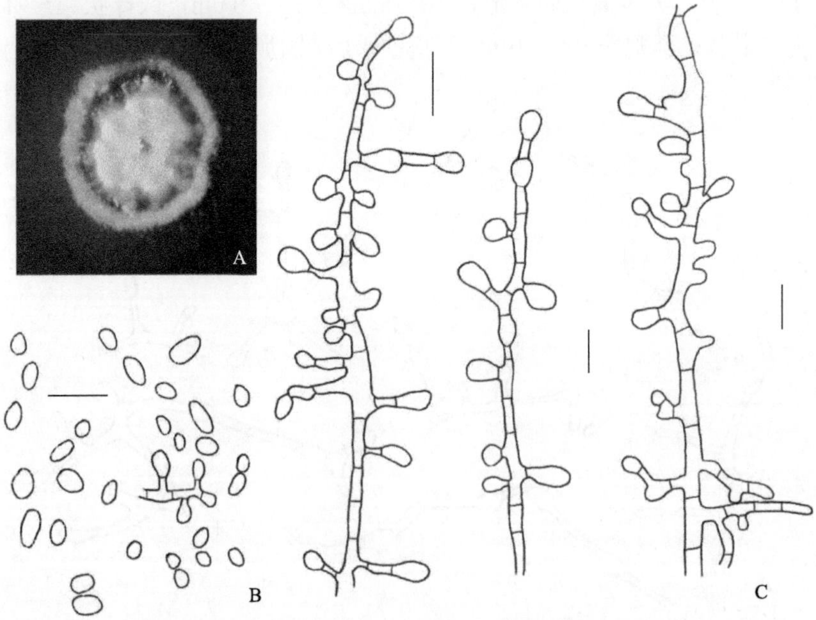

图25　贵州金孢 *Chrysosporium guizhouense* Y.W. Zhang, Y.F. Han & Z.Q. Liang（标本号：GZAC EM14.2002）

A. PDA 培养基上的菌落；B. 分生孢子；C. 产孢结构。标尺：B、C = 10 μm

讨论：该种的近似种有勒克瑙金孢、异味金孢、西格氏金孢和水生金孢。西格氏金孢和水生金孢的分生孢子粗糙；勒克瑙金孢和异味金孢无间生孢子（van Oorschot，1980）。该种的主要鉴别特征是：具球拍状菌丝和间生孢子，分生孢子近球形至倒卵圆形（Zhang *et al.*，2016）。

不规则金孢 图 26，图版 I：2

Chrysosporium irregularum Y.F. Han, W.H. Chen, J.D. Liang & Z.Q. Liang, Phytotaxa 539(1): 8, 2022.

在 PDA 培养基上，26℃培养 14 天，菌落直径 35～36 mm，中心微隆，近边缘处灰色，稀疏短绒状，边缘不规则，半透明；背面白色至浅黄色。菌丝有隔，光滑，无色至半透明，宽 0.5～3.0 μm，菌丝有膨大结构，无球拍状菌丝。分生孢子单生，表面光滑，无色，着生于菌丝或短柄或短突上，单胞或双胞，柱状、3.5～9.5 × 1～2.5 μm，或梨形、3.5～5 × 1.5～3.0 μm，或不规则肾形、3～5 × 1.5～3.5 μm，两端平截；基痕宽 0.5～1 μm；间生孢子单生或成串，椭圆形，2～15 × 1～4 μm。

研究菌株：甘肃：玉门关市，沙地植物根系土样，2017 年 7 月 13 日，王健健，干模式标本 GZAC J1.1（主模式），GZU J1.1，GZU J102。

世界分布：中国。

讨论：不规则金孢与绒毛金孢相似，但绒毛金孢的分生孢子为卵形、梨形或棒状，不规则金孢的分生孢子为柱状、梨形或不规则肾形，且菌落边缘不规则（Han *et al.*，2022）。

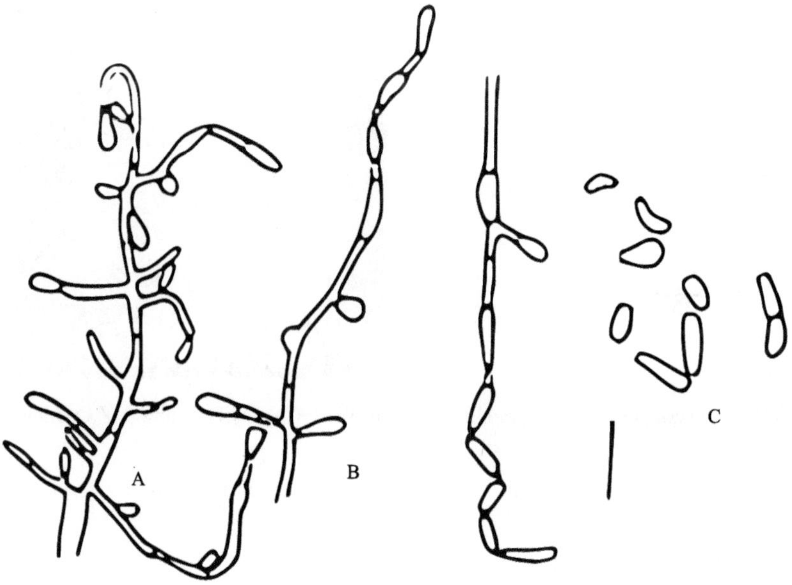

图 26 不规则金孢 *Chrysosporium irregularum* Y.F. Han, W.H. Chen, J.D. Liang & Z.Q. Liang（标本号：GZAC J1.1）

A. 产孢结构；B. 间生孢子；C. 分生孢子。标尺：A～C = 20 μm

江苏金孢 图 27

Chrysosporium jiangsuense Y.F. Han, W.H. Chen, J.D. Liang & Z.Q. Liang, Phytotaxa 539(1): 9, 2022.

在 PDA 培养基上，26℃培养 14 天，菌落直径 40～42 mm，白色，短绒状，具 3 圈轮纹，边缘较规则；背面淡黄色。菌丝分隔，光滑，无色，宽 1.0～3.5 μm，无球拍状菌丝。分生孢子单生，粗糙，侧生或顶生于菌丝上，倒卵形、3.5～6.0 × 1.5～2.5 μm，或椭圆形、1.5～3 × 1.5～2.5 μm，基部平截；基痕宽 0.5～1 μm。

研究菌株：江苏：扬州市，火车站土样，2017 年 8 月 3 日，韩燕峰，GZAC I10（主模式）；山西：晋中市，祁县，绿化地土样，2017 年 8 月 10 日，韩燕峰，GZU D14.3。

世界分布：中国。

讨论：江苏金孢在系统发育上与印度金孢和临汾金孢的亲缘关系较近，但印度金孢和临汾金孢具球拍状菌丝（Liang *et al.*，2009a；van Oorschot，1980），江苏金孢的主要鉴别特征是分生孢子粗糙，倒卵形或椭圆形，无球拍状菌丝和间生孢子（Han *et al.*，2022）。

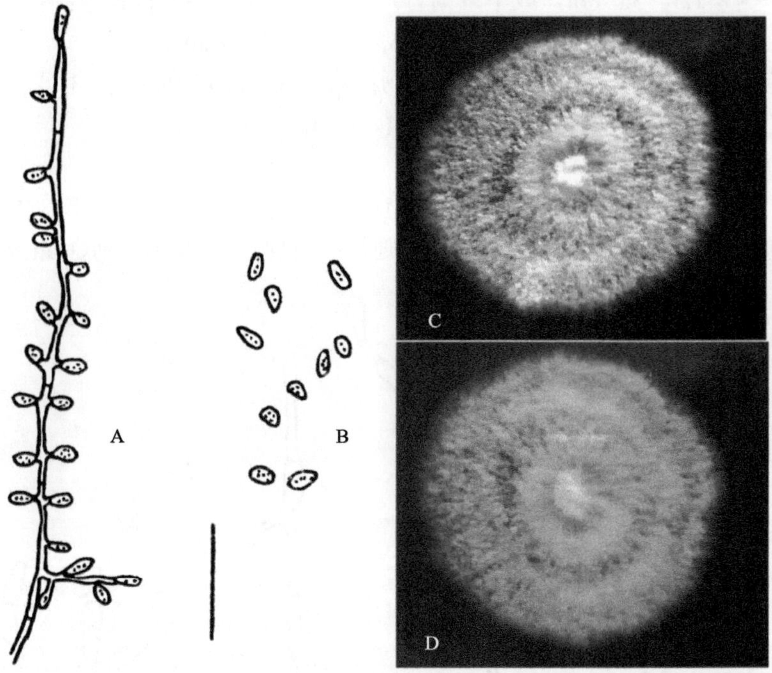

图 27　江苏金孢 *Chrysosporium jiangsuense* Y.F. Han, W.H. Chen, J.D. Liang & Z.Q. Liang（标本号：GZAC I10）

A. 产孢结构；B. 分生孢子；C、D. PDA 培养基上的菌落。标尺：A、B = 20 μm

荆州金孢　图 28

Chrysosporium jingzhouense Y.W. Zhang, Y.F. Han & Z.Q. Liang, Phytotaxa 303(2): 175, 2017.

在 PDA 培养基上，25℃培养 14 天，菌落直径达 25 mm，白色，绒毛状，圆形，边缘规则；背面浅黄色。菌丝分隔，无色，光滑，1.6～4.3 μm；有球拍状菌丝，8.6～15 × 3.2～6.5 μm。分生孢子侧生或顶生，大多数着生于短突起或短侧枝上，大多单生，少数成短链，无色，光滑，单胞，偶见双胞，长倒卵圆形至长椭圆形，4.3～16.2 × 3.2～8.6 μm；有时棒状，8.6～25.9 × 3.2～10.8 μm；基痕宽 0.8～5.4 μm。间生孢子单生，桶状至长棒状，4.3～32.4 × 2.2～7.6 μm。

研究菌株：湖北：荆州市，农田土壤，2015 年 7 月 10 日，王玉荣，GZUIFR EB1303M（主模式），GZUIFR EB1301M。

世界分布：中国。

讨论：荆州金孢在系统发育上与节状金孢和嗜毛金孢亲缘关系相近；但在形态上，节状金孢菌落毯状，分生孢子 5.8～17 × 2.5～5.5 μm，有时轻微厚壁（van Oorschot，1980）；嗜毛金孢菌落绒毛状至毡状，分生孢子光滑或具刺（van Oorschot，1980）。荆州金孢的主要鉴别特征是绒毛状菌落，分生孢子光滑，具球拍状菌丝和间生孢子（Zhang *et al.*，2017a）。

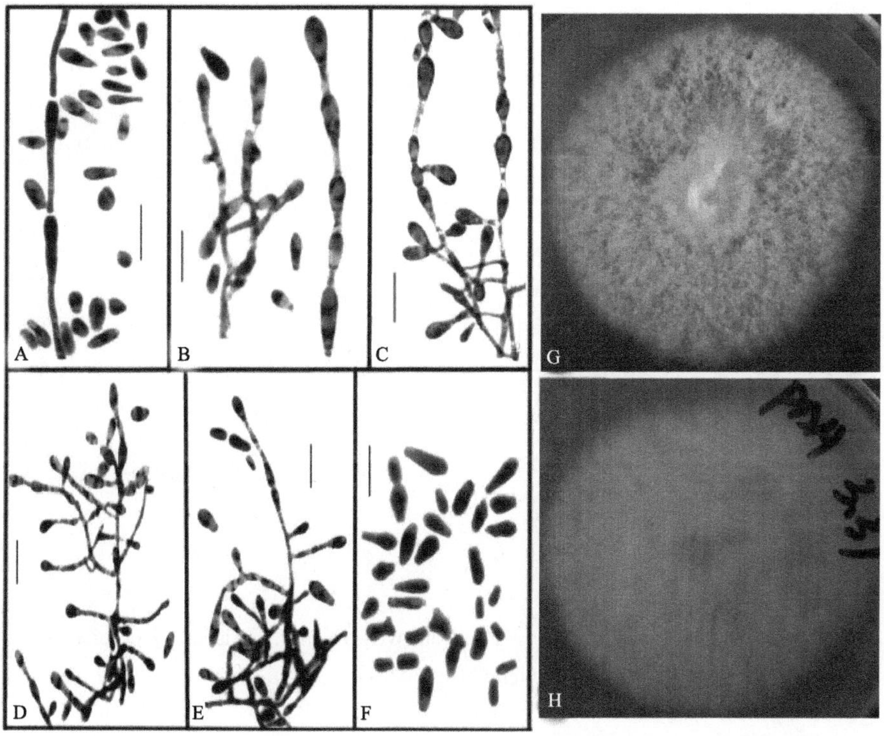

图 28　荆州金孢 *Chrysosporium jingzhouense* Y.W. Zhang, Y.F. Han & Z.Q. Liang（标本号：GZUIFR EB1303M）

A、D、E. 产孢结构；B. 球拍状菌丝；C. 间生孢子；F. 顶侧生孢子；G、H. PDA 培养基上的菌落。标尺：A～F = 10 μm

开阳金孢　图 29

Chrysosporium kaiyangense Y.F. Han, W.H. Chen, J.D. Liang & Z.Q. Liang, Phytotaxa 539(1): 9, 2022.

在查氏培养基上，26℃培养 14 天，菌落直径达 45 mm，稀疏绒毛状，白色，边缘不规则；背面淡黄色。

在 PDA 培养基上，26℃培养 14 天，菌落直径达 35 mm，毡状到绒状，白色，中间微隆，边缘较规则；背面浅黄色到奶油色。菌丝分隔，光滑，无色至半透明，宽 2～3 μm；无球拍状菌丝。分生孢子单生，光滑，无色，顶生或侧生于菌丝上或短突起或短柄上，分隔，1～3 个细胞，倒卵圆形、2～3.5 × 1～2.5 μm，或柱状至棒状、4～10.5 × 2～3 μm，两端平截；基痕宽 1.5～2 μm；无间生孢子，厚垣孢子未见。

研究菌株：贵州：贵阳市，开阳县，堕秧村，大叶黄杨根系土，2012 年 2 月 10 日，王玉荣，GZAC EB0702M（主模式），GZU EB0402M。

世界分布：中国。

讨论：开阳金孢在系统发育上与卵孢金孢在亲缘关系上相近，但卵孢金孢在 PDA 培养基上菌落局限，背面沙棕色，分生孢子较大（Li *et al.*，2019）。开阳金孢的主要鉴别特征是分生孢子倒卵圆形或柱状至棒状，无球拍状菌丝和间生孢子（Han *et al.*，2022）。

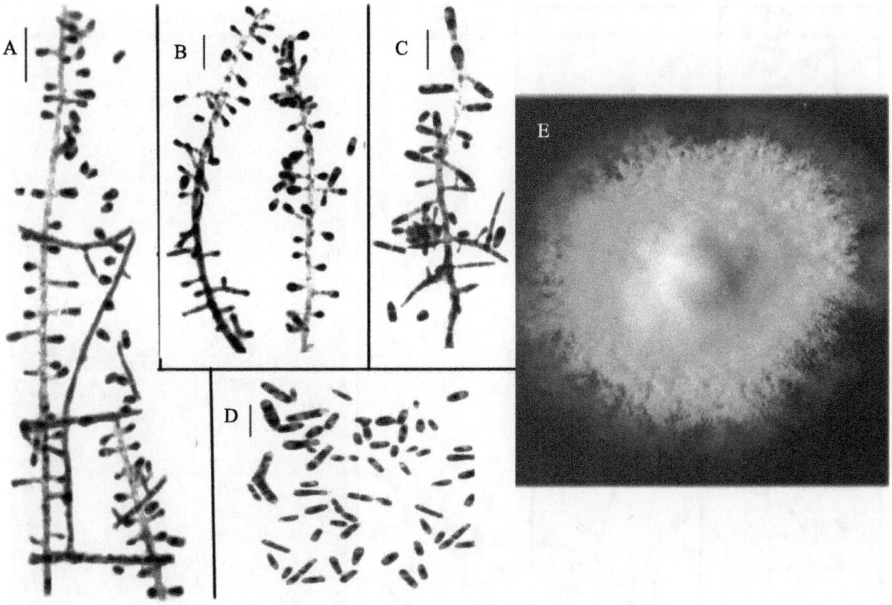

图 29 开阳金孢 *Chrysosporium kaiyangense* Y.F. Han, W.H. Chen, J.D. Liang & Z.Q. Liang（标本号：GZAC EB0702M）

A～C. 产孢结构；D. 分生孢子；E. PDA 培养基上的菌落。标尺：A～D = 10 μm

嗜毛金孢　图 30

Chrysosporium keratinophilum D. Frey ex J.W. Carmich., Can. J. Bot. 40: 1157, 1962.

Geng, A survey on soil dematiaceous hyphomycetes at species level from Tibetan Plateau p. 36, 2008.

在 PDA 培养基上，28℃培养 7 天，菌落直径 24～29 mm，淡黄色或灰褐色，绒毛状至毡状，平展，边缘不规则；背面淡黄色。菌丝有隔，光滑，无色，宽 0.7～2.8 μm，无球拍状菌丝。分生孢子单生，光滑或具刺，无色或淡褐色至中等褐色，着生于菌丝或短柄上，单胞，偶见双胞，梨形，5～17.1 × 3.1～6.8 μm；间生孢子梨形，4.5～10.1 × 3.1～6.8 μm；基痕宽 3.1～4.6 μm。

研究菌株：贵州：遵义市，湄潭县，茄子根际土土样，2017 年 7 月 5 日，韩燕峰，E20-2；西藏：尼木县，桃树下森林土，2008 年，耿月华，HSAUP072550；重庆：北碚区，中医院绿地土，2019 年 7 月 20 日，张芝元，D4.6。

世界分布：中国、日本、印度、法国、捷克、西班牙、尼日利亚。

讨论：嗜毛金孢的相近种是热带金孢。但嗜毛金孢具有较大的分生孢子，偶见间生孢子，分生孢子常具刺（Carmichael，1962；耿月华，2008），该特征能与热带金孢相区别。Hubálek 和 Hornich（1977）将嗜毛金孢腹腔接种小白鼠。2 个月后，该真菌可再次被分离，不同菌株均可引起小白鼠明显的脾肿大、肝脏和大网膜结节，以及肠脓肿。组织切片可观察到菌丝、分生孢子和出芽细胞。结果表明，嗜毛金孢对小白鼠具有潜在的致病性。Krempl-Lamprecht（1965）报道了人类指甲上偶尔自发感染，也能从腐烂的蹄子和鸡的鸡冠上分离到。

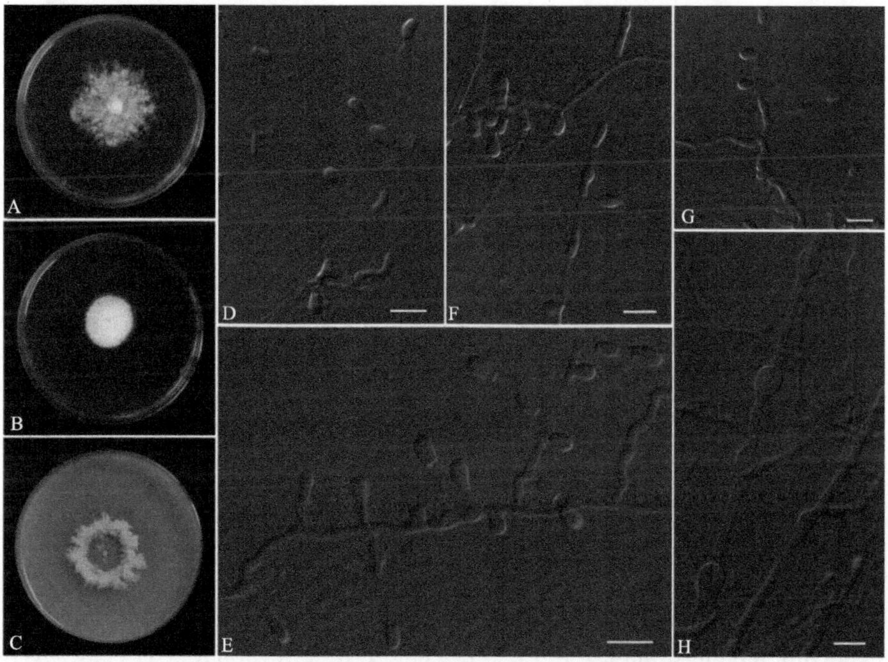

图 30　嗜毛金孢 *Chrysosporium keratinophilum* D. Frey ex J.W. Carmich.（标本号：E20-2）

A～C. PDA、MEA 和 OA 培养基上的菌落；D～G. 产孢结构；H. 间生孢子。标尺：D～H = 10 μm

侧孢金孢　图 31

Chrysosporium laterisporum Z. Li, Y.W. Zhang, W.H. Chen & Y.F. Han, Phytotaxa 400(5): 260, 2019.

在 PDA 培养基上，25℃培养 7 天，菌落直径 32 mm，中央浅黄色至白色，绒状，平展，中央隆起，边缘不整齐；背面奶酪黄色。菌丝具隔膜，无色，表面光滑；无球拍状菌丝。顶生和侧生的分生孢子着生于短柄上或侧枝上，侧生孢子较多，单生，无色，光滑，倒梨形、椭圆至圆形，5～12.5 × 2.5～10 μm；基痕宽 0.8～1.5 μm。

研究菌株：福建：福州市，森林公园，植物根系土，2014 年 11 月 7 日，罗韵，GZUIFR G310（主模式），GZUIFR G310.1，GZUIFR G310.2。

世界分布：中国。

讨论：该种形态特征近似于翼手金孢（Vidal *et al.*，1996）、乔治金孢（van Oorschot，

1980）、大孢金孢（Crous *et al.*，2013）和海岩金孢（Crous *et al.*，2013），但翼手金孢和乔治金孢有明显的球拍状菌丝（Vidal *et al.*，1996；van Oorschot，1980），大孢金孢的分生孢子较大（10～27 × 7～12 μm），卵孢金孢的分生孢子为卵圆形（Vidal *et al.*，1996；Han *et al.*，2022）；海岩金孢具间生孢子（Stchigel *et al.*，2013）。侧孢金孢主要以无球拍状菌丝和无间生孢子与其近似种相区别（Li *et al.*，2019）。

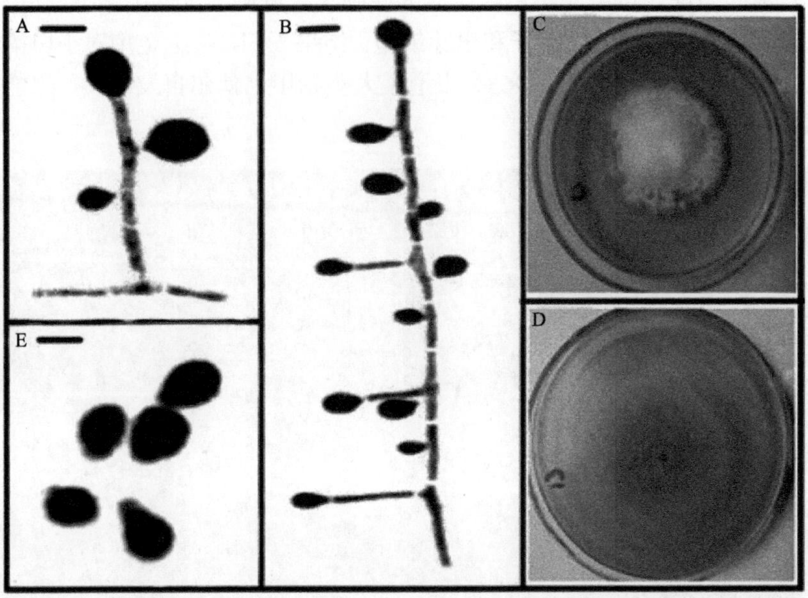

图 31　侧孢金孢 *Chrysosporium laterisporum* Z. Li, Y.W. Zhang, W.H. Chen & Y.F. Han（标本号：GZUIFR G310）

A、B. 产孢结构；C、D. PDA 培养基上的菌落；E.分生孢子。标尺：A、B、E = 10 μm

雷公山金孢　图 32

Chrysosporium leigongshanense Z. Li, G.P. Zeng, J. Ren, X. Zou & Y.F. Han, Mycosphere 8(8): 1033, 2017.

在 PDA 培养基上，40℃培养 14 天，菌落直径达 37～42 mm，白色，致密绒状，中央微隆，具轮纹，边缘规则；背面黄色至浅黄色。菌丝具隔膜，无色，光滑，宽 0.5～1.5 μm；具球拍状菌丝，5.4～10.8 × 2～4.5 μm。分生孢子顶生、侧生，着生于短突、短柄或侧枝上，大多数单生，少数双生或成链，无色，光滑，大多数柱状至棒状，有的稍弯曲、5.4～19.4 × 1～3.2 μm，少数卵圆形、5.4～6.5 × 2.2～3.2 μm；基痕宽 0.5～2 μm。

研究标本：贵州：雷山县，雷公山国家森林公园，植物根系土壤，海拔 1700 m，2013 年 6 月 8 日，邹晓，干模式标本 HMAS 255243（主模式），GZUIFR EB2702H = CGMCC 3.18621。

世界分布：中国。

讨论：该种在基于 ITS-5.8S rDNA 序列构建的系统发育树中与三亚金孢相近，但在形态上，三亚金孢产生近球形至倒卵圆形的分生孢子和大量的间生孢子（Zhang *et al.*，

2013）。该种的主要鉴别特征是具球拍状菌丝，分生孢子大多数柱状至棒状，有时稍弯曲，未见间生孢子和厚垣孢子（Li *et al.*，2017a）。

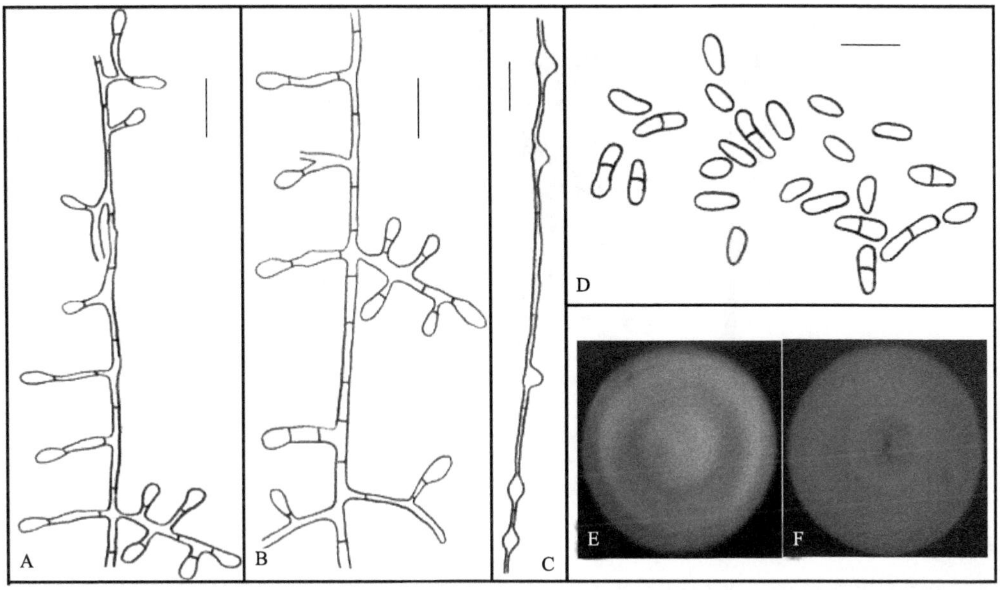

图 32　雷公山金孢 *Chrysosporium leigongshanense* Z. Li, G.P. Zeng, J. Ren, X. Zou & Y.F. Han（标本号：HMAS 255243）

A、B. 产孢结构；C. 球拍状菌丝；D. 分生孢子；E、F. PDA 培养基上的菌落。标尺：A～D = 10 μm

裂叶金孢　图 33

Chrysosporium lobatum Scharapov, Nov. Sist. Niz. Rast., 15: 144, 1978. Han *et al*., Journal of Fungal Research 4(1): 54, 2006.

在查氏培养基上，25℃培养 14 天，菌落平展，致密毡状，直径 15～20 mm，粉红色，夹有黄色；背面深红色，有放射状沟纹。菌丝无色，分隔，宽 0.5～1.8 μm。可育菌丝可见许多短的正交分枝。分生孢子着生于侧枝顶端或侧生于多数 3 个轮生或对生的短侧枝上，偶有双胞；分生孢子光滑或粗糙，无色，单生、顶生、侧生、间生，多数侧生于 1.5～3.5 μm 的短柄上，梨形，2.5～3.5 × 1.8 μm；卵形，1.5～4.5 × 1.5～2.0 μm；近球形，1.8～2.7 μm。间生孢子桶状，1.5～4.5 × 1.5～1.8 μm；厚垣孢子半透明，光滑或厚壁，近球形至球形，5～8 μm。

研究菌株：新疆：克拉玛依，辣椒根际土样，2004 年 8 月 9 日，韩燕峰，GZUIFR XJ22021。

世界分布：中国、印度、捷克斯洛伐克、罗马尼亚、意大利、加拿大。

讨论：裂叶金孢与粪生金孢、乔治金孢的菌落均具有红色色调。但裂叶金孢的可育菌丝可见许多短的正交分枝；粪生金孢具球拍状菌丝，分生孢子倒卵圆形至近球形，基痕较宽（van Oorschot，1980）；乔治金孢无球拍状菌丝，分生孢子倒卵圆形、棒状或不规则椭圆形，间生孢子未见（van Oorscho，1980）。裂叶金孢的主要特征是菌落粉红色，可育菌丝正交分枝多，分生孢子梨形、卵形和近球形，具间生孢子和厚垣孢子（韩燕峰等，2006）。

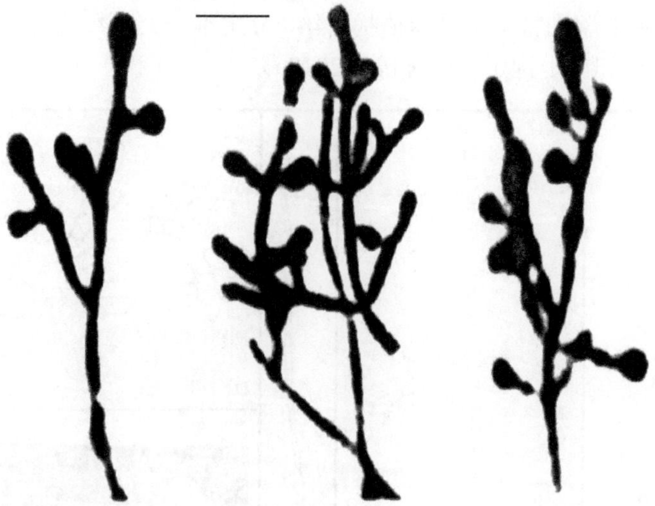

图 33　裂叶金孢 *Chrysosporium lobatum* Scharapov（标本号：GZUIFR XJ22021）

标尺：10 μm

粪生金孢　图 34

Chrysosporium merdarium (Ehrenb.) J.W. Carmich., Can. J. Bot. 40: 1160, 1962. Song *et al*., Mycosystema 27(1): 20, 2008.

≡ *Sporotrichum merdarium* Ehrenb., Sylv. Mycol. Berol. (Berlin): 10, 1818.

= *Chrysosporium corii* Corda, Deutschl. Fl., 3 Abt. (Pilze Deutschl.) 3(13): 85, 1833.

= *Chrysosporium merdarium* var. *roseum* W. Gams & Domsch, Nova Hedwigia 18: 7, 1969.

= *Chrysosporium verruculatum* Scharapov, Nov. Sist. Niz. Rast. 15: 143, 1978.

在查氏培养基上，25℃培养 14 天，菌落直径 30～35 mm，菌落平展，初白色，后变亮黄色或粉色或绿色，绒毛状、毡状至粉状，中部隆起，有时具放射状纵沟；边缘规则，局限；背面中部亮黄色，边缘淡黄色。菌丝无色，分隔，宽 1.3～2 μm；具球拍状菌丝。顶侧生分生孢子直接着生于菌丝上或着生于短突起或侧枝上，单生，半透明至淡黄色，光滑或轻微具细刺，厚壁，倒卵形至近球形，4～10 × 3～6 μm；具少量间生孢子，单生或 2 个串生，柱状至桶状，5～12 × 3～6 μm；基痕宽 1.5～3 μm；有时可见厚垣孢子，光滑或厚壁，半透明，球形至近球形，直径 11 μm 左右。

研究菌株：山西：祁县，绿化地土样，2009 年 1 月 21 日，韩燕峰，GZM0901。

世界分布：中国、英国、美国等，世界性分布。

讨论：粪生金孢的相近种为裂叶金孢和乔治金孢，但乔治金孢的分生孢子倒卵圆形、棒状或不规则椭圆形，间生孢子未见（van Oorscho，1980）；裂叶金孢的可育菌丝正交分枝多，分生孢子近球形至倒卵圆形，具间生孢子和厚垣孢子。粪生金孢的主要鉴别特征是菌落初白色，后变亮黄色或粉色或绿色；具球拍状菌丝；分生孢子倒卵圆形至近球形，具间生孢子和厚垣孢子。大多数粪生金孢的菌株倾向于失去它们最初的颜色，并在重复的传代培养后变成白色。有些菌株甚至停止正常产孢而发育成厚垣孢子。粪生金孢 CBS 475.76 和 477.76 菌株在 30℃条件下生长较差，其他菌株在 25℃条件下均不能生长（van Oorschot，1980）。该菌适应性强，分布广（高云超等，2001；陈曦，2010；

慕东艳，2012；Song *et al.*，2008；宋伟，2008；王娜，2012）。

图 34　粪生金孢 *Chrysosporium merdarium* (Ehrenb.) J.W. Carmich.（引自 Song *et al.*，2008）
标尺：10 μm

多形金孢　图 35

Chrysosporium multiforme Y.F. Han, W.H. Chen, J.D. Liang & Z.Q. Liang, Phytotaxa 539(1): 10, 2022.

在 PDA 培养基上，26℃培养 14 天，菌落直径 45～50 mm，中心淡黄色隆起，中部白色，稀疏绒毛状，可见孢子粉团，圆形，边缘不规则；背面淡黄色。菌丝分隔，光滑，无色，宽 2～3μm；未见球拍状菌丝。顶侧生分生孢子单生，表面粗糙或光滑，无色，着生于菌丝或短柄上，梨形、9.5～15.5 × 7～8 μm，球形至椭圆形、8.5～10 × 7.5～9 μm；基痕宽 1.5～3 μm。间生孢子长卵形，单生或 2～4 个串生，7.5～13.5 × 4～8 μm；梭形，9～24.5 × 5.5～8 μm。

研究菌株：甘肃：兰州市，兰州大学，雪松根际土样，2017 年 7 月 13 日，王健健，GZAC U3（主模式），GZU U3。干模及其模式菌株 GZU U3 均保存于贵州大学真菌资源研究所。

世界分布：中国。

讨论：基于 ITS-5.8S rDNA 序列构建的系统发育树，显示多形金孢与嗜毛金孢的亲缘关系相近，嗜毛金孢的分生孢子无色或淡褐色至中等褐色，梨形，偶见双胞（van Oorschot，1980）；而多形金孢无球拍状菌丝和具有多形的顶侧生分生孢子（Han *et al.*，2022）。

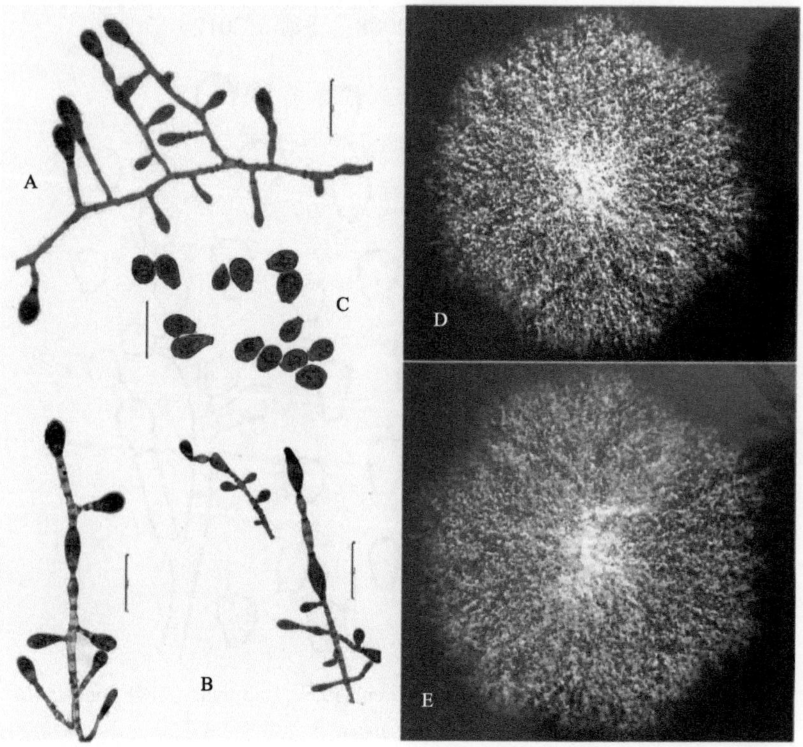

图 35　多形金孢 *Chrysosporium multiforme* Y.F. Han, W.H. Chen, J.D. Liang & Z.Q. Liang（标本号：GZAC U3）

A. 产孢结构；B. 间生孢子；C. 分生孢子；D、E. PDA 培养基上的菌落。标尺：A～C = 20 μm

卵孢金孢　图 36

Chrysosporium ovalisporum Z. Li, Y.W. Zhang, W.H. Chen & Y.F. Han, Phytotaxa 400(5): 261, 2019.

在 PDA 培养基上，25℃培养 14 天，菌落直径 28 mm，白色至淡黄色，粉状，平展，边缘不规则；背面沙棕色。菌丝分隔，无色，光滑，宽 2.5～3.5 μm。分生孢子单生、顶生或侧生，着生于长的或短的小柄上，小柄分隔或不分隔，有的中间膨大，小柄垂直于菌丝；大多数单胞，偶见双胞，分生孢子表面光滑，无色，柱状或广卵形至倒卵球形，5～15 × 2.5～7 μm；基痕宽 0.8～2.5 μm。

研究菌株：福建：福州市，福建动物园，秃鹰毛土，2013 年 9 月 12 日，罗韵，GZUIFR G446（主模式）。

世界分布：中国。

讨论：金孢属已知种中，仅硫磺金孢 *C. sulphureum* (Fiedl.) Oorschot & Samson 和同步金孢的菌落呈白色至淡黄色，无球拍状菌丝，无间生孢子和厚垣孢子，与本种形态上相似。但本种孢子较窄，表面光滑，而硫磺金孢和同步金孢孢子较宽，表面都光滑至粗糙，偶具刺（表 2）。

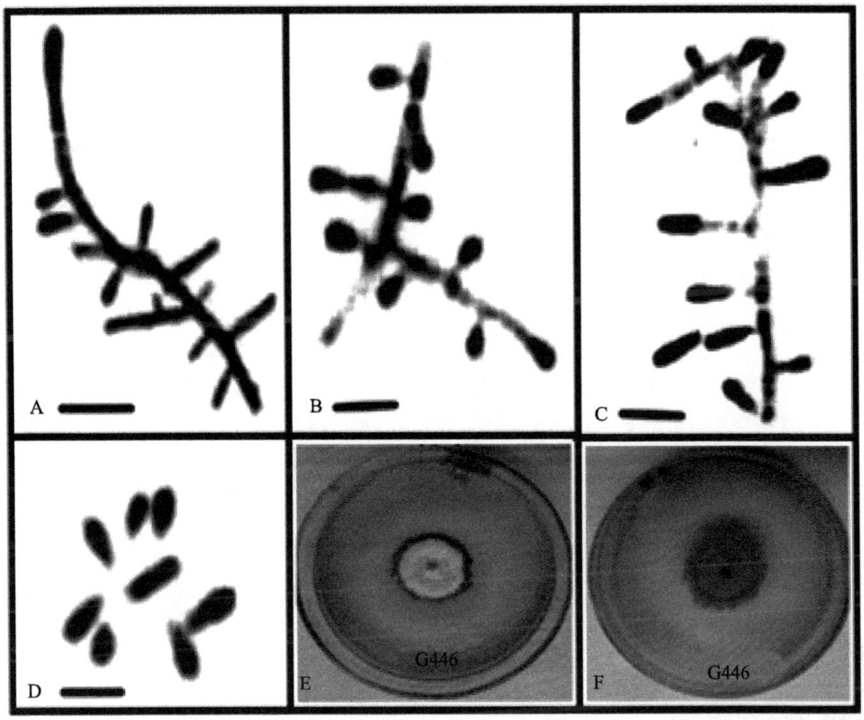

图 36　卵孢金孢 *Chrysosporium ovalisporum* Z. Li, Y.W. Zhang, W.H. Chen & Y.F. Han（标本号：GZUIFR G446）

A～C. 产孢结构；D. 分生孢子；E、F. PDA 培养基上的菌落。标尺：A～D = 10 μm

表 2　卵孢金孢与近似种形态比较

菌名	菌落颜色	菌落质地	孢子着生方式	孢子形状	孢子大小（μm）	孢子表面
硫磺金孢	黄（淡）	毡-粉状	无柄-锥形突起；单胞	近球形-拟卵形-棒状	3～7 × 3～8	粗糙-光滑
同步金孢	白	绒-毡状	锥形突起-侧枝；单胞	拟卵形-梭状	7.5～11 × 4～5.5	粗糙-光滑-少刺
卵孢金孢 GZUIFR G446	白-淡黄	粉状	无柄-短柄-侧枝；单胞-多胞	柱状-广卵形-倒卵球形	5～15 × 2.5～7	光滑

三亚金孢　图 37

Chrysosporium sanyaense Y.W. Zhang, Y.F. Han, J.D. Liang & Z.Q. Liang, Mycosystema 32(4): 613, 2013.

在 PDA 培养基上，25℃培养 7 天，菌落直径 30～35 mm，白色至黄绿色，粉状，平展，边缘整齐，圆形；背面棕色。菌丝具隔膜，无色，表面光滑或粗糙，菌丝宽 1.0～2.1 μm；具球拍状菌丝，4.5～7.2 × 4.5～6.4 μm。顶生和侧生孢子无柄或着生于短柄上或侧枝上，4.5～15 × 1.0～1.8 μm，单生，无色，光滑或粗糙，近球形至倒卵圆形，3.5～9（～18）× 2.5～4.5（～9）μm；间生孢子丰富，单生或成短链，桶状至椭圆形，4.5～6.5 × 2.8～4.5 μm。

研究菌株：海南：三亚市，棕榈树根际土样，2010 年 3 月 5 日，梁宗琦，GZUIFR A10222（主模式），分离活菌株 GZU A10222。

世界分布：中国。

讨论：该种形态特征与粪生金孢、拟粪生金孢、同步金孢和线状金孢较为近似，但粪生金孢的顶侧生孢子具稀少的刺状结构；拟粪生金孢菌丝的侧枝呈树枝状结构；同步金孢无球拍状菌丝（van Oorschot，1980）；线状金孢的侧枝呈直角分枝，并且有长丝状孢子（1～3 × 5～40 μm）（Sigler *et al.*，1982）。该种的主要鉴别特征是具有球拍状菌丝和间生孢子（Zhang *et al.*，2013）。

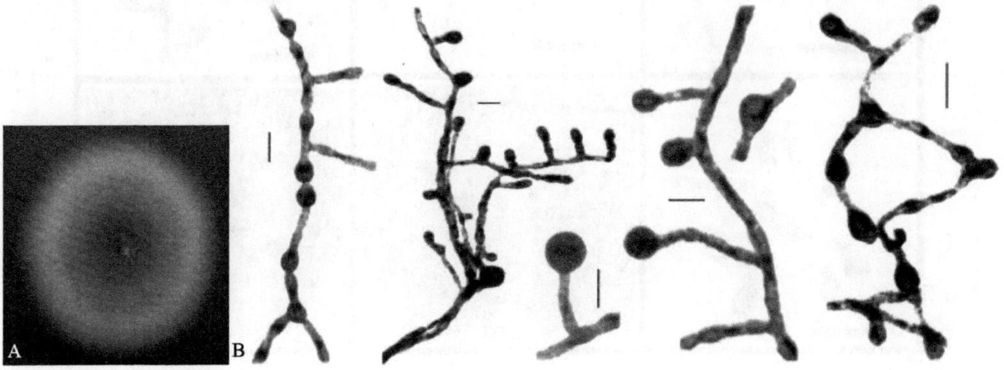

图 37　三亚金孢 *Chrysosporium sanyaense* Y.W. Zhang, Y.F. Han, J.D. Liang & Z.Q. Liang（标本号：GZUIFR A10222）

A. PDA 培养基上的菌落；B. 产孢结构。标尺：B = 10 μm

山西金孢　图 38

Chrysosporium shanxiense Y.W. Zhang, W.H. Chen, X. Zou, Y.F. Han & Z.Q. Liang, Mycosystema 35(11): 1341, 2016.

在 PDA 培养基上，26℃培养 14 天，菌落直径约 25 mm，白色，绒毛状至粉状，边缘较规则；背面浅黄色。菌丝分隔，无色，光滑，2～4.3 μm；有球拍状菌丝，10.8～16.2 × 5.4～6.5 μm。分生孢子侧生或顶生，着生于菌丝上或短的突起或短柄上，单生或 2～3 个串生，大多数单胞，偶见双胞；大多数长倒卵形，6.5～11.9 × 3.2～6.5 μm；少数棒状至柱状，8.6～17.3 × 2.2～4.3 μm；极少数不规则肾形，10.8～18.4 × 2.2～5.4 μm；基痕宽 1.9～4.3 μm；具间生孢子，单生，不规则桶状至肾形，8.6～13 × 3.2～6.5 μm；产孢细胞与分生孢子连接处具有独特的“领状”结构。

研究菌株：山西：祁县，固邑村，麻雀窝，2015 年 3 月 1 日，韩燕峰，GZUIFR EB1601M.1（主模式），GZUIFR EB1601M.2，GZUIFR EB1601M.3。

世界分布：中国。

讨论：山西金孢与热带金孢系统发育上亲缘关系较近，但在形态特征上，热带金孢的分生孢子半透明，且在产孢细胞与分生孢子连接处无独特的“领状”结构（van Oorschot，1980），这些特征能与本种明显区别。山西金孢的主要鉴别特征是具球拍状菌丝，分生孢子大多数长倒卵形，产孢细胞与分生孢子连接处具有独特的“领状”结构（张延威等，2016）。

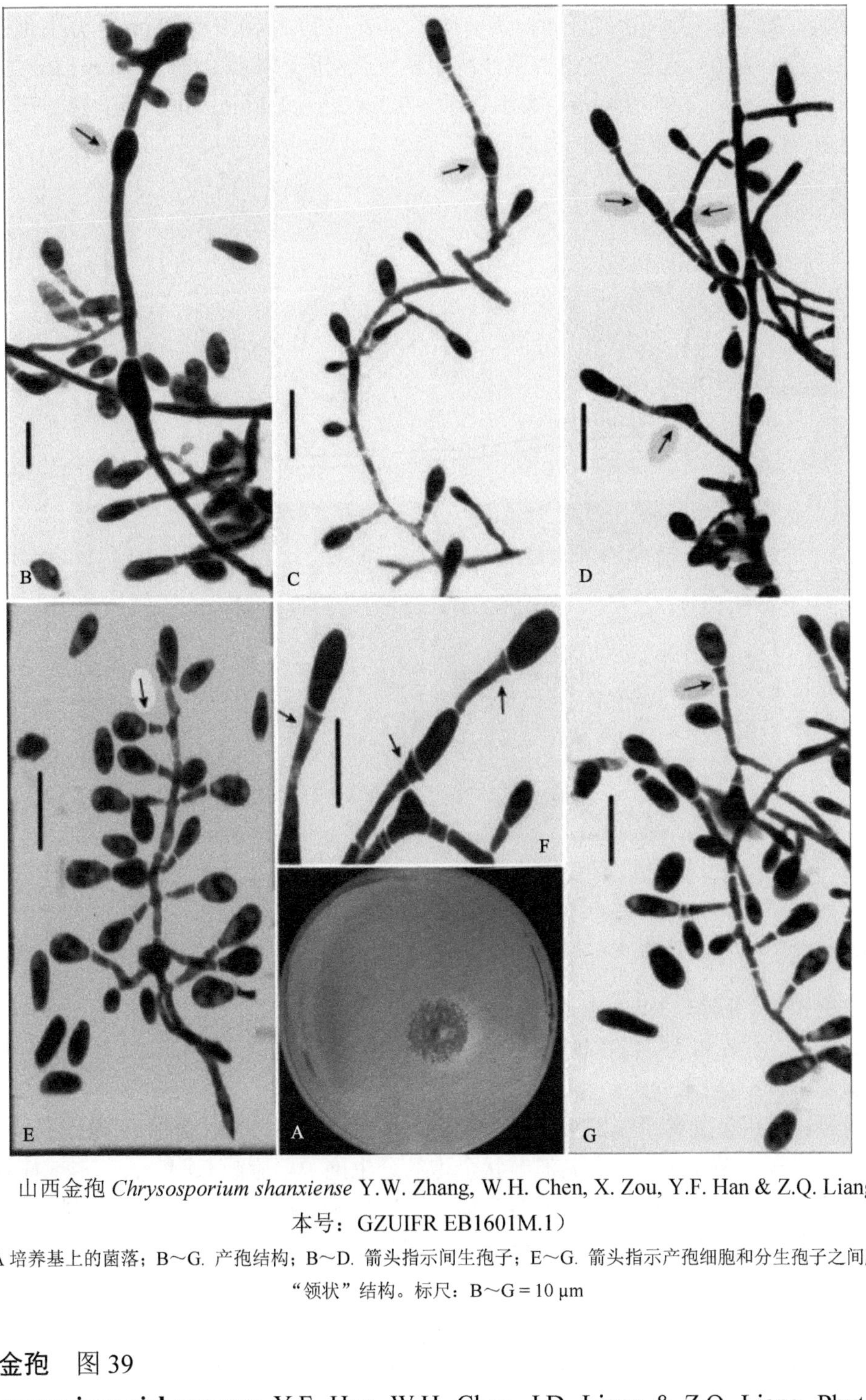

图 38　山西金孢 *Chrysosporium shanxiense* Y.W. Zhang, W.H. Chen, X. Zou, Y.F. Han & Z.Q. Liang（标本号：GZUIFR EB1601M.1）

A. PDA 培养基上的菌落；B～G. 产孢结构；B～D. 箭头指示间生孢子；E～G. 箭头指示产孢细胞和分生孢子之间膨大的“领状”结构。标尺：B～G = 10 μm

四川金孢　图 39

Chrysosporium sichuanense Y.F. Han, W.H. Chen, J.D. Liang & Z.Q. Liang, Phytotaxa 539(1): 11, 2022.

在 PDA 培养基上，28℃培养 7 天，菌落直径 30～31 mm，中心隆起白色绒状，外部稀疏绒状，边缘规则；背面淡黄色。菌丝分隔，表面光滑，无色或半透明，菌丝上有膨大结构，宽 1.4～3.4 μm；具球拍状菌丝，34.2～129.2 × 4.5～6.9 μm。分生孢子单生或串生，表面光滑，无色，着生于菌丝或短柄上，双胞，棒状、6.7～8.4 × 1.9～2.5 μm，倒卵形、4.3～6.3 × 1.9～2.6 μm，梨形、4.9～9.3 × 2.6～4.7 μm；间生孢子椭圆形，4.4～5.5 × 2.2～4.5 μm；基痕宽 1.1～1.7 μm。

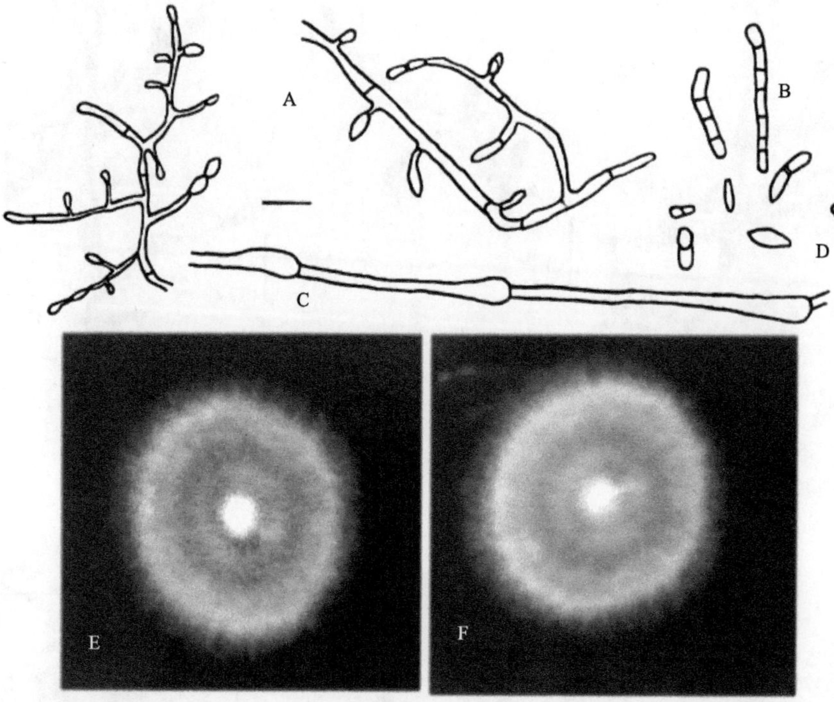

图 39　四川金孢 *Chrysosporium sichuanense* Y.F. Han, W.H. Chen, J.D. Liang & Z.Q. Liang（标本号：GZAC FX8）

A. 产孢结构；B. 节孢子；C. 球拍状菌丝；D. 分生孢子；E、F. PDA 培养基上的菌落。标尺：A～D = 20 μm

研究菌株：四川：巴中市，香樟树土样，2017 年 8 月 1 日，韩燕峰，GZAC FX8（主模式）；山西：祁县，绿化地土样，2017 年 8 月 10 日，韩燕峰，GZAC I17，GZAC I18。

世界分布：中国。

讨论：四川金孢在系统发育上与金孢属其他种能明显区分，在形态学上，该种具有明显的节孢子与节状金孢相似，但节状金孢的分生孢子单胞或双胞，长卵形或棒状，具长卵形的间生孢子。四川金孢的主要鉴别特征是具球拍状菌丝，分生孢子棒状或倒卵形或梨形，具椭圆形的间生孢子（Han *et al.*，2022）。

中国金孢　图 40

Chrysosporium sinense Z.Q. Liang, Acta Mycol. Sin. 10(1): 51, 1991.

在查氏培养基上，18℃培养 20 天，菌落直径 30 mm，毡状，灰白色至蚌肉白色，具放射状沟纹和钻状的孢梗束，高约 2 mm；延长培养时间菌落表面可见粒状的菌丝球；菌落背面中部淡笋皮棕色或咖啡色。菌丝无色，壁薄，粗，（0.5～）0.9～1.5 μm。顶生和侧生的分生孢子无柄，着生于短突起或侧枝上，单生，无色，光滑，单胞，梨形或倒卵形，2.4～3.8 × 4～5.7 μm；间生孢子多，单生或被胞间连体细胞分开成对，柱状、卵形或桶状，无色，光滑，（1.2～）1.4～3.7 × 3.7～7（～10）μm。

研究菌株：四川：理县，冬虫夏草内菌核，1991 年 6 月 10 日，梁宗琦，CGAC80-501。

世界分布：中国。

讨论：在形态上，中国金孢与昆士兰金孢在间生孢子大量产生、单生和不成链等方面相似，但昆士兰金孢的生长最适温度为 25℃，菌落粉状，不形成孢梗束，而中国金孢喜低温环境，最适温度低于 20℃，菌落毡状，形成孢梗束（梁宗琦，1991）。

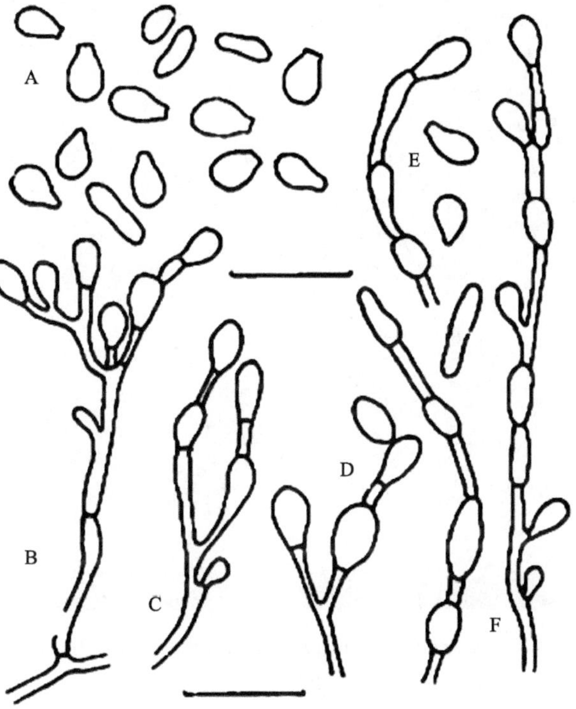

图 40　中国金孢 *Chrysosporium sinense* Z.Q. Liang（标本号：CGAC80-501）

A. 分生孢子；B～F. 产孢结构和间生孢子。标尺：A～F = 10 μm

热带金孢　图 41

Chrysosporium tropicum J.W. Carmich., Can. J. Bot. 40: 1170, 1962. Liang *et al*., Mycosystema 27(3): 448, 2008.

在 PDA 培养基上，37℃培养 14 天，菌落直径 45 mm；26℃培养 14 天，菌落直径 80 mm，薄，平展，粉状，中央致密，边缘稀疏，乳白色；背面淡黄色。菌丝半透明，表面光滑，宽 1.3～3.6 μm；有球拍状菌丝，8 × 4.5 μm。分生孢子多，顶生、侧生，或着生于短的突起、长柄（12～27 μm）或侧分枝上，单生，偶见双生或 3 个链生，半透明，表面光滑，大多倒卵形，3.6～7.2 × 2.0～4.5 μm；基痕宽 1.8～2.8 μm；间生孢子很少，柱状。偶见厚垣孢子，球形。

研究菌株：甘肃：格尔木，骆驼刺根际土样，2006 年 8 月 6 日，梁建东和韩燕峰，GZDXIFR M314-2；贵州：贵阳市，黔灵山公园，土样，2013 年 6 月 10 日，韩燕峰，

GZAC O.1.2。

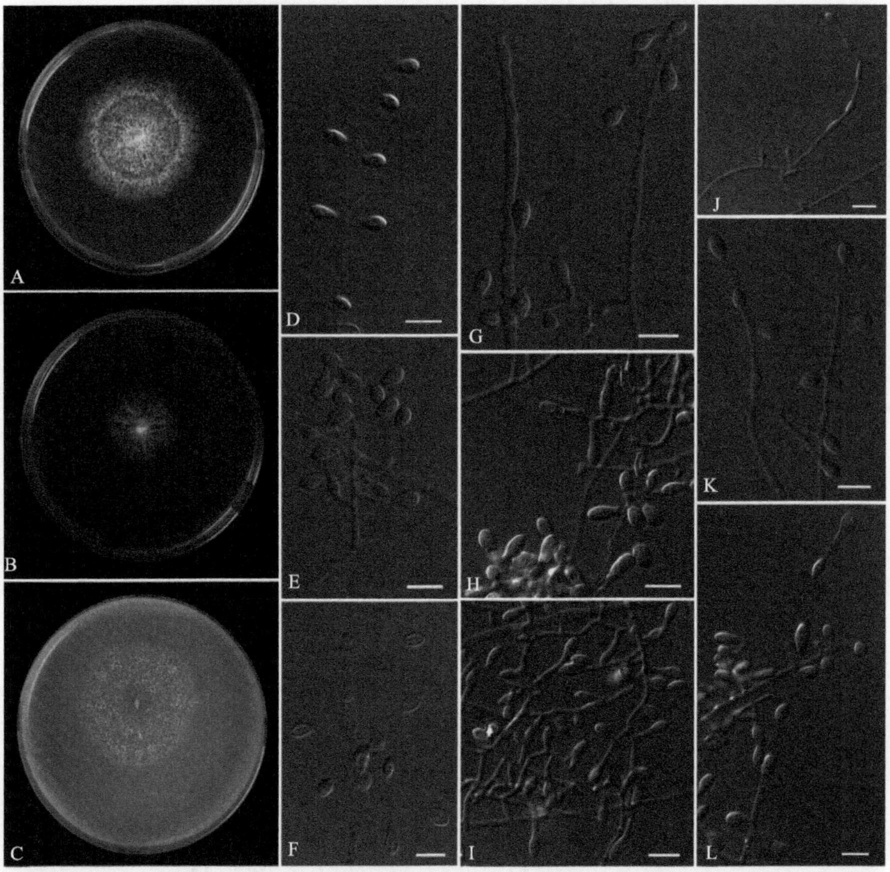

图 41　热带金孢 *Chrysosporium tropicum* J.W. Carmich. (标本号：GZDXIFR M314-2)

A～C. PDA、MEA 和 OA 培养基上的菌落；D～L. 产孢结构。标尺：D～L = 10 μm

世界分布：中国、印度、所罗门群岛、法国、瑞典、埃及、荷兰。

讨论：该种与嗜毛金孢、节状金孢、热带金孢形态特征较为相似，但嗜毛金孢和节状金孢的分生孢子明显较大（表 3）。热带金孢的分生孢子较小，间生孢子少（梁建东等，2008）。

表 3　热带金孢和其相关种的主要特征的比较

菌名	菌落颜色	球拍状菌丝	孢子着生方式	孢子形状	孢子大小（μm）	孢子表面	间生孢子
嗜毛金孢	白-黄	有	无柄-短突-短柄	倒卵形-棒状	11～22 × 3.5	光滑	有
节状金孢	白色	有	无柄-短突-侧枝	倒卵形-棒状-近椭圆	6～9.5 × 3～3.5	光滑	有（多）
热带金孢	白色	有	无柄-短突-侧枝	倒卵形-棒状	4.5～7.5 × 3～3.5	光滑	有（少）

绒毛金孢 图 42

Chrysosporium villiforme Y.F. Han, W.H. Chen, J.D. Liang & Z.Q. Liang, Phytotaxa 539(1): 12, 2022.

在 PDA 培养基上，26℃培养 7 天，菌落直径约 40 mm，淡黄色，毡状，中心白色微隆，边缘稀疏长绒毛状，不规则；背面白色至淡黄色。菌丝分隔，光滑或粗糙，无色，宽 1.5～3 μm，无球拍状菌丝。分生孢子单生或 2～3 个串生，表面光滑，无色，顶生或侧生于菌丝或短侧枝上，卵形、4～7.5 × 2～2.5 μm，或梨形、4～5.5 × 2.5～3 μm，或棒状、4.5～7 × 2～2.5 μm；间生孢子椭圆形，2～3 × 1.0～2.5 μm；基痕宽 0.5～1.5 μm；厚垣孢子未见。

研究菌株：山西：临汾市，吉县，生活垃圾土土样，2017 年 8 月 2 日，梁建东、韩燕峰，GZAC L19.4（主模式），GZU L19.4；湖北：宜昌市，土样，2017 年 8 月 15 日，张芝元，GZAC A1，菌株 GZU A1。

世界分布：中国。

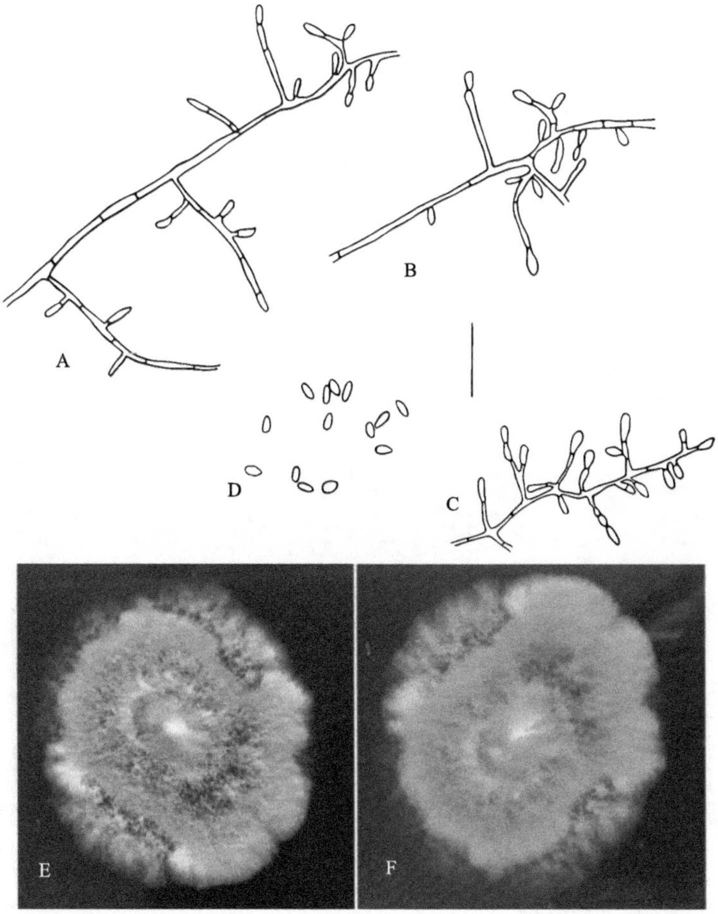

图 42 绒毛金孢 *Chrysosporium villiforme* Y.F. Han, W.H. Chen, J.D. Liang & Z.Q. Liang（标本号：GZAC L19.4）

A～C. 产孢结构；D. 分生孢子；E、F. PDA 培养基上的菌落。标尺：A～D = 20 μm

讨论：绒毛金孢与青海金孢形态相似，但青海金孢的分生孢子丰富，椭圆、棒状至柱状，偶见双胞，柱状，具椭圆形的间生孢子，产生厚垣孢子（韩燕峰等，2013），但青海金孢已被移入角蛋白癣菌属中，二者亲缘关系较远；而绒毛金孢的分生孢子形状多样，卵形或梨形或棒状，具间生孢子，不产生厚垣孢子（Han *et al.*，2022）。

轮带金孢　图 43

Chrysosporium zonatum Al-Musallam & C.S. Tan, Persoonia 14(1): 69, 1989. Han *et al.*, Microbiology China 46(9): 2211, 2019.

在 PDA 培养基上，26℃培养 14 天，菌落直径 52～53 mm，白色，絮状，中心淡黄色，粉状，近边缘部有一圈白色轮纹，边缘整齐，半透明色；背面白色，中心淡黄色。菌丝分隔，光滑，无色，宽 0.6～1.6 μm；未见球拍状菌丝。分生孢子单生，表面粗糙或光滑，无色，着生于菌丝短柄上，梨形、2.7～3.8 × 1.2～2.2 μm，倒卵形、2.6～3.1 × 1～1.9 μm；间生孢子未见；厚垣孢子 4.2～5.3 μm；基痕宽 0.6～1.3 μm。

研究菌株：云南：腾冲市，土样，2017 年 8 月 25 日，韩燕峰，HMAS 255385，GZAC O.1。

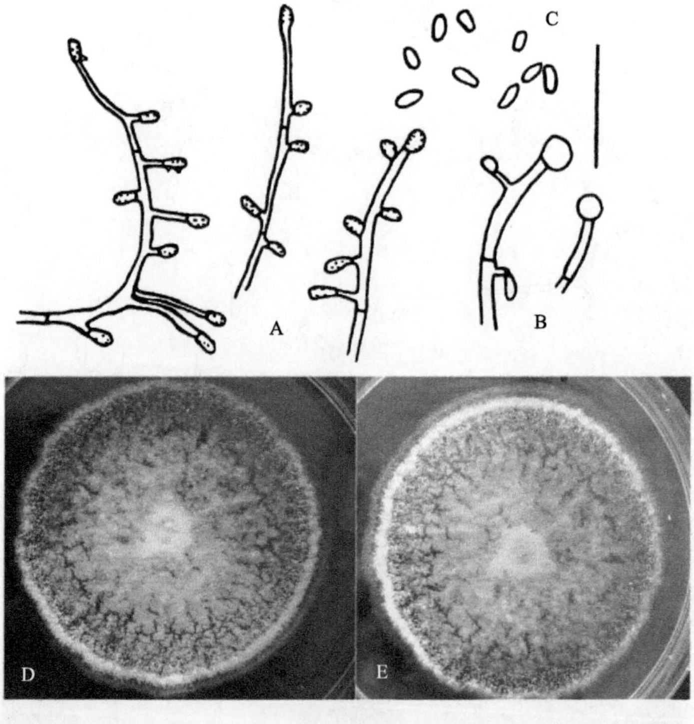

图 43　轮带金孢 *Chrysosporium zonatum* Al-Musallam & C.S. Tan（标本号：HMAS 255385）

A. 产孢结构；B. 厚垣孢子；C. 分生孢子；D、E. PDA 培养基上的菌落。标尺：A～C = 20 μm

世界分布：中国、科威特。

讨论：轮带金孢在系统发育上为单独的一个分支。在形态学上，与节状金孢和昆士兰金孢相似，但它们在孢子大小上明显不同，节状金孢为 5.8～17 × 2.5～5.5 μm（van Oorschot，1980），昆士兰金孢为 7～8 × 3～4 μm（van Oorschot，1980）；而轮带金孢的分生孢子较小，5～7.5 × 3～5 μm（Al-Musallam & Tan，1989）、2.7～3.8 × 1.2～2.2 μm（韩燕峰等，2019）。

作者未观察的种

法斯特金孢　图 44

Chrysosporium fastidium Pitt, Trans. Br. Mycol. Soc. 49(3): 467, 1966. Wu, Taxonomic studies on soil dematiaceous Hyphomycetes from north of North China p. 81, 2006.

讨论：据该种的描述（吴悦明，2006），在 PDA 培养基上，25℃培养 14 天，菌落平展，微绒毛状，直径 40～50 mm，灰色至灰褐色。分生孢子直接生长在膨大的产孢瓶体顶端，近球形、梨形或椭球形，淡金黄色到淡褐色，无隔，基部平截有时具柄，3～5 × 4～7 μm。目前仅在内蒙古阿尔山有报道。

世界分布：中国、印度、澳大利亚。

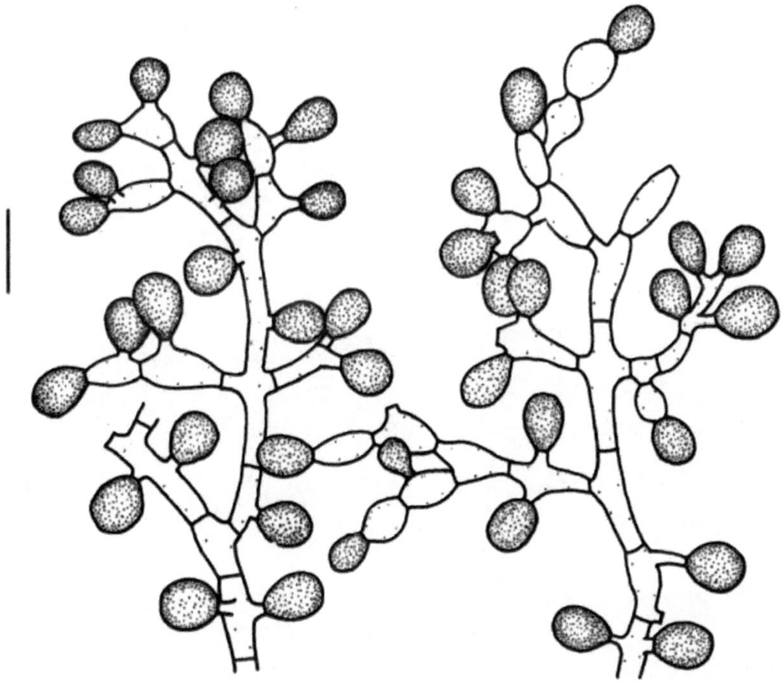

图 44　法斯特金孢 *Chrysosporium fastidium* Pitt（标本号：HSAUP041087；引自吴悦明，2006）
标尺：10 μm

苍白金孢　图 45

Chrysosporium pallidum Z.F. Zhang & L. Cai, Fungal Diversity 106: 64, 2021a.

讨论：根据原始描述（Zhang *et al.*，2021a），该种的显著特征如下：在 PDA 培养基上，25℃培养 28 天，菌落直径 28～34 mm，平展，毯状，具轮纹。子囊果大量产生，球形，初白色，后变为淡黄色。子囊具 8 个子囊孢子，倒梨形、近球形或球形，无色，8.0～13.0 × 7.5～10.5 μm。子囊孢子扁球形、球形，直径 2.5～3.5 μm。关节状分生孢子大量产生，间生、侧生或顶生，单生，无色。该种分离于喀斯特洞穴的动物粪便中。

世界分布：中国。

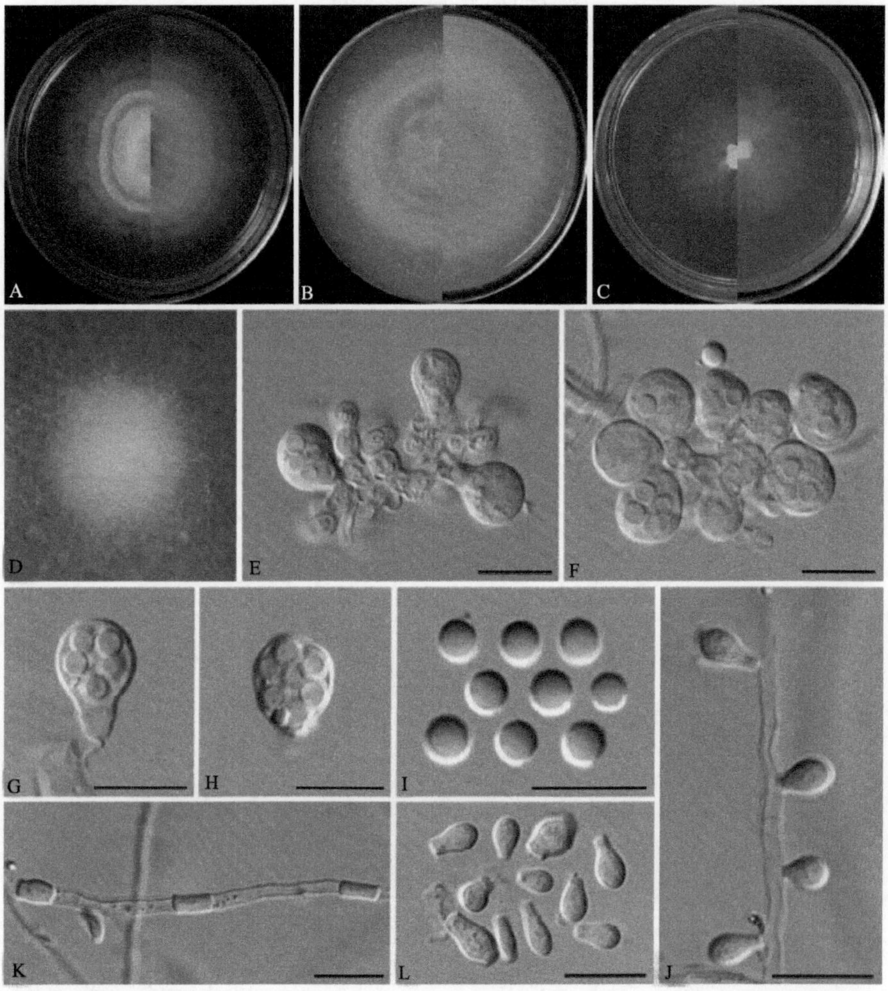

图 45　苍白金孢 *Chrysosporium pallidum* Z.F. Zhang & L. Cai（标本号：HMAS 247992；引自 Zhang *et al.*，2021a）

A～C. PDA、OA 和合成低营养琼脂（SNA）培养基上的菌落；D. 子囊果；E～H. 子囊；I. 子囊孢子；J～L. 关节状分生孢子。标尺：E～L = 10 μm

栉霉属 **Ctenomyces** Eidam

Beitr. Biol. Pfl. 3: 274, 1880

闭囊壳橙褐色，球形。包被由两层菌丝组织组成：内层为紧密排列的拟薄壁组织细胞，细胞壁薄、表面光滑；外层稀疏疏松，由菌丝构成，其细胞壁厚、呈橙褐色，表面具细疣至小刺状突起，多数隔膜处细胞膨大形成节状突起。附属物从外层产生顶端分枝或侧分枝；附属物由 1～3 伸长的基部细胞和 5～11 个较短的、厚壁的、粗糙的细胞组成。无性阶段的特征为菌落绒状至粉状，大多圆形；菌丝无色，光滑，分隔，分枝；顶生或侧生分生孢子着生于菌丝、短突、侧枝上。分生孢子单生或 2～3 个成短链，椭圆形、近球形、卵圆形，表面光滑或粗糙；有时可见间生孢子。

模式种：锯齿栉霉 *Ctenomyces serratus* Eidam。

中国栉霉属分种检索表

1. 间生孢子存在，近球形至椭圆形…………………………………………白色栉霉 ***C. albus***
1. 无间生孢子……………………………………………………………………………………2
  2. 常单个分生孢子着生于安瓿形膨大结构上……………………青灰栉霉 ***C. peltricolor***
  2. 常 1～2 个分生孢子着生于安瓿形膨大结构上………………………………………………3
3. 分生孢子最大可超过 20 μm ……………………………………………锯齿栉霉 ***C. serratus***
3. 分生孢子最大不超过 20 μm……………………………………………………………………4
  4. 分生孢子倒卵形或椭圆形………………………………………倒卵形栉霉 ***C. obovatus***
  4. 分生孢子椭圆形至梨形……………………………………………棉毛栉霉 ***C. vellereus***

白色栉霉　图 46，图版 I：3

Ctenomyces albus Y.F. Han, Z.Q. Liang & Zhi Y. Zhang, MycoKeys 47: 8, 2019.

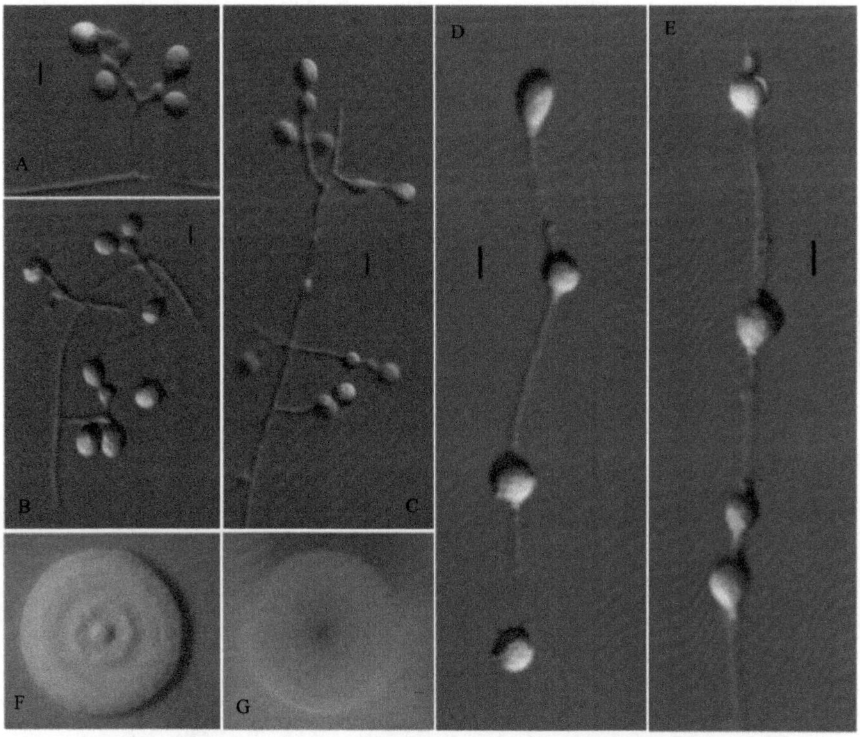

图 46　白色栉霉 *Ctenomyces albus* Y.F. Han, Z.Q. Liang & Zhi Y. Zhang（标本号：HMAS 255389）
A～C. 产孢结构与分生孢子；D、E. 间生孢子；F、G. PDA 培养基上的菌落。标尺：A～E = 10 μm

在 PDA 培养基上，25℃培养 14 天，菌落直径为 32 mm，短绒状至粉状，具环纹，圆形，边缘规则；背面淡黄色。菌丝无色，光滑，分隔，分枝，宽 1.1～2.4 μm；无球拍状菌丝。顶生、侧生分生孢子着生于菌丝、短突、侧枝，或安瓿形膨大结构上。分生孢子单生，或 2～3 个由短而纤细的菌丝连接成链，椭圆形，表面光滑或粗糙，疣状，12.8～18.6 × 10.8～14.7 μm；间生孢子近球形至椭圆形，光滑或粗糙，13.1～16.9 × 11.2～14.4 μm。

研究菌株：贵州：土壤，2016 年 9 月 9 日，张芝元，HMAS 255389（主模式），HMAS 255442，HMAS 255443，CGMCC 3.19232，CGMCC 3.18631 = GZUIFR QL17.11 和 CGMCC 3.18632 = GZUIFR QL17.12。

世界分布：中国。

讨论：形态上，白色栉霉区别于栉霉属中的其他物种在于其能产生间生孢子（Zhang *et al.*，2019）。此外，基于 ITS 序列的系统发育分析，白色栉霉与锯齿栉霉的系统发育关系较近，但其相似性较低，为 92%（ITS 序列：508 bp 中差异碱基达 37 bp）。

倒卵形栉霉　图 47

Ctenomyces obovatus Y.F. Han, Z.Q. Liang & Zhi Y. Zhang, MycoKeys 47: 8, 2019.

在 PDA 培养基上，25℃培养 14 天，菌落直径为 14～15 mm，淡黄色，边缘白色，绒毛状，圆形，边缘规则；背面棕色。菌丝无色，光滑，分隔，大量分枝，宽 1.2～2.4 μm；无球拍状菌丝。顶生、侧生分生孢子着生于菌丝、短突、侧枝，或安瓿形膨大结构上。分生孢子单生或 2 个链状生长，椭圆形、倒卵形，光滑或粗糙，疣状，刺状，10.3～17.3 × 9.7～10.5 μm。

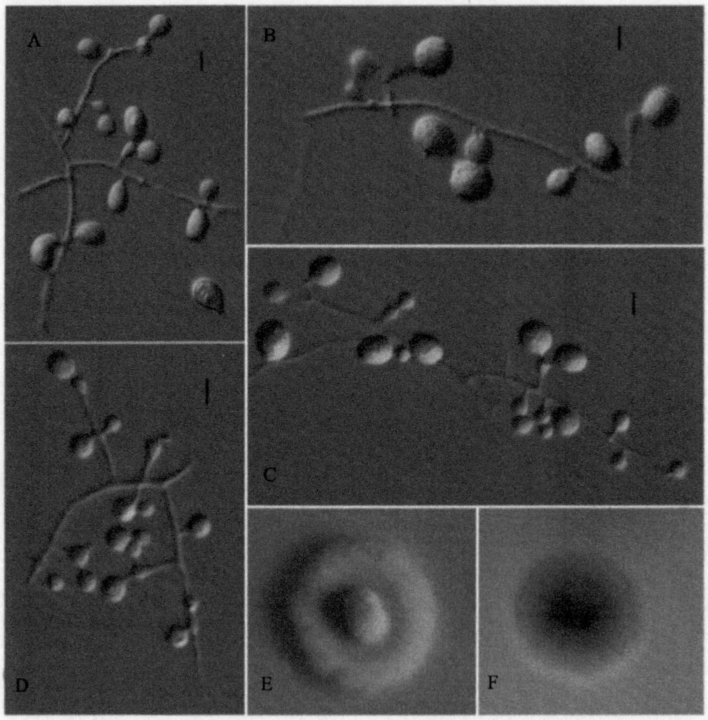

图 47　倒卵形栉霉 *Ctenomyces obovatus* Y.F. Han, Z.Q. Liang & Zhi Y. Zhang（标本号：HMAS 255446）
A～D. 产孢结构与分生孢子；E、F. PDA 培养基上的菌落。标尺：A～D = 10 μm

研究菌株：山西：土壤，2017 年 11 月 2 日，张芝元，HMAS 255446（主模式），

HMAS 255447，HMAS 255448，CGMCC 3.19225。

世界分布：中国。

讨论：形态上，倒卵形栉霉与棉毛栉霉较为相似，这两个物种的孢子大小及产孢细胞非常相近。然而，倒卵形栉霉是栉霉属中唯一孢子呈倒卵形的物种。此外，基于 ITS 和 ITS + *EF1A* + *RPB2* 的系统发育树，倒卵形栉霉的菌株聚为一支，并与其他物种菌株形成的支序相分离（Zhang *et al.*，2019）。

青灰栉霉　图 48

Ctenomyces peltricolor Y.F. Han, Z.Q. Liang & Zhi Y. Zhang, MycoKeys 47: 10, 2019.

在 PDA 培养基上，25℃培养 14 天，菌落直径为 12 mm，中部青灰色，边缘白色，中部粉状至絮状，边缘绒毛状，具环纹，近圆形，边缘规则；背面中部棕色，边缘淡黄色。菌丝无色，光滑，分隔，分枝，宽 1.2～3.3 μm；无球拍状菌丝。顶生、侧生分生孢子着生于菌丝、短突、侧枝，或安瓿形膨大结构上。分生孢子常单生于安瓿形膨大结构上，近球形至球形，光滑或粗糙，疣状，刺状，8.3～20.2 μm。

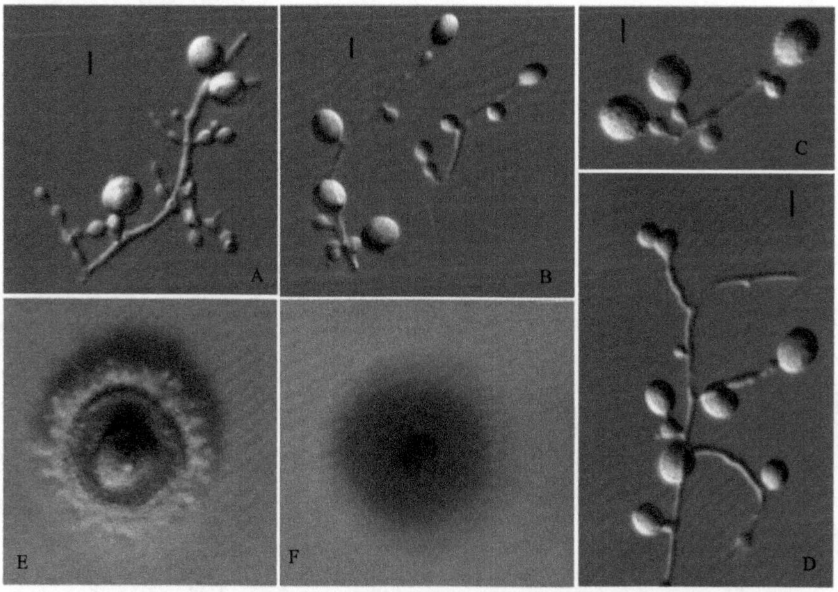

图 48　青灰栉霉 *Ctenomyces peltricolor* Y.F. Han, Z.Q. Liang & Zhi Y. Zhang（标本号：HMAS 255387）
A～D. 产孢结构与分生孢子；E、F. PDA 培养基上的菌落。标尺：A～D = 10 μm

研究菌株：甘肃：土壤，2017 年 11 月 20 日，张芝元，HMAS 255387（主模式），HMAS 255439，HMAS 255440，CGMCC 3.19229，CGMCC 3.19230 = GZUIFR C03011 和 CGMCC 3.19231 = GZUIFR C03012。

世界分布：中国。

讨论：青灰栉霉因孢子单生于安瓿形膨大结构及菌落青灰色而与栉霉属的其他物种相区别。基于 ITS 和 ITS + *EF1A* + *RPB2* 的系统发育树，青灰栉霉的菌株聚为一支，并与其他物种菌株形成的支序相分离（Zhang *et al.*，2019）。

锯齿栉霉 图 49，图版 I：4

Ctenomyces serratus Eidam, Beitr. Biol. Pfl. 3: 274, 1880. Zhang *et al*., MycoKeys 47: 11, 2019.

在 PDA 培养基上，25℃培养 14 天，菌落直径为 30 mm，棕色，边缘白色，中部絮状，其余短绒状，具环纹，圆形，边缘规则，分界明显；背面淡黄色。菌丝无色，光滑，分隔，分枝，宽 0.9～3.3 μm；无球拍状菌丝。顶生、侧生分生孢子着生于菌丝、短突、侧枝，或安瓿形膨大结构上；安瓿形膨大结构单生或 2 个成链。分生孢子单生，或 2～3 个由短而纤细的菌丝连接成链，常椭圆形，偶尔近球形，光滑或粗糙，疣状、刺状，11.5～21.9 × 8～15.2 μm。

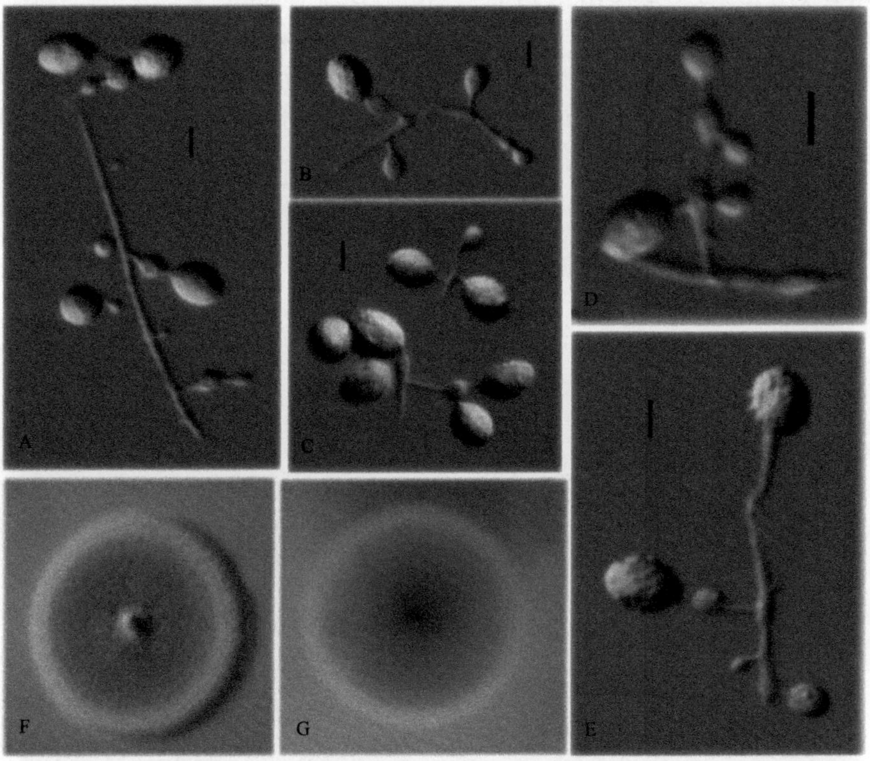

图 49 锯齿栉霉 *Ctenomyces serratus* Eidam（标本号：HMAS 255390）

A～E. 产孢结构与分生孢子；F、G. PDA 培养基上的菌落。标尺：A～E = 10 μm

研究菌株：贵州：土壤，2016 年 9 月 2 日，张芝元，HMAS 255390，HMAS 255444，HMAS 255445，CGMCC 3.18622 = GZUIFR S37.1、CGMCC 3.18623 = GZUIFR S37.2，CGMCC 3.18624 = GZUIFR S37.3。

世界分布：中国、印度、德国、英格兰、捷克斯洛伐克、阿根廷、澳大利亚。

讨论：形态上，锯齿栉霉因不产生间生孢子，常 1～2 个分生孢子着生于安瓿形膨大结构而与倒卵形栉霉和棉毛栉霉相似，但它们的分生孢子大小不同：锯齿栉霉的分生孢子最大可超过 20 μm（Zhang *et al.*，2019），而其他 2 个种则小于 20 μm（van Oorschot，1980）。

棉毛栉霉 图 50

Ctenomyces vellereus (Sacc. & Speg.) P.M. Kirk, Index Fungorum 120: 1, 2014.

≡ *Sporotrichum vellereum* Sacc. & Speg., Michelia 2 (no. 7): 287, 1881.

≡ *Myceliophthora vellerea* (Sacc. & Speg.) Oorschot, Stud. Mycol. 20: 47, 1980. Han *et al*., Microbiology China 43(9): 1963, 2016.

在 PDA 培养基上，25℃培养 14 天，菌落直径达 54 mm，白色至浅棕色，绒毛状，微隆，圆形，边缘较规则；背面浅黄色。菌丝分隔，光滑，无色，宽 0.5～2 μm；无球拍状菌丝。分生孢子顶生、侧生，直接着生于菌丝上或短的突起或膨大的支撑细胞上，大多数单生，厚壁，表面光滑或粗糙，无色至半透明，椭圆形至梨形，4～16 × 2～8.5 μm；基痕宽 1～2 μm；间生孢子单生或成链，梨形、桶状、梭形、柱状、椭圆形等，粗糙或光滑，4～24 × 2～8.5 μm。

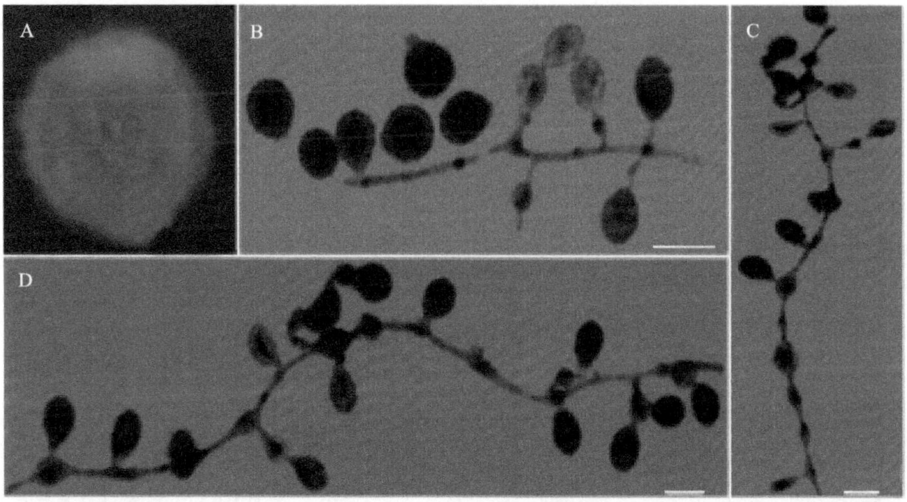

图 50 棉毛栉霉 *Ctenomyces vellereus* (Sacc. & Speg.) P.M. Kirk（标本号：GZUIFR EB6301M）

A. PDA 培养基上的菌落；B～D. 产孢结构。标尺：B～D = 10 μm

研究菌株：陕西：合阳县，茄子地土样，2013 年 6 月 29 日，韩燕峰，GZUIFR EB6301M。

世界分布：中国、俄罗斯、德国、美国、巴西等，世界性分布。

讨论：形态上，棉毛栉霉与倒卵形栉霉较为相似，通常 1～2 个分生孢子着生于安瓿形膨大结构上，且分生孢子直径不超过 20 μm。但棉毛栉霉的分生孢子为椭圆形至梨形（韩燕峰等，2016），而倒卵形栉霉的分生孢子倒卵形或椭圆形。

地丝霉属 Geomyces Traaen

Nytt Mag. Natur. 52: 28, 1914

菌落略微扩展，表生，初白色，后变为浅褐色、深褐色、灰色、淡红色或黄色，毡状或粉状。菌丝无色或淡黄色，壁薄，窄，不育菌丝的分枝近乎垂直生长，常与支撑菌丝的分隔不相关，可育菌丝分枝频繁，呈锐角，常一轮或两轮上具有 2～4 个分枝，二级或三级分枝分别直接着生于最初级或是第二级分枝的隔膜下。菌丝型孢子常着生于轮生分枝末端，中间着生间生孢子或着生于短突起或侧枝的侧面或单生，半透明，浅黄色或淡绿色，壁光滑或具小刺，壁薄或稍加厚，楔形、倒卵形、椭圆形或棒状，单胞，宽的基痕呈截形。间生孢子着生于轮生菌丝的外层分枝上，交替着生，由短、比其宽的可育菌丝分隔，等分，2～4 个相串联，半透明，淡绿色或淡黄色，表面光滑或具小刺，壁薄或稍加厚，桶形，单胞，比支撑菌丝宽，两端呈截形。不耐热，偶喜角质。

模式种：金黄色地丝霉 *Geomyces auratus* Traaen。

中国地丝霉属分种检索表

1. 菌落产生脂溶性黑色素……………………………………………………**贵阳地丝霉 *G. guiyangensis***
1. 菌落无色素产生……………………………………………………………………………………2
   2. 无间生孢子………………………………………………………………**福建地丝霉 *G. fujianensis***
   2. 间生孢子丰富………………………………………………………………………………………3
3. 间生孢子长椭圆形或筒形…………………………………………………………**光滑地丝霉 *G. laevis***
3. 间生孢子非长椭圆形………………………………………………………**倒卵形地丝霉 *G. obovatus***

福建地丝霉　图 51

Geomyces fujianensis W.H. Chen, G.P. Zeng, Y. Luo, Z.Q. Liang & Y.F. Han, Mycosphere 8(1): 39, 2017.

在 PDA 培养基上，25℃培养 14 天，菌落直径可达 43 mm，短绒毛状，起初为白色，逐渐中部变为淡粉色，周围为淡灰色，边缘呈白色，不规则；菌落背面中部呈深褐色，外部区域为淡褐色。菌丝分隔，无色，光滑，宽 1.1～3.2 μm，分生孢子梗大量存在，其上着生呈锐角的一级和二级分枝，轮生或对生。分生孢子（粉孢子）顶生，光滑，半透明，倒卵形至近球形，少数棍棒状，2.5～7.5 × 2.5～5 μm；基痕宽 1.1～2.2 μm；厚垣孢子未见。

研究菌株：福建：福州市，鼓山风景区，某树洞土，2015 年 7 月 15 日，罗韵，GZDXIFR G242.1（主模式），GZDXIFR G242。

世界分布：中国。

讨论：本种与贵阳地丝霉和光滑地丝霉的形态较为相似，但本种分生孢子倒卵形至近球形，无间生孢子，无色素产生（Chen *et al.*，2017），明显不同于这两个近似种（表 4）。

表 4　福建地丝霉及其近似种的形态特征比较

种名	孢子（μm）	间生孢子	色素
贵阳地丝霉	倒卵形 5.0～7.5 × 2.5～5.0	无	黑色
光滑地丝霉	梨形、楔形至倒卵形 1.5～2.5 × 3.5（～4）	有	无
福建地丝霉	倒卵形至近球形 2.5～7.5 × 2.5～5	无	无

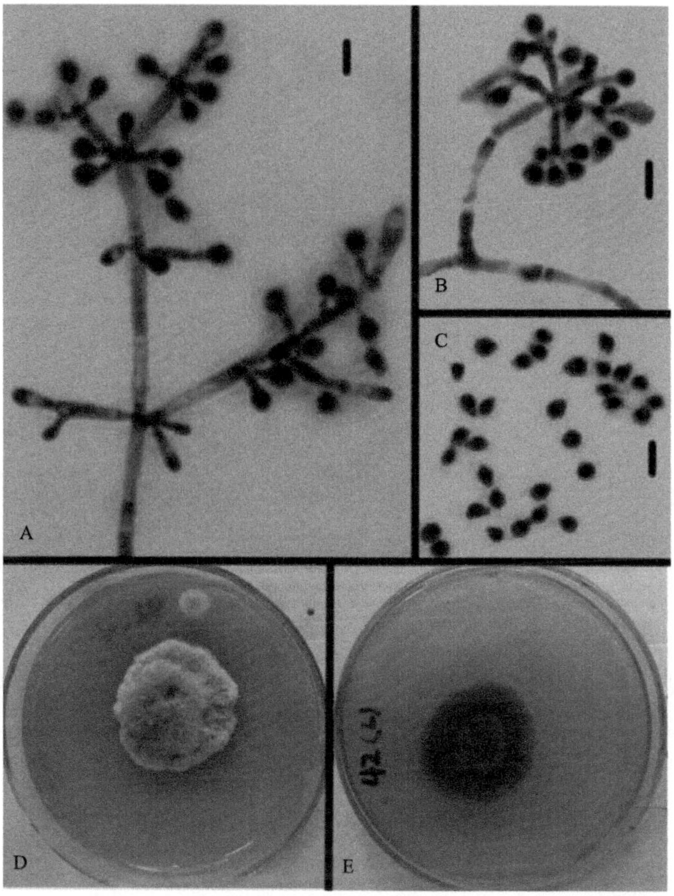

图 51 福建地丝霉 *Geomyces fujianensis* W.H. Chen, G.P. Zeng, Y. Luo, Z.Q. Liang & Y.F. Han（标本号：GZDXIFR G242.1）

A、B. 产孢结构；C. 分生孢子；D、E. PDA 培养基上的菌落。标尺：A～C = 10 μm

贵阳地丝霉 图 52

Geomyces guiyangensis Y.F. Han, Y. Luo & Z.Q. Liang, Mycosystema 35: 125, 2016.

在查氏培养基上，25℃培养 14 天，菌落直径达 35 mm，白色至淡黄色，平展，毡状，边缘整齐；背面淡黄色。菌丝无色，分隔，光滑，宽 1.5～2.5 μm。分生孢子梗大量存在，其上着生呈锐角的一级和二级分枝，轮生或对生；分生孢子（粉孢子）顶生，单生，光滑，半透明，倒卵形，5.0～7.5 × 2.5～5.0 μm；无厚垣孢子。

在 PDA 培养基上，25℃培养 14 天，菌落直径达 30 mm，菌落中心淡褐色，边缘白色，毡状至绒毛状，具轮纹，并可见淡褐色角变。

研究菌株：贵州：贵阳市，贵州大学，4℃冰箱空气，2014 年 5 月 12 日，罗韵，G014512.1（主模式），G014512。

世界分布：中国。

讨论：本种形态特征与光滑地丝霉较为相似，但本种分生孢子明显大于后者。此外，本种能在 CA、PDA 和 PYE 培养基中产生明显的黑色素（罗韵等，2016），而光滑地丝霉不产生色素。

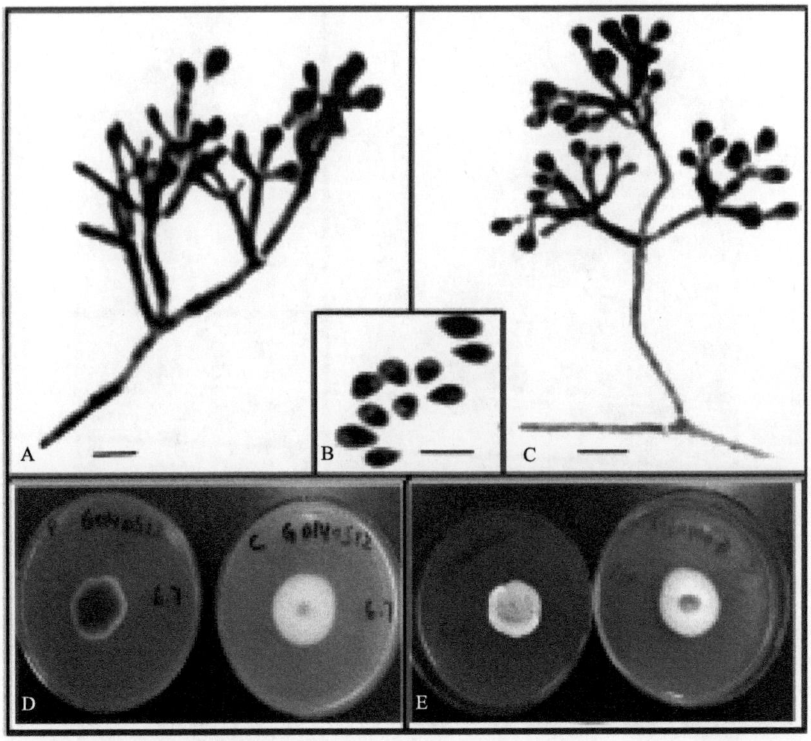

图 52　贵阳地丝霉 *Geomyces guiyangensis* Y.F. Han, Y. Luo & Z.Q. Liang （标本号：G014512）

A、C. 产孢结构；B. 分生孢子；D. PDA 培养基上的菌落；E. 查氏培养基上的菌落。标尺：A～C = 10 μm

倒卵形地丝霉　图 53

Geomyces obovatus Zhi Y. Zhang, Y.F. Han & Z.Q. Liang, Frontiers in Microbiology 11(572596): 5, 2020.

在 PDA 培养基上，25℃培养 14 天，菌落直径达 28～30 mm，中部卡其色，边缘白色，毡状至绒毛状，气生菌丝常具有离散的簇，中央厚达 2 mm，有时有放射状的脊，边缘局限，规则；背面淡褐色。气生菌丝茂盛，光滑，薄壁，无色，分隔，宽 0.5～2 μm；无球拍状菌丝。分生孢子大量产生，具顶生或间生孢子，常 1～4 个串生，少数单生于短柄上，3.0～5 × 2.5～4 μm；顶生孢子大多倒卵形，少数近球形；间生孢子梨形、桶状或不规则形。有性型未知。

研究菌株：贵州：贵阳市，黔灵山公园，土样，2017 年 8 月 8 日，张芝元，GZUIFR QL20.8.1 = CGMCC 3.18491，GZUIFR QL20.8.2 = CGMCC 3.18492。

世界分布：中国。

讨论：本种在形态上与金黄色地丝霉较为相似，但本种菌落卡其色至白色，顶生分生孢子倒卵形，偶尔近球形（Zhang *et al.*，2020），而明显不同于金黄色地丝霉。

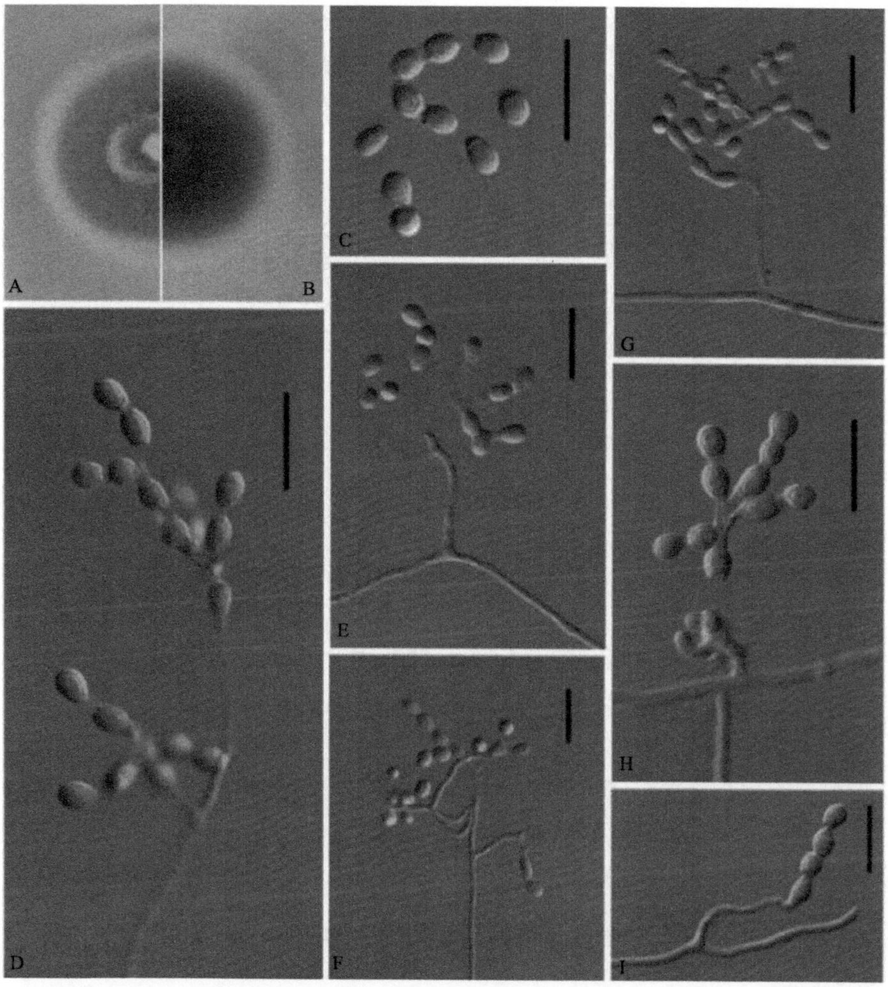

图 53 倒卵形地丝霉 *Geomyces obovatus* Zhi Y. Zhang, Y.F. Han & Z.Q. Liang（标本号：CGMCC 3.18491）

A、B. PDA 培养基上的菌落；C～I. 产孢结构和分生孢子。标尺：C～I = 10 μm

作者未观察的种

光滑地丝霉 图 54

Geomyces laevis Zhong Q. Li & C.Q. Cui, Acta Mycol. Sin. 8(1): 48, 1989.

讨论：据 Li 和 Cui（1989）报道，该种曾在北京发现。分生孢子梗直，10～25（～35）× 1～1.5（～2）μm，轮状或在主轴的顶端形成锐角分枝或反向分枝。可育菌丝分枝，宽 2～4 × 1～1.5 μm。孢子向基式，梨形、楔形至倒卵形，1.5～2.5 × 3.5（～4）μm；间生孢子长椭圆形或筒形，1.5～2.5 × 3.5（～4）μm，光滑，壁薄，无色或略微淡黄色。本种与毡状地丝霉酒红变种较为相似，均产生小而光滑的分生孢子，但本种菌落为白色，后者菌落则为淡红棕色至红色。

世界分布：中国。

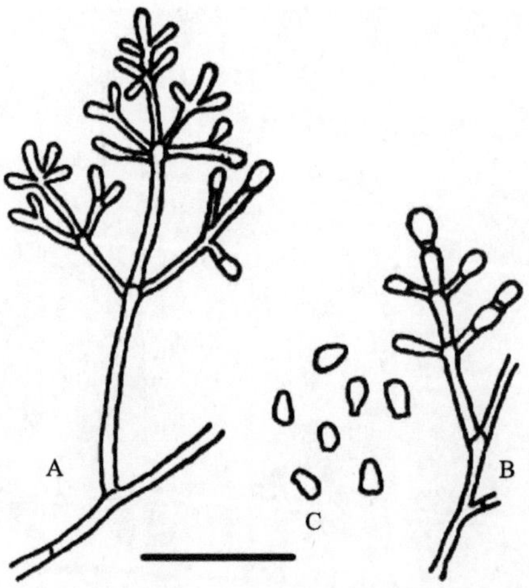

图 54 光滑地丝霉 *Geomyces laevis* Zhong Q. Li & C.Q. Cui（标本号：HMAS 54875；引自 Li & Cui，1989）

A、B. 分生孢子梗及其分枝；C. 分生孢子。标尺：A～C = 15 μm

角蛋白癣菌属 **Keratinophyton** H.S. Randhawa & R.S. Sandhu Sabouraudia 3: 252, 1964

角蛋白癣菌属是基于分子系统学研究，发现其在金孢属中聚为单系而建立，属内种大多来自于金孢属，因此形态特征与金孢属相似，其属级特征可参考金孢属的属级特征。

模式种：棕灰角蛋白癣菌 *Keratinophyton terreum* H.S. Randhawa & R.S. Sandhu。

中国角蛋白癣菌属分种检索表

1\. 间生孢子缺或少 ………………………………………… 2
1\. 间生孢子丰富 ………………………………………… 6
　2. 间生孢子少 ………………………………………… 3
　2. 间生孢子无 ………………………………………… 4
3\. 顶侧生孢子单生或成链，单胞 …………………… **具刺角蛋白癣菌 *K. echinulatum***
3\. 顶侧生孢子仅单生，单胞，偶见双胞 ……………… **河泥角蛋白癣菌 *K. fluviale***
　4. 分生孢子非棒状，倒卵圆至椭圆形 ……………… **湖北角蛋白癣菌 *K. hubeiense***
　4. 分生孢子棒状 ………………………………………… 5
5\. 菌落奇特，有深裂纹，稀疏，绒毛状 …………… **棒孢角蛋白癣菌 *K. clavisporum***
5\. 菌落破布状，致密，短绒状至粉状 ……………… **临汾角蛋白癣菌 *K. linfenense***
　6. 厚垣孢子可见 ………………………………………… 7
　6. 厚垣孢子未见 ………………………………………… 8
7\. 无球拍状菌丝 ……………………………………… **青海角蛋白癣菌 *K. qinghaiense***
7\. 具球拍状菌丝 ……………………………………… **水生角蛋白癣菌 *K. submersum***
　8. 无间生孢子 ………………………………………… **四川角蛋白癣菌 *K. sichuanense***
　8. 具间生孢子 ………………………………………… 9
9\. 分生孢子光滑至轻微粗糙 ………………………… **重庆角蛋白癣菌 *K. chongqingense***
9\. 分生孢子光滑 ………………………………………… 10

10. 间生孢子梨形 ·· 印度角蛋白癣菌 ***K. indicum***
10. 间生孢子形状多样 ·· 阿尔维亚角蛋白癣菌 ***K. alvearium***

阿尔维亚角蛋白癣菌　图 55

Keratinophyton alvearium (F. Liu & L. Cai) Kandemir & de Hoog, Fungal Diversity 115: 21, 2022.

≡ *Chrysosporium alvearium* F. Liu & L. Cai, Mycosphere, 9(6):1096, 2018.

在 PDA 培养基上，25℃培养 7 天，菌落直径 14～17 mm，平展，中心微隆，粉状，白色，边缘不规则；背面白色。菌丝无色，光滑，分枝，宽 1.5～2.5 μm。顶侧生分生孢子无柄或直接生长在或长或短的垂直侧突起上，单生或成可达 10 个孢子的链，无色，光滑，球形、梨形、棒状或倒卵圆形，少数柱状，单胞，4～9 × 2～7.5 μm；间生孢子大量存在，单生或成链，光滑，球形、桶状、椭圆形至倒卵圆形，4～11 × 3.5～9.5 μm。

研究菌株：湖北：蜜蜂蜂巢中的花粉，2017 年，蔡磊，HMAS 247780（主模式），CGMCC 3.18783；贵州：遵义市，遵义医学院附属医院，绿地土，2020 年 8 月 8 日，张芝元，GZAC 2020，CGMCC 3.20866。

世界分布：中国。

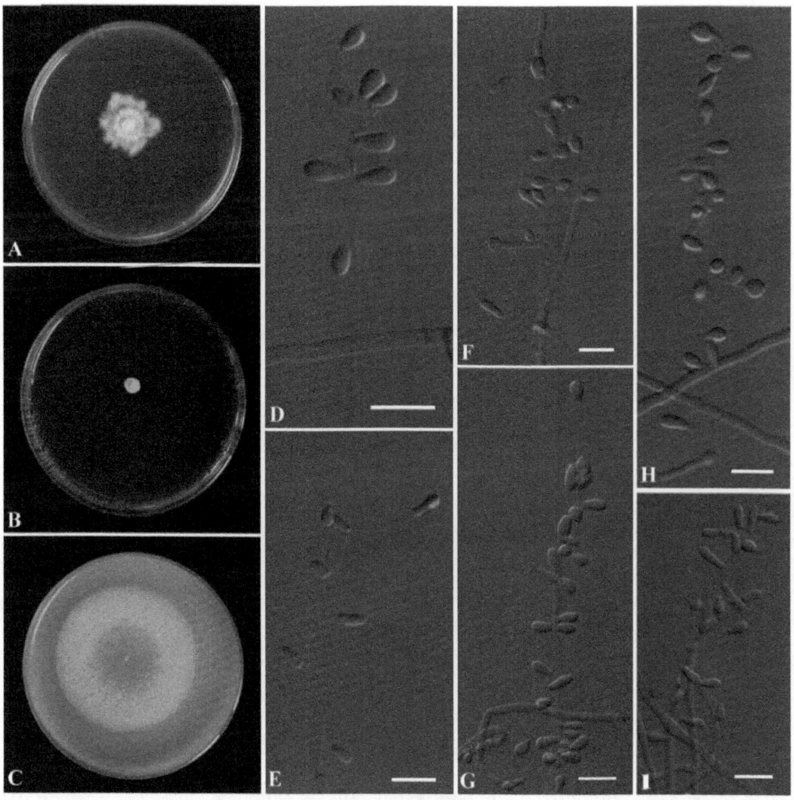

图 55　阿尔维亚角蛋白癣菌 *Keratinophyton alvearium* (F. Liu & L. Cai) Kandemir & de Hoog（标本号：CGMCC 3.18783）

A～C. PDA、MEA 和 OA 培养基上的菌落；D～I. 产孢结构及分生孢子。标尺：D～I = 10 μm

讨论：阿尔维亚角蛋白癣菌与水生角蛋白癣菌和湖北角蛋白癣菌在形态上相似，但在 PDA 培养基上生长缓慢，分生孢子单胞（Zhao *et al.*，2018），能与水生角蛋白癣菌相区别；该种产生的分生孢子比湖北角蛋白癣菌更长（4～9 × 2～7.5 μm vs. 2.2～4.3 × 1.6～3.2 μm）（Zhang *et al.*，2016）。

重庆角蛋白癣菌 图 56

Keratinophyton chongqingense Xin Li, Zhi. Y. Zhang, W.H. Chen, J.D. Liang, Z.Q. Liang & Y.F. Han, Int. J. Syst. Evol. Microbiol. 72(8, no. 005468): 2, 2022.

在 PDA 培养基上，25℃培养 14 天，菌落直径 51～55 mm，白色，绒毛状，中部致密，微隆，边缘不整齐；背面白色。37℃培养 14 天，菌落直径 62～67 mm，白色至奶酪色，绒毛状，边缘整齐；背面白色至奶酪色。菌丝具隔膜，无色，表面光滑，菌丝宽 1.0～5.5 μm；具球拍状菌丝。分生孢子（粉孢子）无色，薄壁，光滑至轻微粗糙，顶生和侧生于主干菌丝或侧枝上，无柄或具短突起，单生，柱状至棒状，有时弯曲，4～10 × 2～2.5 μm；倒卵圆形、近球形至球形、椭圆形，3～8 × 2.5～4.5 μm。间生孢子量大，光滑，椭圆形，14～18.5 × 4～6 μm。厚垣孢子未见。该菌生长的最低和最高温分为 8℃和 40℃。

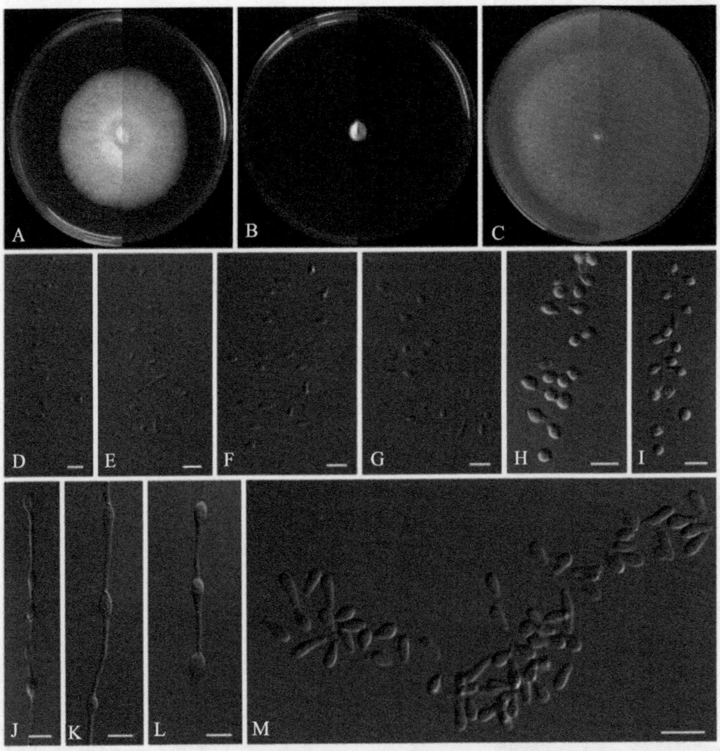

图 56　重庆角蛋白癣菌 *Keratinophyton chongqingense* Xin Li, Zhi. Y. Zhang, W.H. Chen, J.D. Liang, Z.Q. Liang & Y.F. Han（标本号：HMAS 351886）

A～C. PDA、MEA 和 OA 培养基上的菌落；D～I. 产孢结构；J. 间生孢子；K、L. 球拍状菌丝；M. 顶侧生分生孢子。标尺：D～M = 10 μm

研究菌株：重庆：北碚区中医院，绿地土，29.8056°N，106.4114°E，2016 年 7 月 8 日，张芝元，HMAS 351886（干模式），CGMCC 3.20867 = GZUIFR 22.006（模式菌株）；贵州：贵阳市，黔灵山公园，绿地土，2017 年 8 月 8 日，张芝元，GZUIFR 22.030，GZUIFR 22.031。

世界分布：中国。

讨论：在角蛋白癣菌属中，重庆角蛋白癣菌与阿尔维亚角蛋白癣菌亲缘关系很近，但该种以无分生孢子链和无厚垣孢子等形态特征与后者相区别（Li *et al.*，2022a）。

棒孢角蛋白癣菌 图 57

Keratinophyton clavisporum (Yan W. Zhang, Y.F. Han & Z.Q. Liang) Labuda & Bernreiter, IMA Fungus 12(17): 6, 2021.

≡ *Chrysosporium clavisporum* Yan W. Zhang, Y.F. Han & Z.Q. Liang, Phytotaxa 303(2): 177, 2017.

在 PDA 培养基上，25℃培养 14 天，菌落直径达 53 mm，白色，稀疏，绒毛状，中部较密，边缘稀疏，圆形，不规则，具深裂纹；背面中部红棕色，边缘浅黄色。菌丝分隔，无色，光滑，1.5～3.5 μm；有球拍状菌丝，7.5～15 × 5～7.5 μm。分生孢子侧生或顶生，大多着生于短突起或短侧枝上，大多单生，少数成短链，无色，光滑，单胞，棒状至长椭圆形，5～10 × 2.5～5 μm；基痕宽 2.5～5.0 μm；间生孢子和厚垣孢子未见。

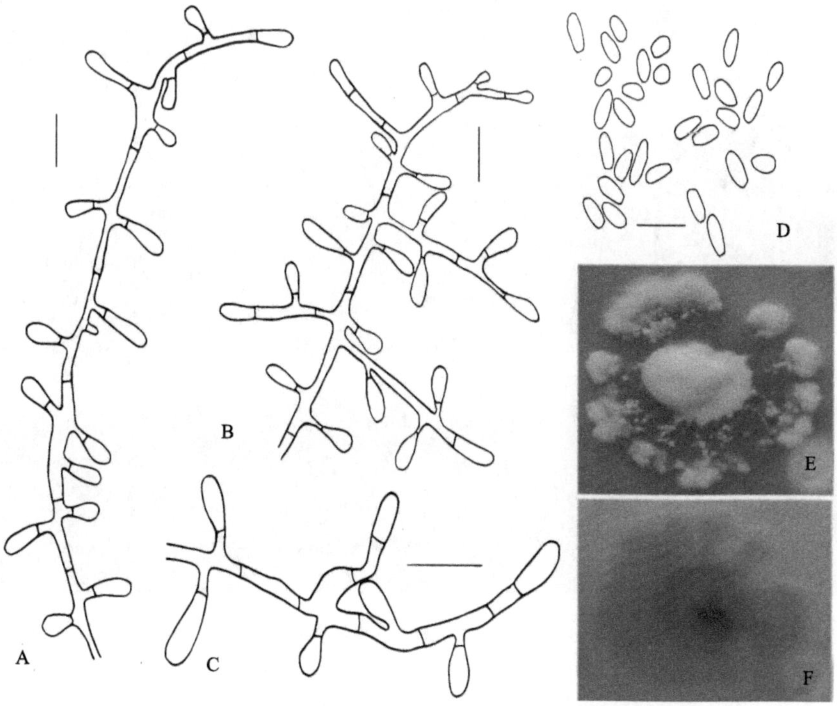

图 57 棒孢角蛋白癣菌 *Keratinophyton clavisporum* (Yan W. Zhang, Y.F. Han & Z.Q. Liang) Labuda & Bernreiter（标本号：GZUIFR G80）

A～C. 产孢结构；D. 分生孢子；E、F. PDA 培养基上的菌落。标尺：A～D = 10 μm

研究菌株：广西：贵港市，马草江，植物根际土样，2012 年 8 月 10 日，王玉荣，GZUIFR G80（主模式），GZUIFR G80.1，GZUIFR G80.2。

世界分布：中国。

讨论：棒孢角蛋白癣菌与青海角蛋白癣菌、印度角蛋白癣菌和临汾角蛋白癣菌在基于 ITS rDNA 的系统发育上相近，但青海角蛋白癣菌为粉质菌落，并缺乏球拍状菌丝；印度角蛋白癣菌的分生孢子较小，能产生间生孢子；临汾角蛋白癣菌有小的椭圆形至棒状的分生孢子；棒孢角蛋白癣菌的主要鉴别特征是具有较大的棒状至长椭圆形的分生孢子，产生球拍状菌丝，不产生间生孢子（Zhang *et al.*，2017a）。

具刺角蛋白癣菌 图 58

Keratinophyton echinulatum (Hubka, Mallátová, Čmoková & M. Kolařík) Labuda & Bernreiter, IMA Fungus 12(17): 6, 2021.

≡ *Chrysosporium echinulatum* Hubka, Mallátová, Čmoková & M. Kolařík, Persoonia 36: 411, 2016.

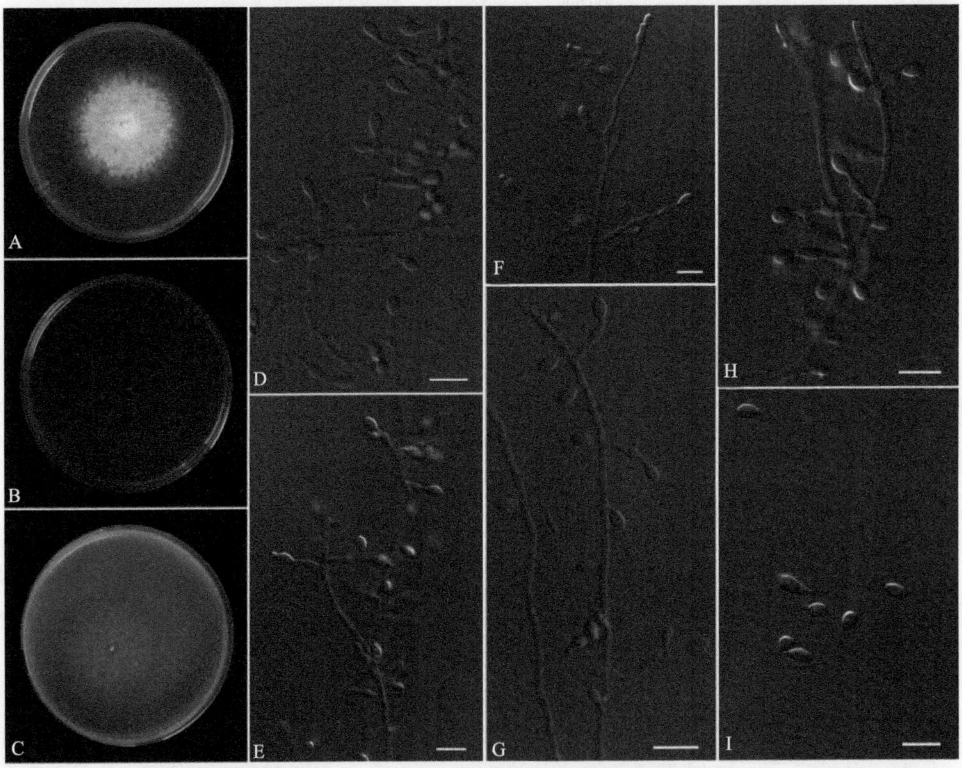

图 58 具刺角蛋白癣菌 *Keratinophyton echinulatum* (Hubka, Mallátová, Čmoková & M. Kolařík) Labuda & Bernreiter（标本号：CGMCC 3.20868）

A～C. PDA、MEA 和 OA 培养基上的菌落；D～I. 产孢结构及分生孢子。标尺：D～I = 10 μm

在 PDA 培养基上，25℃培养 14 天，菌落直径 28～45 mm，致密绒毛状，平展或中心微隆，边缘到中心由白色至淡黄色再至橘黄色；背面橘黄色。菌丝无色，分隔，光滑，1.2～5 μm，直立，分枝少；具球拍状菌丝。分生孢子无色，成团时淡黄色，顶生或侧生孢子着生于无柄或短直角状突起或不同长度的柄上，单生或成链，初光滑，成熟时具刺，倒卵圆形至棒状，单胞，4.5～7 × 2.5～4 μm；间生孢子单生或成短链，光滑，桶状至椭圆形。

研究菌株：贵州：遵义市，遵义医学院附属医院，绿地土，2020 年 8 月 8 日，张芝元，CGMCC 3.20868。

世界分布：中国、捷克。

讨论：具刺角蛋白癣菌与毡状地丝霉形态相似，但它在 35℃条件下不能生长且孢子较小（Crous *et al.*，2016）。

河泥角蛋白癣菌　图 59

Keratinophyton fluviale (P. Vidal & Guarro) Labuda & Bernreiter, IMA Fungus 12(17): 6, 2021.

≡ *Chrysosporium fluviale* P. Vidal & Guarro, Mycol. Res. 104(2): 245, 2000.

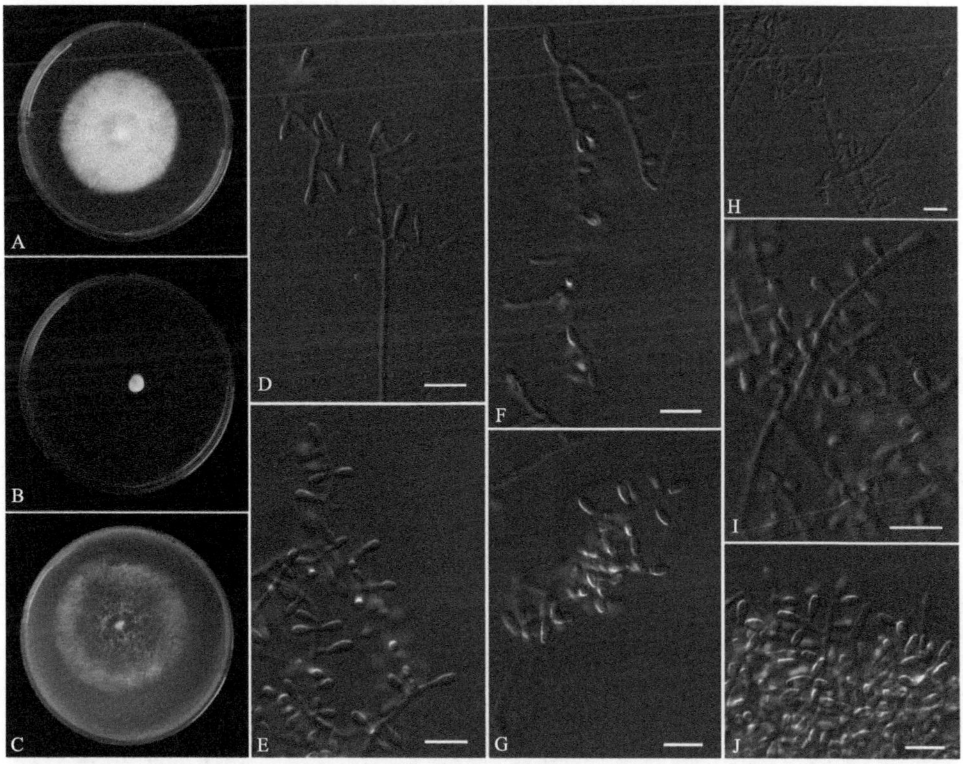

图 59　河泥角蛋白癣菌 *Keratinophyton fluviale* (P. Vidal & Guarro) Labuda & Bernreiter（标本号：CGMCC 3.20869）

A～C. PDA、MEA 和 OA 培养基上的菌落；D～J. 产孢结构及分生孢子。标尺：D～J = 10 μm

在 PDA 培养基上，25℃培养 7 天，菌落直径 55～60 mm，绒毛状，白色至奶酪色，边缘整齐，圆形；背面淡黄色；菌丝无色，光滑，分隔，宽 1.5～3 μm，分枝；具球拍状菌丝。分生孢子无色至半透明，成团时奶酪色，光滑，成熟时表面粗糙，侧生和顶生孢子直接着生于菌丝或与主菌丝成直角的短突或有限长度的侧枝上，大量产生，单生，倒卵圆形至棒状，有时近椭圆形，单胞，4～7（～12）× 2～3 μm；偶见双胞，7～12 × 2～3 μm；基痕窄，1～2 μm。间生孢子少，单生，柱状至桶状，单胞，4～15 × 2～3 μm；偶见厚垣孢子，半透明，球形至近球形，直径 5～6 μm。

研究菌株：贵州：遵义市，遵义医学院附属医院，绿地土，2020 年 8 月 8 日，CGMCC 3.20869。

世界分布：中国、荷兰、西班牙、英国、比利时。

讨论：该种与具刺角蛋白癣菌在形态上很相似，都具球拍状菌丝，光滑至粗糙的分生孢子及有间生孢子（Vidal *et al.*，2000）。二者不同之处在于本种分生孢子单生，偶见双胞，而具刺角蛋白癣菌为单胞，分生孢子可成短链。

湖北角蛋白癣菌　图 60

Keratinophyton hubeiense (Yan W. Zhang, Y.F. Han & Z.Q. Liang) Labuda & Bernreiter, IMA Fungus 12(17): 8, 2021.

≡ *Chrysosporium hubeiense* Yan W. Zhang, Y.F. Han & Z.Q. Liang, Phytotaxa, 270(3): 213, 2016.

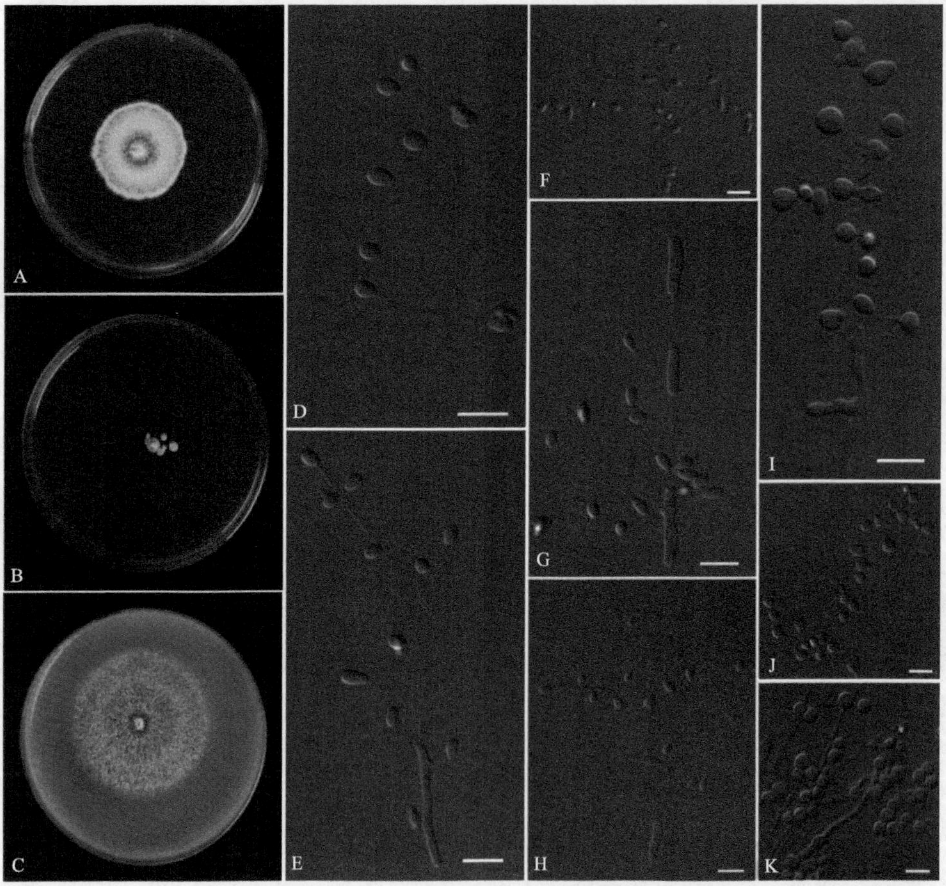

图 60　湖北角蛋白癣菌 *Keratinophyton hubeiense* (Yan W. Zhang, Y.F. Han & Z.Q. Liang) Labuda & Bernreiter（标本号：GZUIFR EM66601）

A～C. PDA、MEA 和 OA 培养基上的菌落；D～K. 产孢结构及分生孢子。标尺：D～K = 10 μm

在查氏培养基上，25℃培养 14 天，菌落直径达 35～39 mm，白色，粉状，边缘不规则。在 PDA 培养基上，菌落直径达 65～67 mm，灰白色至白色，平展，粉状，中部致密，边缘稀疏绒毛状；背面浅黄色。菌丝分隔，无色，光滑，1.1～2.2 μm；有球拍状菌丝，5.4～7.6 × 2.2～3.2 μm。分生孢子侧生或顶生，着生于垂直于菌丝的或短或长的突起或短柄上，单生，无色，光滑，倒卵圆形至椭圆形，2.2～4.3 × 1.6～3.2 μm；基痕宽 2.2～3.2 μm；间生孢子和厚垣孢子未见。

研究菌株：湖北：松滋市，鸡毛土，2013 年 6 月 10 日，王玉荣，GZUIFR EM66601（主模式）；四川：成都市，华西医院，绿地土，2017 年 8 月 15 日，张芝元，CGMCC 3.20870。

世界分布：中国。

讨论：湖北角蛋白癣菌的近似种有水生角蛋白癣菌。水生角蛋白癣菌产生间生孢子，分生孢子粗糙（van Oorschot，1980）。该种的主要鉴别特征是：具球拍状菌丝，无间生孢子，分生孢子为倒卵圆形至椭圆形。

印度角蛋白癣菌　图 61

Keratinophyton indicum (Garg) Kandemir & de Hoog, Fungal Diversity 115: 21, 2022.

≡ *Chrysosporium indicum* Garg, Sabouraudia, 4(4): 262,1966. Han *et al*., Journal of Fungal Research 4(1): 53, 2006.

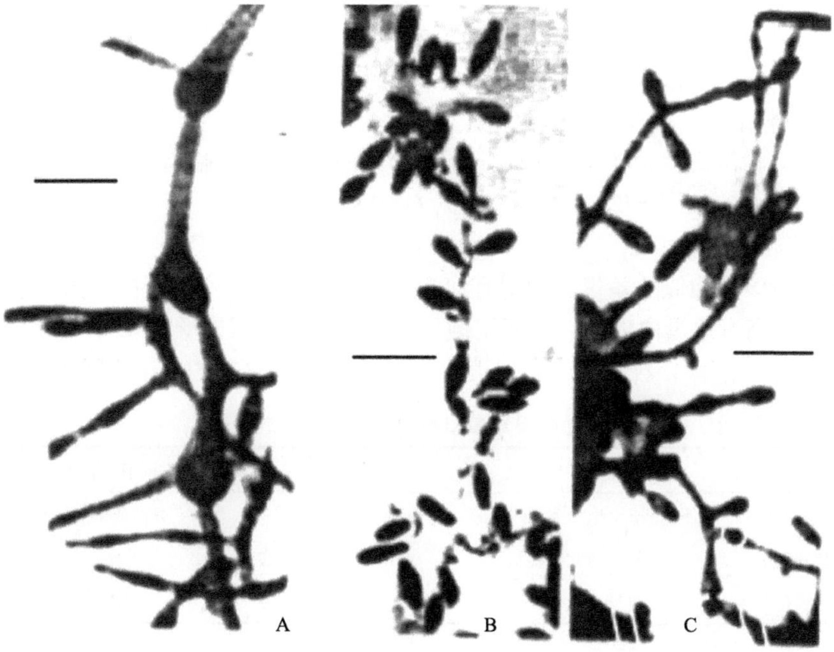

图 61　印度角蛋白癣菌 *Keratinophyton indicum* (Garg) Kandemir & de Hoog（标本号：GZDXIFR 324）

A. 球拍状菌丝；B、C. 侧生和顶生分生孢子。标尺：A～C = 5 μm

在查氏培养基上，25℃培养 14 天，菌落平展，稀疏绒毛状，白色，生长局限，直径 20～25 mm；背面白色。菌丝无色，分隔，宽约 1.3 μm；有球拍状菌丝。分生孢子无色，光滑，椭圆至棒状、3.6～8.1 × 1.8～2.2 μm，或柱状、7.2～8.1 × 1.3～2.7 μm，大多单生，有时可串生或簇生于菌丝顶端或侧生于菌丝，大多着生在长 1.8～2.9 μm 的短柄或短侧枝上；间生孢子大量，梨形，7.2～11.5 × 3.6～4.5 μm。

研究菌株：山西：吉县，土样，2005 年 8 月 3 日，梁建东，GZDXIFR 324。

世界分布：中国、印度。

讨论：印度角蛋白癣菌起初放在发癣菌属中，由于发癣菌属中的其他种未见这一结构，因此将其移入金孢属，后转入角蛋白癣菌属中。该种与其他近似种的区别主要是产生球拍状菌丝，具椭圆至棒状的分生孢子和梨形的间生孢子（van Oorschot，1980；韩燕峰等，2006）。

临汾角蛋白癣菌　图 62

Keratinophyton linfenense (Z.Q. Liang, J.D. Liang & Y.F. Han) Labuda & Bernreiter, IMA Fungus 12(17): 12, 2021.

≡ *Chrysosporium linfenense* Z.Q. Liang, J.D. Liang & Y.F. Han, Mycotaxon 110: 67, 2009.

在 PDA 培养基上，30℃培养 14 天，菌落直径 72 mm，中央微突，致密，边缘稀疏，不整，破布状，白色，短绒状至粉状；背面淡黄色。菌丝具隔膜，半透明，表面光滑，菌丝宽 0.8～1.5（～2.3）μm；有球拍状菌丝，球拍直径常 9 μm，菌丝直径常 4.5 μm。顶生和侧生分生孢子无柄，或生于短柄上，单生或 2～3 个成链或 2 个簇生，半透明，表面光滑，大多椭圆形、3.2～5.4 × 1.4～2.2 μm，少数棒状、4.2～6.5 × 1.6～2.5 μm；基痕宽 1～2 μm，产孢丰富。

研究菌株：山西：临汾市，雪松根际土，2005 年 8 月 4 日，梁建东，GZDXIFR H-31-1（主模式）；临汾市，大叶冬青根际土，GZDXIFR H-25-2；临汾市，法国梧桐根际土，GZDXIFR H-26-3；临汾市，黄楝根际土，GZDXIFR H-29-2。

世界分布：中国。

讨论：与该种相似的角蛋白癣菌有河泥角蛋白癣菌和印度角蛋白癣菌，它们的主要区别见表 5。临汾角蛋白癣菌以无间生孢子，分生孢子表面光滑、椭圆形至棒状等特征能与近似种相区别；该种最适生长温度为 30℃，大于 40℃不生长；可分解角蛋白。

表 5　临汾角蛋白癣菌与其近似种的区别

种名	分生孢子形状	分生孢子大小（μm）	分生孢子表面特征	间生孢子
河泥角蛋白癣菌	倒卵圆形、棒状至近椭圆形	4～7（～12）× 2～3	光滑或粗糙	稀少
印度角蛋白癣菌	椭圆至棒状或柱状	3.6～8.1 × 1.3～2.7	光滑	大量
临汾角蛋白癣菌	椭圆形至棒状	3.2～6.5 × 1.4～2.5	光滑	无

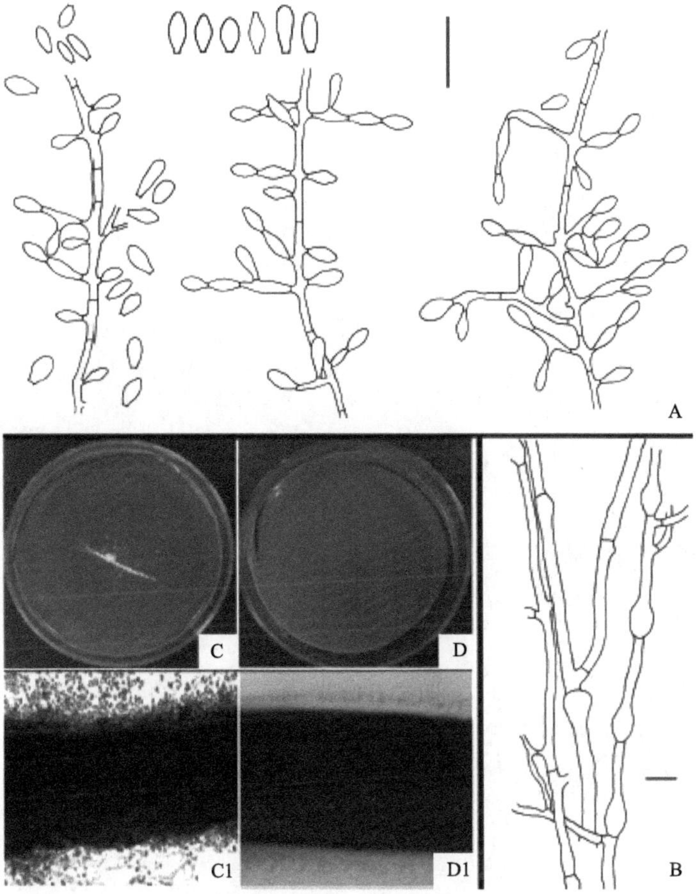

图 62　临汾角蛋白癣菌 *Keratinophyton linfenense* (Z.Q. Liang, J.D. Liang & Y.F. Han) Labuda & Bernreiter（标本号：GZDXIFR H-31-1）

A. 产孢结构和成熟的分生孢子；B. 球拍状菌丝；C. 在无碳氮源的查氏培养基上培养后可见沿着头发生长的临汾角蛋白癣菌；D. 在无碳氮源的查氏培养基上未培养时的头发；C1. 培养 14 天后被临汾角蛋白癣菌降解的头发（×400）；D1. 未被降解时的头发（×400）。标尺：A、B = 10 μm

青海角蛋白癣菌　图 63

Keratinophyton qinghaiense (Y.F. Han, J.D. Liang & Z.Q. Liang) Labuda & Bernreiter, IMA Fungus 12(17): 14, 2021.

≡ *Chrysosporium qinghaiense* Y.F. Han, J.D. Liang & Z.Q. Liang, Mycosystema 32(4): 607, 2013.

在 PDA 培养基上，25℃培养 7 天，菌落直径 30 mm，白色至浅黄色，粉状，平展，薄，边缘不整齐；背面淡黄色。菌丝具隔膜，无色，表面光滑，菌丝宽 0.9～1.5 μm；无球拍状菌丝。分生孢子顶生和侧生于短柄上或侧枝上，侧生孢子较多，单生，无色，光滑，椭圆、棒状至柱状，3.6～9 × 1.8～3.6 μm；基痕宽 1～2 μm；有双胞分生孢子，柱状，有的稍弯，9～13 × 1.8～2.7 μm。间生孢子多出现在长的侧分枝中，椭圆形，3.6～9 × 1.8～3.6 μm；厚垣孢子近球形至椭圆形，4.5～5.5 × 3.6～4.5 μm。

研究菌株：青海：格尔木，农田土样，2005 年 8 月 11 日，韩燕峰，GZUIFR Chry11

（主模式）；重庆：北碚区中医院，绿地土，2018 年 8 月 10 日，张芝元，CGMCC 3.20872。

世界分布：中国。

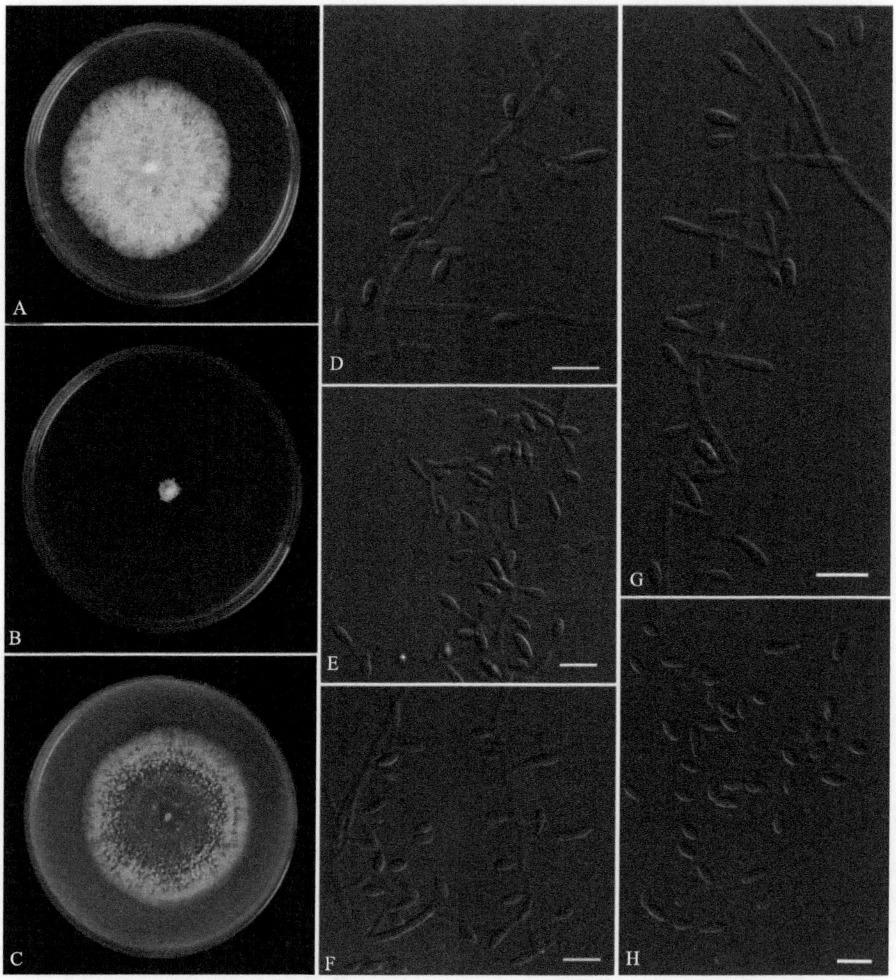

图 63　青海角蛋白癣菌 *Keratinophyton qinghaiense* (Y.F. Han, J.D. Liang & Z.Q. Liang) Labuda & Bernreiter（标本号：GZUIFR Chry11）

A～C. PDA、MEA 和 OA 培养基上的菌落；D～H. 产孢结构及分生孢子。标尺：D～H = 10 μm

讨论：在角蛋白癣菌属中，该种能以无球拍状菌丝、具厚垣孢子和分生孢子光滑等特征与其他种相区别。该种能分解角蛋白（韩燕峰等，2013）。

四川角蛋白癣菌　图 64

Keratinophyton sichuanense Xin Li, Zhi Y. Zhang, W.H. Chen, J.D. Liang, Z.Q. Liang & Y.F. Han, Int. J. Syst. Evol. Microbiol. 72(8, no.005468): 6, 2022.

在 PDA 培养基上，25℃培养 14 天，菌落直径 50～54 mm，白色，绒毛状，可见分生孢子团，微隆，中部轻微凹陷，边缘不整齐；背面中部奶酪色，其余白色。37℃培养 14 天，菌落直径 60～63 mm，白色至奶酪色，绒毛状，边缘不规则；背面白色。菌丝分隔，无色，光滑，分枝，菌丝宽 1～4 μm；具球拍状菌丝。分生孢子（粉孢子）无色，薄壁，光滑至轻微粗糙，顶生和侧生于主干菌丝或长度不一的侧枝上，无柄或具短突起，单生，倒卵圆形至棒状，有时弯曲，4～8.5 × 2～4 μm。该菌生长的最低温和最高温分别为 10℃和 42℃。

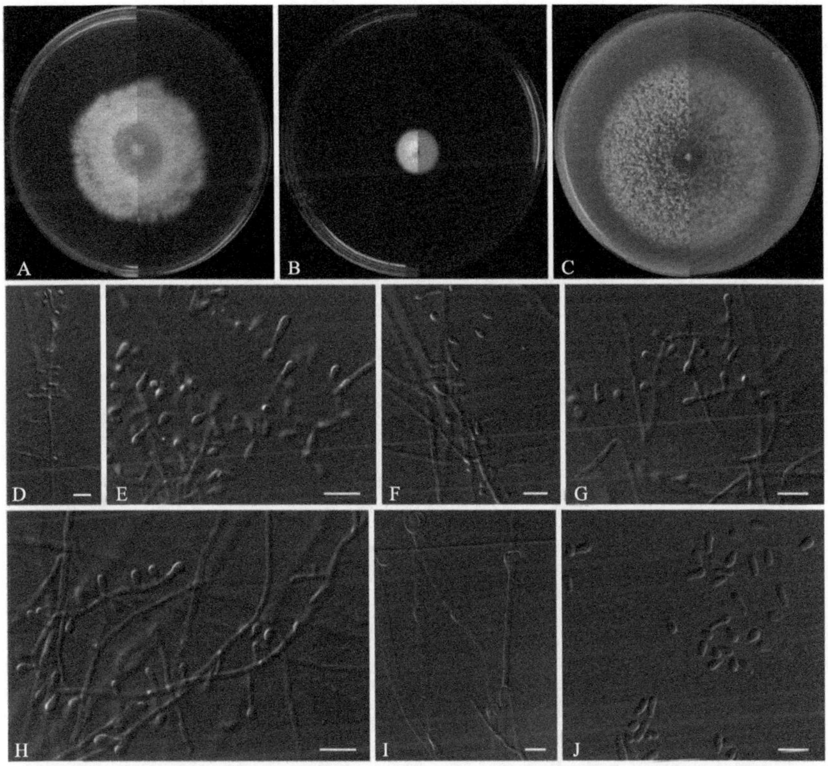

图 64　四川角蛋白癣菌 *Keratinophyton sichuanense* Xin Li, Zhi Y. Zhang, W.H. Chen, J.D. Liang, Z.Q. Liang & Y.F. Han（标本号：HMAS 351888）

A～C. PDA、MEA 和 OA 培养基上的菌落；D～H. 产孢结构；I. 球拍状菌丝；J. 分生孢子。标尺：D～J = 10 μm

研究菌株：四川：成都市，华西医院，绿地土，30.6417°N，104.0610°E，2016 年 8 月 8 日，张芝元，HMAS 351888（主模式），CGMCC 3.20871 = GZUIFR 22.008；贵州：贵阳市，贵州中医药大学第一附属医院，绿地土，26.59°N，106.71°E，2016 年 9 月 10 日，张芝元，GZUIFR 22.030。

世界分布：中国。

讨论：角蛋白癣菌属中，四川角蛋白癣菌与湖北角蛋白癣菌亲缘关系较近，但该种的顶侧生孢子比后者的大（4～8.5 × 2～4 μm vs. 2.2～4.3 × 1.6～3.2 μm）（Li *et al.*, 2022a）。

水生角蛋白癣菌　图 65

Keratinophyton submersum (P. Vidal & Guarro) Labuda & Bernreiter, IMA Fungus 12(17): 12, 2021.

≡ *Chrysosporium submersum* P. Vidal & Guarro, Stud. Mycol. 47: 200, 2002. Han *et al.*, Microbiology 46(9): 2210, 2019.

在 PDA 培养基上，25℃培养 14 天，菌落直径 50～60 mm，浅黄色，中部粉质，边缘绒毛状，圆形，边缘规则；背面浅黄色。菌丝分隔，光滑，壁薄，分枝，宽 2～3.5 μm；具球拍状菌丝。分生孢子起初无色，光滑，薄壁，成团后变为半透明至淡黄色或厚壁，有时具疣状突起。分生孢子顶生和侧生于短柄、侧枝或膨大的突出上，大多单生，有时成由 2～4 个孢子组成的短链，主要为棒状，有时为梨形、卵圆或近球形，单胞或 2～3 个细胞，4.5～30 × 2.5～4.5 μm；基痕宽 1.2～2.5 μm；延长培养后出现间生孢子，柱状，一端膨大或桶状，6～38 × 2～5 μm；厚垣孢子较少，无色，单生，球形至近球形或不规则，光滑，厚壁，5～12 μm。

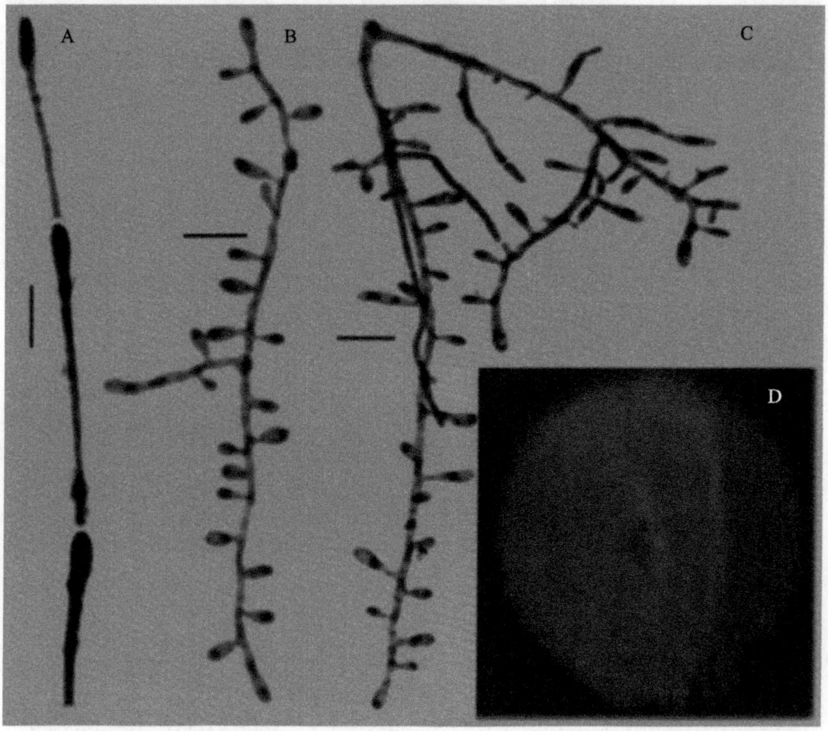

图 65　水生角蛋白癣菌 *Keratinophyton submersum* (P. Vidal & Guarro) Labuda & Bernreiter（标本号：GZAC EB8803M）

A. 球拍状菌丝；B、C. 产孢结构；D. PDA 培养基上的菌落。标尺：A～C = 10 μm

研究菌株：云南：昆明市，植物根际土，2006 年 8 月 14 日，王玉荣，GZAC EB8803M，GZU EB8803M。

世界分布：中国、西班牙。

讨论：水生角蛋白癣菌在系统发育上与阿尔维亚角蛋白癣菌的亲缘关系相近，但在形态上，后者不产生球拍状菌丝和厚垣孢子，分生孢子单生或成可达 10 个孢子的链，光滑，球形、梨形、棒状或倒卵圆形（Zhao *et al.*，2018）；而水生角蛋白癣菌具球拍状菌丝，分生孢子具疣状突起（Vidal *et al.*，2002；韩燕峰等，2019）。

冠癣菌属 **Lophophyton** Matr. & Dassonv.

Rev. Gén. Bot. 11: 432 bis, 1899

菌落粉状或天鹅绒状，产褐色到红色色素。大分生孢子松散簇生，大，可达 60 μm 长，壁厚，粗糙，多隔。产生小分生孢子。不同基因型的菌丝配接后形成有性型，节皮样。

模式种：鸡禽冠癣菌 *Lophophyton gallinae* (Mégnin) Matr. & Dassonv.。

鸡禽冠癣菌　图 66

Lophophyton gallinae (Mégnin) Matr. & Dassonv., Rev. Gén. Bot. 11: 429, 1899.

≡ *Epidermophyton gallinae* Mégnin, C. r. Soc. Biol., 33: 404, 1881.

≡ *Microsporum gallinae* (Mégnin) Grigoraki, Annals Derm. Syph., 5 Série 10: 42, 1929. Wang, Medical Mycology-Guide to Laboratory Examination p. 159, 2005.

≡ *Trichophyton gallinae* (Mégnin) Marg. Silva & Benham, Mycologia 44: 486, 1952.

在沙氏培养基上，菌落中等速度生长，直径约 45 mm，绒毛状，白色至粉色，25℃时呈白色，30℃时转淡红色；背面深红色，可溶性色素弥散至培养基中；菌丝分隔，光滑，宽 2～4 μm。大分生孢子丰富，长椭圆形至棒状，常弯曲或在顶部变窄，光滑或棘状突起，有 2～9 隔，45～60 × 15～20 μm；小分生孢子丰富，卵圆形至梨形，10～20 × 2～4 μm。

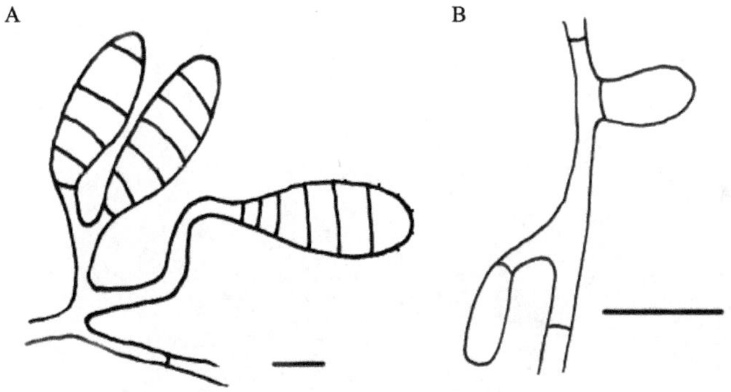

图 66　鸡禽冠癣菌 *Lophophyton gallinae* (Mégnin) Matr. & Dassonv.（标本号：GZAC A1720）

A. 大分生孢子；B. 小分生孢子。标尺：10 μm

研究菌株：贵州：贵阳市，某鸡场，土样，韩燕峰，2016 年 9 月 10 日，GZAC A1720。

分布：中国、日本、美国。

讨论：该种为亲动物性真菌（de Hoog *et al.*，2017；王端礼，2005），主要对家禽具有致病性。

小孢子菌属 **Microsporum** Gruby

C. r. hebd. Séanc. Acad. Sci., Paris 17: 302, 1843

菌落绒毛状、粉状。具有营养菌丝和气生菌丝两种类型，菌丝一般有隔，表面光滑，有大、小孢子两种类型，大孢子侧生或顶生于菌丝，表面光滑或粗糙，分隔，大多为梭形或船形；小孢子丰富，着生于菌丝或短柄，形状多样，一般为梨形或倒卵形或圆形或棒状等。

模式种：奥杜盎小孢子菌 *Microsporum audouinii* Gruby。

中国小孢子菌属分种检索表

1. 厚垣孢子丰富……2
1. 厚垣孢子无或罕见……3
  2. 缺少典型的大、小分生孢子……**铁锈色小孢子菌 *M. ferrugineum***
  2. 大孢子梭形，表面具疣状突起；小孢子棒状……**奥杜盎小孢子菌 *M. audouinii***
3. 大分生孢子梭形至纺锤形，大小变化较大……**犬小孢子菌 *M. canis***
3. 大分生孢子椭圆形至梭形……4
  4. 小分生孢子梨形或椭圆形或棒状……**布拉尔小孢子菌 *M. boullardii***
  4. 小分生孢子棒状至柱状……**贵州小孢子菌 *M. guizhouense***

布拉尔小孢子菌　图 67

Microsporum boullardii Dominik & Majchr. [as '*Microsporon*'], Ekon. Pol., Ser. A 13: 428, 1965. Shao *et al*., Journal of Mountain Agriculture and Biology 38(3): 18, 2019.

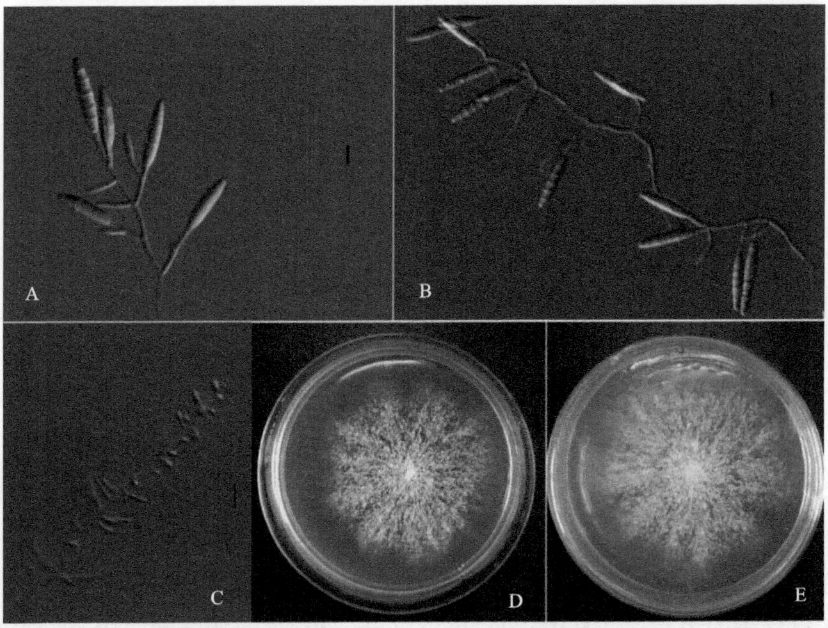

图 67　布拉尔小孢子菌 *Microsporum boullardii* Dominik & Majchr.（标本号：HMAS 255405）

A、B. 大分生孢子；C. 小分生孢子；D、E. PDA 培养基上的菌落。标尺：A～C = 20 μm

在 PDA 培养基上，28℃培养 14 天，菌落直径 64～69 mm；中部乳白色、粉状，边缘灰白色，短绒毛状；背面中部乳白色，边缘灰白色。菌丝分隔，表面光滑，无球拍状菌丝和瘤状菌丝，宽 1.6～5.3 μm。有两种类型孢子，大分生孢子侧生或顶生于锐角分枝上，光滑，分 2～6 隔，长椭圆形或梭形，不弯曲，32.7～78.1 × 8.0～11.1 μm；小分生孢子少，着生于菌丝或短柄，梨形或椭圆形或棒状，3.2～22.9 × 1.8～2.0 μm。厚垣孢子未见。

研究菌株：山西：临汾市，吉县，生活垃圾土，2017 年 8 月 3 日，梁建东，HMAS 255405，L19.3；甘肃：敦煌市，莫高窟，沙土，2017 年 7 月 15 日，张芝元，S61。

世界分布：中国、埃及、几内亚、波兰。

讨论：形态上，布拉尔小孢子菌与奇妙小孢子菌（现名奇妙帕拉癣菌）较为相似，两者均无厚垣孢子，大分生孢子以梭形孢子为主，且不弯曲，无球拍状菌丝，小分生孢子稀少。但二者的主要区别在于：布拉尔小孢子菌的小分生孢子梨形或椭圆形或棒状，3.2～22.9 × 1.8～2.0 μm（邵秋雨等，2019）；而奇妙帕拉癣菌的小分生孢子梨形或倒卵形，2.4～5.9 × 1.5～2 μm；此外，二者隶属于同属时，其在系统发育上的亲缘关系也较远。

贵州小孢子菌　图 68

Microsporum guizhouense Zhi Y. Zhang, X. Zou, Y.F. Han & Z.Q. Liang, Mycosystema 36(5): 537, 2017.

在查氏培养基上，26℃培养 14 天，菌落直径达 40 mm，灰白色至乳白色，绒状，气生菌丝稀疏，边缘较规则；背面乳白色。

在 PDA 培养基上，25℃培养 14 天，菌落直径达 49 mm，白色至浅棕色，绒状，中央微隆，边缘不规则；背面褐色到棕色，有浅棕色色素扩散。菌丝分隔，无色，光滑，宽 2～4 μm；无球拍状菌丝。分生孢子侧生或顶生，大分生孢子较丰富，无色，光滑或粗糙，具疣状突起，2～10 个隔，梭形至椭圆形，16～56 × 5～8.5 μm；小分生孢子丰富，着生于菌丝或短柄上，无色，光滑，棒状至柱状，3.2～11 × 1～2 μm。

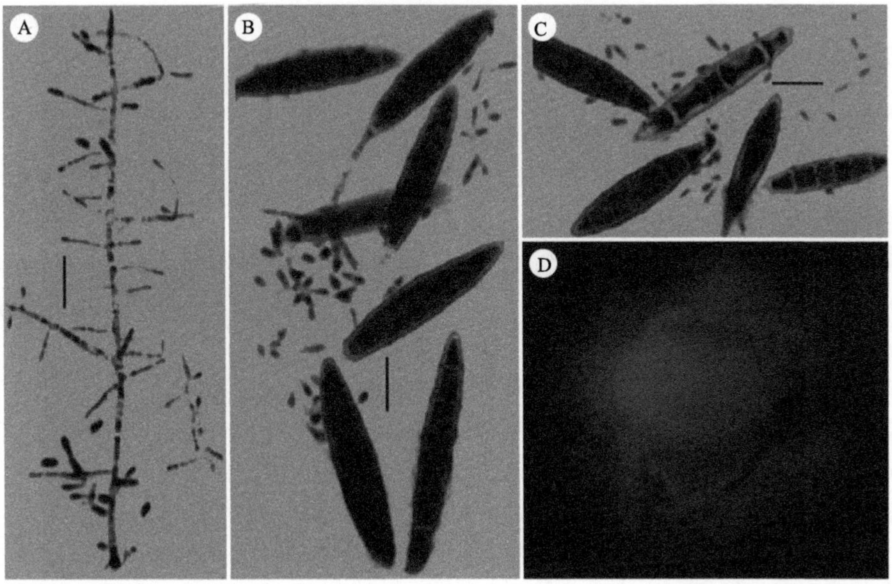

图 68　贵州小孢子菌 *Microsporum guizhouense* Zhi Y. Zhang, X. Zou, Y.F. Han & Z.Q. Liang（标本号：GZAC EB2001M）

A. 小分生孢子；B、C. 大分生孢子和小分生孢子；D. 查氏培养基上的菌落。标尺：A～C = 10 μm

研究菌株：贵州：开阳县，烟地土，2012 年 8 月 10 日，GZAC EB2001M（干模式），模式菌株 GZUIFR EB2001M。

世界分布：中国。

讨论：本种的典型特征为：无球拍状菌丝；分生孢子侧生或顶生，有大小孢子两种类型，大分生孢子较丰富，无色，光滑或粗糙，分 2～10 隔，梭形至椭圆形，16～56 × 5～8.5 μm；小分生孢子同样较丰富，着生于菌丝上或短柄上，棒状至柱状，3.2～11 × 1～2 μm（张芝元等，2017c），而近似种布拉尔小孢子菌的小孢子为梨形或椭圆形或棒状，明显不同于本种。

作者未观察的种

奥杜盎小孢子菌

Microsporum audouinii Gruby, C. r. hebd. Séanc. Acad. Sci., Paris 17: 302, 1843. Wang, Medical Mycology-Guide to Laboratory Examination p. 152, 2005.

≡ *Sporotrichum audouinii* (Gruby) Sacc., Syll. Fung. 4: 101, 1886.

≡ *Sabouraudites audouinii* (Gruby) M. Ota & Langeron, Annls Parasit. Hum. Comp. 1: 327, 1923.

≡ *Closteroaleurosporia audouinii* (Gruby) Grigoraki, Annls Sci. Nat., Bot., Sér. 10, 7: 412, 1925.

≡ *Veronaia audouinii* (Gruby) Benedek, Mycopath. Mycol. Appl. 14: 115, 1961.

= *Sabouraudites langeronii* Vanbreus., Annls Parasit. Hum. Comp. 25: 516, 1950.

讨论：王端礼（2005）记载，在沙氏培养基上，菌落中度快速生长，毡状至绒毛状，灰色至米黄色；背面橙红色或苍白色。菌丝分隔，有时具梳状菌丝或球拍状菌丝，常存在厚垣孢子，端生或间生。小分生孢子罕见，棒状；大分生孢子罕见，梭形，表面具疣状突起。该种可引起儿童头癣。

世界分布：中国、法国、德国、葡萄牙、美国。

犬小孢子菌

Microsporum canis E. Bodin ex Guég., Les Champignons parasites de l'homme et des animaux domestiques 137, 1902. Wang, Medical Mycology-Guide to Laboratory Examination p. 141, 2005.

≡ *Microsporum audouinii* var. *canis* E. Bodin, Prat. Derm. 810, 1900.

≡ *Sabouraudites canis* (E. Bodin ex Guég.) Langeron, Précis Mycol. 534, 1945.

= *Nannizzia otae* A. Haseg. & Usui, Jpn. J. Med. Mycol. 16: 151, 1975.

= *Microsporum distortum* di Menna & Marples, Trans. Br. Mycol. Soc. 37: 372, 1954.

讨论：王端礼（2005）记载，该种在沙氏培养基上，菌落具放射状沟纹；大分生孢子丰富，梭形至纺锤形，大小变化较大，粗糙，具疣状突起；小分生孢子在沙氏培养基上较少，在其他培养基上丰富，3.5～8.5 × 1.5～3.5 μm，光滑，单胞，偶见双胞。该种为亲动物性菌，主要为猫狗的病原，常传染给人；据报道还从黑猩猩、栗鼠、狐狸、狮子、猴子、猪、兔子、羊和老虎等上分离到了。

世界分布：中国、法国、英国、美国等。

铁锈色小孢子菌

Microsporum ferrugineum M. Ota, Jap. J. Derm. Urol. 21: 201, 1921. Wang, Medical Mycology-Guide to Laboratory Examination p. 150, 2005.

≡ *Grubyella ferruginea* (M. Ota) M. Ota & Langeron, Annls Parasit. Hum. Comp. 1: 330, 1923.

≡ *Arthrosporia ferruginea* (M. Ota) Grigoraki, Annls Sci. Nat., Bot., Sér. 10, 7: 414, 1925.

≡ *Achorion ferrugineum* (M. Ota) Guiart & Grigoraki, Lyon Méd. 141: 377, 1928.

≡ *Trichophyton ferrugineum* (M. Ota) Langeron & Miloch. ex Talice, Annls Parasit. Hum. Comp. 9: 83, 1931.

讨论：据王端礼（2005）的记载，在沙氏培养基上，菌落生长缓慢，绒毛状；具球拍状菌丝及梳状体；厚垣孢子大量存在，单个或成串，形状不规则。该菌可引起人类和动物的头癣与体癣。

世界分布：中国、日本。

毁丝霉属 **Myceliophthora** Costantin

C. r. hebd. Séanc. Acad. Sci., Paris 114: 849, 1892

菌落平铺，白色、奶油色、淡黄色、浅棕色或深绿色，绒状、毡状、絮状或粉状。菌丝无色，表面光滑，有隔膜，不规则，或多或少直生分枝。芽殖分生孢子，顶生、侧生于菌丝，无柄或生于短的突起或侧分枝上，通常 1～4 个芽殖型分生孢子着生于一个膨大突起上，单生，或 2～4 个成连，半透明或淡黄色、浅棕色或红棕色，表面光滑或具疣状突起，球形、近球形、椭圆形或梨形，单胞，基痕狭窄，无间生孢子，厚垣孢子未见，耐热。

模式种：土黄毁丝霉 *Myceliophthora lutea* Costantin。

土黄毁丝霉　图 69，图版 II：5

Myceliophthora lutea Costantin, C. r. hebd. Séanc. Acad. Sci., Paris 114: 850, 1892.

≡ *Scopulariopsis lutea* (Costantin) Tubaki, Nagaoa 5: 29, 1955.

≡ *Chrysosporium luteum* (Costantin) J.W. Carmich., Canad. J. Bot. 40: 1158, 1962.

= *Sporotrichum carthusioviride* J.N. Rai & Mukerji, Mycopathol. Mycol. Appl. 18: 122, 1962.

在查氏培养基上，25℃培养 14 天，菌落直径 19～21 mm，平展，粉状，乳白色；背面无色。

在 PDA 培养基上，25℃培养 14 天，菌落直径 9～19 mm，白色至淡黄色，褶皱，裂开；形状不规则，分界明显；背面淡黄色，培养基裂开。菌丝无色，光滑，分隔，宽 1.2～3.8 μm；无球拍状菌丝。分生孢子梗分枝，分生孢子常无色，光滑，顶生于菌丝，短柄，短突；单生或2～3个簇生，单胞，表面光滑或粗糙，大多数近球形至球形，直径为16.6～26.3 μm；少数椭圆形，具1平截底部或1窄且不明显的具柄部分，4～7 × 3.5～4 μm。间生孢子未见，厚垣孢子未见。

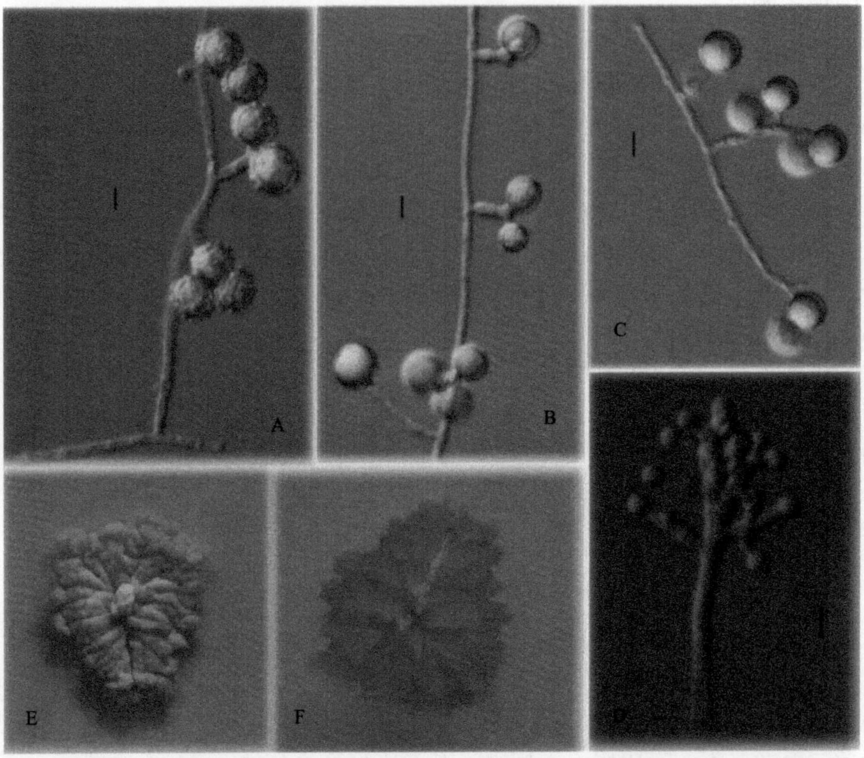

图69　土黄毁丝霉 *Myceliophthora lutea* Costantin（标本号：E0603M）

A～D. 产孢结构和孢子；E、F. PDA培养基上的菌落。标尺：A～D = 10 μm

研究菌株：贵州：修文县，动物园土，2010年9月1日，韩燕峰，E0603M，E0603M.1。

世界分布：中国、法国。

讨论：土黄毁丝霉广泛分离自菌床、土壤、干草、大麦 *Hordeum vulgare*、猪圈的空气和马厩的灰尘中（van Oorschot，1977）。形态上，该种与嗜热疣霉属的无性型相似，但其在琼脂培养基上生长受限，耐热但不嗜热（梁晨和吕国忠，2002）。

奈尼兹皮菌属 **Nannizzia** Stockdale

Sabouraudia 1: 45, 1961

菌落大多棉絮状至粉状，白色至棕色。菌丝薄壁，无色，菌丝型大分生孢子和小分生孢子，如果存在，产生于正交排列的菌丝上。大分生孢子，2或多胞，壁薄，光滑或粗糙，无色，圆柱状，或棒状到雪茄状；小分生孢子壁薄，光滑，无色，单胞，卵圆形、梨状到棒状。菌丝配接后通常产生有性型，节皮样。

该属的形态特征与小孢子菌属很相似，其中部分曾为小孢子菌属的成员，现更名为奈尼兹皮菌属。该属的研究概况可参考本卷小孢子菌属。目前该属包括17个种，国内已报道 5 个种，即粉奈尼兹皮菌 *Nannizzia fulva* Stockdale、石膏样奈尼兹皮菌 *N. gypsea* (Nann.) Stockdale、弯奈尼兹皮菌 *N. incurvata* Stockdale、猪奈尼兹皮菌 *N. nana* (C.A. Fuentes) Y. Gräser & de Hoog 和桃色奈尼兹皮菌 *N. persicolor* Stockdale，这几个种属于亲动物性，均有致病性（de Hoog *et al.*，2017）。

模式种：弯奈尼兹皮菌 *Nannizzia incurvata* Stockdale。

中国奈尼兹皮菌属分种检索表

1. 大分生孢子常弯曲 ………………………………………… **弯奈尼兹皮菌 *N. incurvata***
1. 大分生孢子不弯曲 ………………………………………… 2
  2. 大分生孢子多隔，超过 3 个细胞 ………………………………………… 3
  2. 大分生孢子通常为双胞 ………………………………………… **猪奈尼兹皮菌 *N. nana***
3. 大分生孢子光滑，仅顶部稍粗糙 ………………………………………… **桃色奈尼兹皮菌 *N. persicolor***
3. 大分生孢子粗糙或具瘤 ………………………………………… 4
  4. 大分生孢子梭形至纺锤形，小分生孢子椭圆形至棒状 ………………………… **粉奈尼兹皮菌 *N. fulva***
  4. 大分生孢子椭圆形至梭形，小分生孢子卵圆形至椭圆形 …………… **石膏样奈尼兹皮菌 *N. gypsea***

粉奈尼兹皮菌　图 70

Nannizzia fulva Stockdale, Sabouraudia 3(1): 120, 1963.

≡ *Microsporum fulvum* Uriburu, Argentina Med. 7: 565, 1909.

≡ *Closterosporia fulva* (Uriburu) Grigoraki, Annls. Sci. Nat., Bot. Sér. 10, 7: 411, 1925.

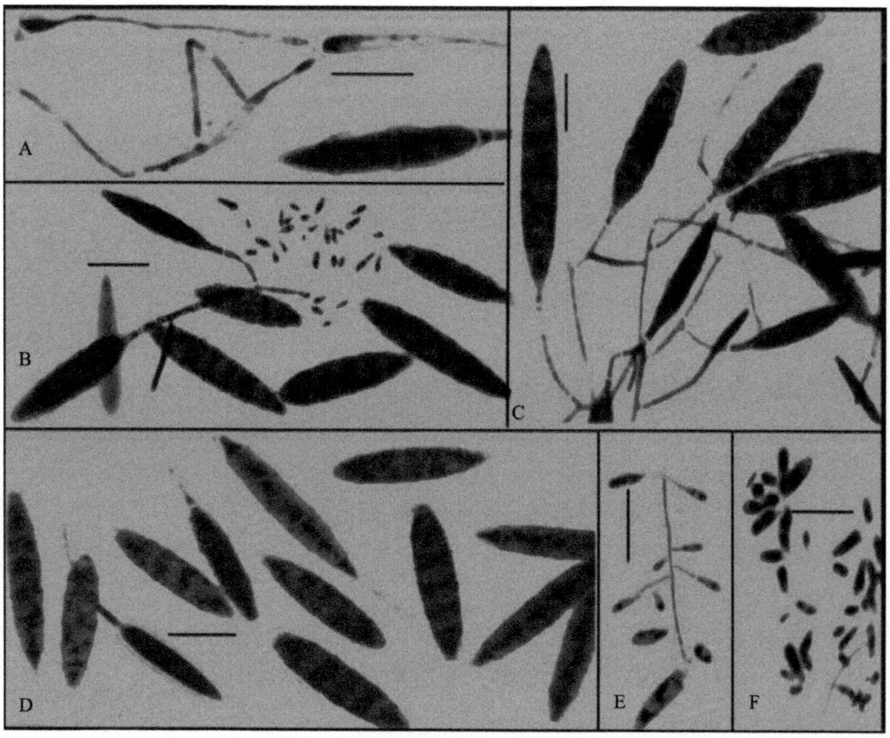

图 70　粉奈尼兹皮菌 *Nannizzia fulva* Stockdale（标本号：GZUIFR EB1401M）

A. 球拍状菌丝；B～D. 大分生孢子；E、F. 小分生孢子。标尺：A～F = 10 μm

在查氏培养基上，25℃培养 14 天，菌落直径 30 mm，乳白色，气生菌丝十分稀疏，极不易分辨；背面灰白色。

在 PDA 培养基上，25℃培养 14 天，菌落直径 40 mm，毡状，淡棕色到灰白色再到淡黄红色，中央密集，具多条环状条纹，边缘规则；背面黄棕色到乳白色。菌丝分隔，无色，光滑，宽 1～5 μm；具球拍状菌丝，21～26 × 4.3～8.6 μm。大分生孢子十分丰富，表面粗糙，一般 2～6 个隔膜，少数 8 个隔膜，梭形至纺锤形，15～58 × 4.3～10.8 μm；小分生孢子较少，多数侧生，着生于菌丝或短柄上，光滑，无色，单胞，偶见双胞，椭圆形至棒状，4.3～8.6 × 1～3.2 μm。厚垣孢子未见。

研究菌株：山西：朔州市，甘庄村，农田土，2012 年 8 月 6 日，张延威，GZUIFR EB1401M；陕西：渭南市，辣椒地土，2013 年 6 月 5 日，韩燕峰，GZUIFR EB6401M。

世界分布：中国、阿根廷、匈牙利、波兰、墨西哥等。

讨论：形态上，粉奈尼兹皮菌与石膏样奈尼兹皮菌因两者均无厚垣孢子，大分生孢子以梭形孢子为主，且不弯曲，存在球拍状菌丝而较为相似。但二者的主要区别在于：粉奈尼兹皮菌的小分生孢子椭圆形至棒状，4.3～8.6 × 1～3.2 μm；而石膏样奈尼兹皮菌的小分生孢子大多数卵圆形至椭圆形，4.3～11 × 3.2～5.4 μm。

石膏样奈尼兹皮菌　图 71

Nannizzia gypsea (Nann.) Stockdale, Sabouraudia 3(1): 119, 1963.

≡ *Arthroderma gypseum* (Nann.) Weitzman, McGinnis, A.A. Padhye & Ajello, Mycotaxon 25: 514, 1986.

= *Microsporum gypseum* (E. Bodin) Guiart & Grigoraki, Lyon Médical 141: 377, 1928.

≡ *Gymnoascus gypseus* Nann., Atti Accad. Fisioscr. Siena Med.-Fis. 2: 93, 1927.

≡ *Arthroderma gypseum* (Nann.) Weitzman, McGinnis, A.A. Padhye & Ajello, Mycotaxon 25: 514, 1986.

= *Microsporum appendiculatum* Bhat & Miriam, in Miriam & Bhat, Kavaka 25: 93, 1998.

在查氏培养基上，25℃培养 14 天，菌落直径 65 mm，绒状，白色，菌丝十分稀疏，边缘不规则，树枝状；背面白色。

在 PDA 培养基上，25℃培养 14 天，菌落直径达 80 mm，绒状到粉状，白色到黄棕色，中央微隆，边缘不规则；背面黄棕色。菌丝分隔，无色，光滑或粗糙，宽 1～5 μm；具球拍状菌丝，光滑或粗糙，大小为 8.6～29 × 5.4～11 μm。大分生孢子十分丰富，无色，表面光滑或粗糙，分隔或不分隔，一般分 3～6 个隔，椭圆形至梭形，15～51 × 7.5～13 μm；小分生孢子较少，顶生、侧生，着生于菌丝上，无色，光滑或粗糙，大多数卵圆形至椭圆形，4.3～11 × 3.2～5.4 μm，少数近球形或棒状。厚垣孢子未见。

研究菌株：陕西：合阳县，茄子地土，2013 年 6 月 6 日，韩燕峰，GZUIFR EB6302M。

世界分布：中国、印度、美国、澳大利亚等。

讨论：形态上，石膏样奈尼兹皮菌的典型特征为：具球拍状菌丝；大分生孢子以梭形孢子为主，且大分生孢子不弯曲；小分生孢子较少，小分生孢子大多数卵圆形至椭圆形，4.3～11 × 3.2～5.4 μm；无厚垣孢子。其近似种粉奈尼兹皮菌的小分生孢子椭圆形至棒状，4.3～8.6 × 1～3.2 μm，明显不同于本种。该种常从土壤中和小啮齿动物的皮毛中分离到。

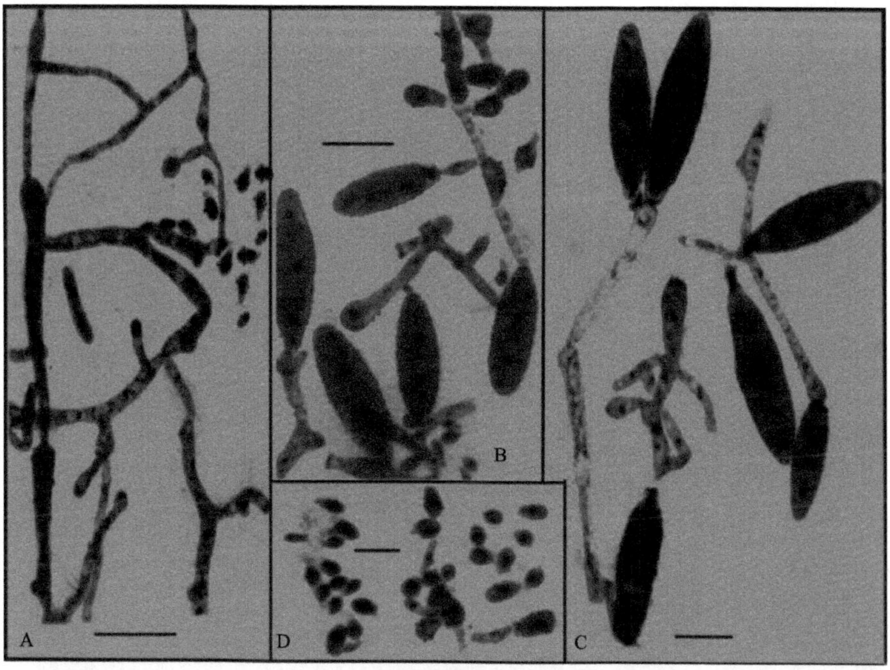

图 71　石膏样奈尼兹皮菌 *Nannizzia gypsea* (Nann.) Stockdale（标本号：GZUIFR EB6302M）

A. 球拍状菌丝；B. 分生孢子；C. 大分生孢子；D. 小分生孢子。标尺：A～D = 10 μm

弯奈尼兹皮菌　图 72，图版 II：6

Nannizzia incurvata Stockdale, Sabouraudia 1: 46, 1961.

≡ *Microsporum incurvatum* (Stockdale) P.L. Sun & Y.M. Ju, Medical Mycol. 52(3): 46, 2014. Shao *et al*., Journal of Mountain Agriculture and Biology, 38(3): 19, 2019.

≡ *Nannizzia gypsea* var. *incurvata* (Stockdale) Apinis, Mycol. Pap. 96: 32, 1964.

≡ *Arthroderma incurvatum* (Stockdale) Weitzman, McGinnis, A.A. Padhye & Ajello, Mycotaxon 25: 514, 1986.

在 PDA 培养基上，28℃培养 7 天，菌落直径 48～51 mm，菌落近圆形，树枝状，粉状，边缘灰白色，中部褐色，边缘不规则，分界不明；背面边缘白色，中部浅褐色。菌丝分隔，表面光滑；无球拍状菌丝和瘤状菌丝，宽 1.2～4.6 μm。大分生孢子顶生或侧生，单生、表面光滑或粗糙，具 2～6 个隔，梭形，有时会弯曲，29.2～50.3 × 7.2～11.9 μm；小孢子丰富，着生于菌丝或短柄，表面光滑，倒卵形或棒状，3.7～5.7 × 1.2～1.7 μm。

研究菌株：甘肃：玉门市，某服务区“羊胡子草”根系土，2017 年 7 月 10 日，张芝元，GZUIFR K82.3。

世界分布：中国、英国。

讨论：早前认为石膏样奈尼兹皮菌是弯奈尼兹皮菌的无性型，虽然这两个无性型形态非常相似，但弯奈尼兹皮菌的大分生孢子梭形，有时会弯曲，具 2～6 隔；小分生孢子倒卵形或棒状而有别于石膏样小孢子菌。此外，基于 ITS rDNA 聚类分析，发现它们并不聚为一支，石膏样奈尼兹皮菌反而与粉奈尼兹皮菌聚在一起，因此弯奈尼兹皮菌是独立的一个种（Sun *et al.*，2014；邵秋雨等，2019）。该种常从人或狗上分离到。

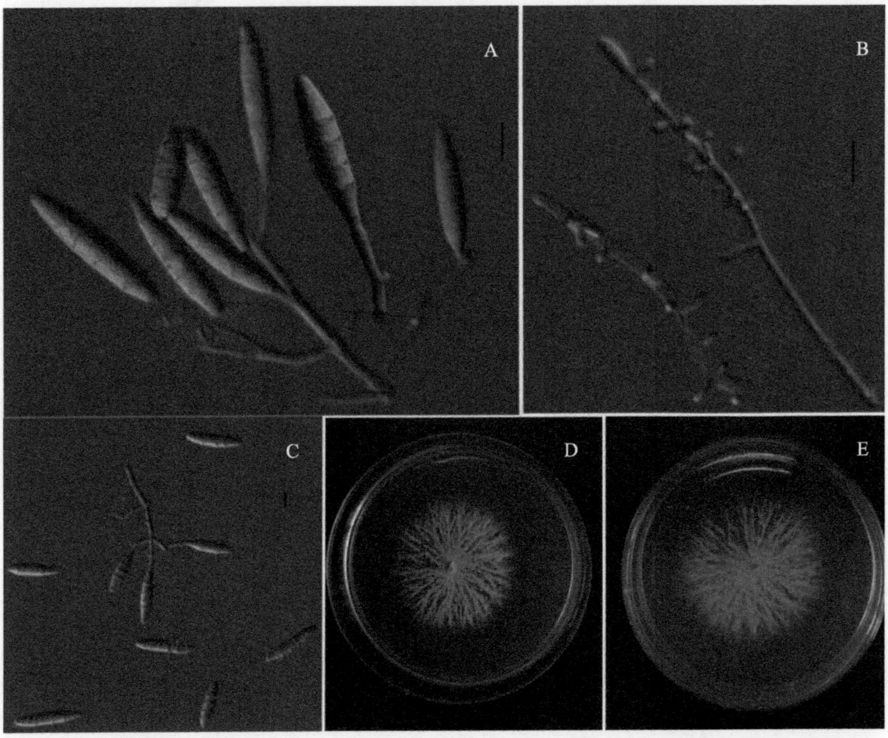

图 72　弯奈尼兹皮菌 *Nannizzia incurvata* Stockdale（标本号：GZUIFR K82.3）

A、C. 大分生孢子；B. 小分生孢子；D、E. PDA 培养基上的菌落。标尺：A～C = 20 μm

作者未观察的种

猪奈尼兹皮菌　图 73

Nannizzia nana (C.A. Fuentes) Y. Gräser & de Hoog, Mycopathologia 182: 23, 2016.

≡ *Microsporum nanum* C.A. Fuentes, Mycologia 48: 614, 1956. Wang, Medical Mycology-Guide to Laboratory Examination p. 149, 2005.

讨论：毛叶红等（2017）记载，该菌在沙氏培养基上，菌落绒毛状，白色至黄色，可见水滴状分泌物；具有两种分生孢子，大分生孢子卵圆形至椭圆形，通常有 2 个细胞，具瘤，10～18 × 8～10 μm；小分生孢子棒状，2.5～5 × 1～1.5 μm。其显著特征是大分生孢子分隔少，具瘤，常具 1 个隔膜；在猪的感染中常见。

世界分布：中国、波兰、古巴、墨西哥、危地马拉。

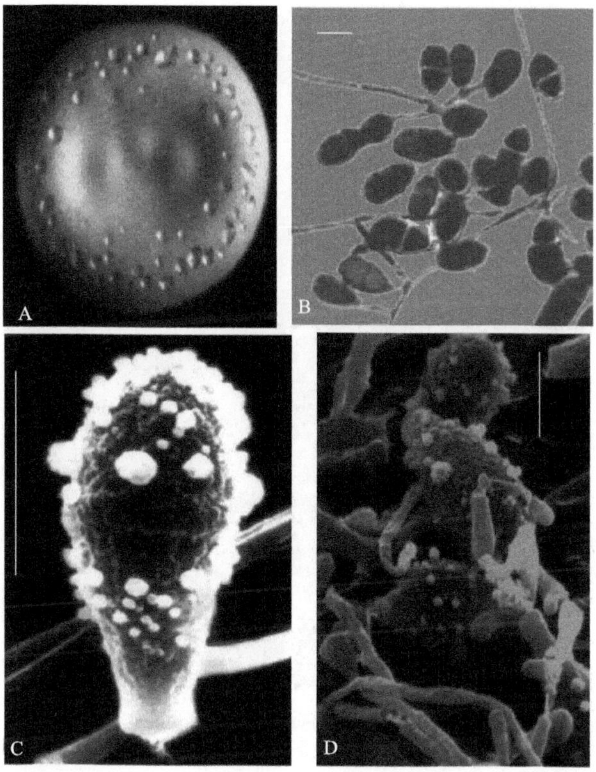

图 73　猪奈尼兹皮菌 *Nannizzia nana* (C.A. Fuentes) Y. Gräser & de Hoog（引自王端礼，2005）

A. 菌落特征；B、C. 大分生孢子；D. 产孢结构。标尺：B～D = 10 μm

桃色奈尼兹皮菌　图 74

Nannizzia persicolor Stockdale, Sabouraudia 5: 357, 1967.

≡ *Microsporum persicolor* (Sabour.) Guiart & Grigoraki, Lyon Médical 141: 377, 1928. Wang, Medical Mycology-Guide to Laboratory Examination p. 157, 2005.

≡ *Arthroderma persicolor* (Stockdale) Weitzman, McGinnis, A.A. Padhye & Ajello, Mycotaxon 25: 514, 1986.

= *Nannizzia quinckeana* Balab. & G. Schick, Derm. Venereol. 9: 35, 1970.

讨论：de Hoog 等（2017）记载，该种在沙氏培养基上，菌落生长快速，绒毛状至粉状，黄色至玫瑰红色；常见螺旋菌丝；大分生孢子梭形或子弹形，光滑，顶部稍粗糙，常有 5 隔，75～80 × 12～16 μm；小分生孢子丰富，棒状至球状，常有蒂，2.5～5 × 1～1.5 μm。该种的显著特征是产生螺旋菌丝，两型分生孢子，大分生孢子的顶部粗糙。该种为亲动物性，主要宿主有人类（寄生于头皮、皮肤和足）、啮齿动物、蝙蝠和狗（王端礼，2005）。

世界分布：中国、英国、保加利亚。

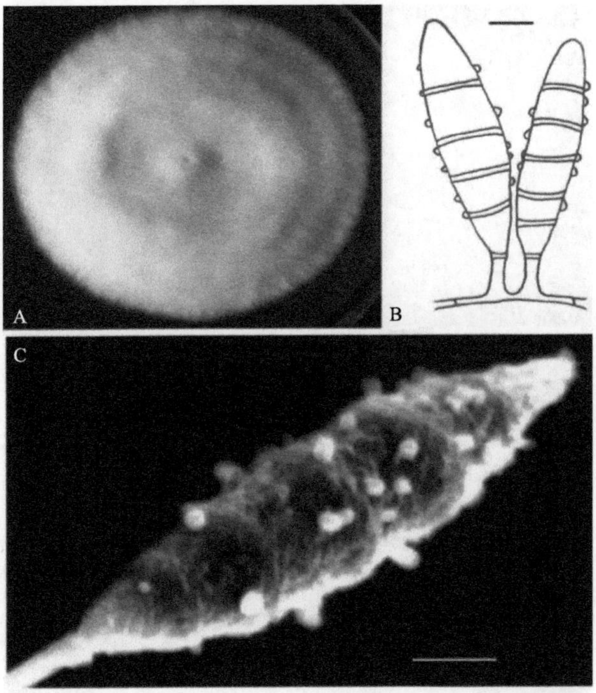

图 74　桃色奈尼兹皮菌 *Nannizzia persicolor* Stockdale（引自王端礼，2005）

A. 菌落特征；B、C. 大分生孢子。标尺：B、C = 10 μm

帕拉癣菌属 **Paraphyton** Y. Gräser, Dukik & de Hoog

Mycopathologia 182: 24, 2017

菌落大多可见颗粒状，褐色；背面棕色。菌丝薄壁，无色。菌丝型大分生孢子和小分生孢子，如果存在，产生于正交排列的菌丝上。大分生孢子，多胞，壁厚，粗糙，近无色，棒状或雪茄状。小分生孢子壁薄，光滑，无色，单胞，棒状。菌丝配接后产生有性型，节皮样。

模式种：库柯帕拉癣菌 *Paraphyton cookei* (Ajello) Y. Gräser, Dukik & de Hoog。

中国帕拉癣菌属分种检索表

1. 大分生孢子粗糙，小分生孢子丰富 ………………………………………………… **库柯帕拉癣菌 *P. cookei***
1. 大分生孢子光滑，小分生孢子少 ………………………………………………… **奇妙帕拉癣菌 *P. mirabile***

库柯帕拉癣菌　图 75

Paraphyton cookei (Ajello) Y. Gräser, Dukik & de Hoog, Mycopathologia 182: 24, 2017.

≡ *Microsporum cookei* Ajello, Mycologia 51: 71, 1959. Wang, Medical Mycology-Guide to Laboratory Examination p. 159, 2005.

= *Arthroderma cajetani* (Ajello) Ajello, Weitzman, McGinnis & A.A. Padhye, Mycotaxon 25: 513, 1986.

讨论：王端礼（2005）记载，该种在沙氏培养基上，菌落中等速度生长，致密短绒毛状至粉状，白色至黄色，可见明显的放射状沟纹；大分生孢子丰富，厚壁，粗糙，具6～13 个隔膜，梭形，50～90 × 12～25 μm；小分生孢子丰富，光滑，卵圆形，3～5.5 × 1.5～2.0 μm。该种为亲土性真菌，与奇妙帕拉癣菌的主要区别是大分生孢子比较大，粗糙；而奇妙帕拉癣菌具有节孢子状菌丝，且大分生孢子比前者小，光滑。该种主要从土壤中发现，有时能从猫、狗和啮齿动物中分离到。

世界分布：中国、美国等，世界性分布。

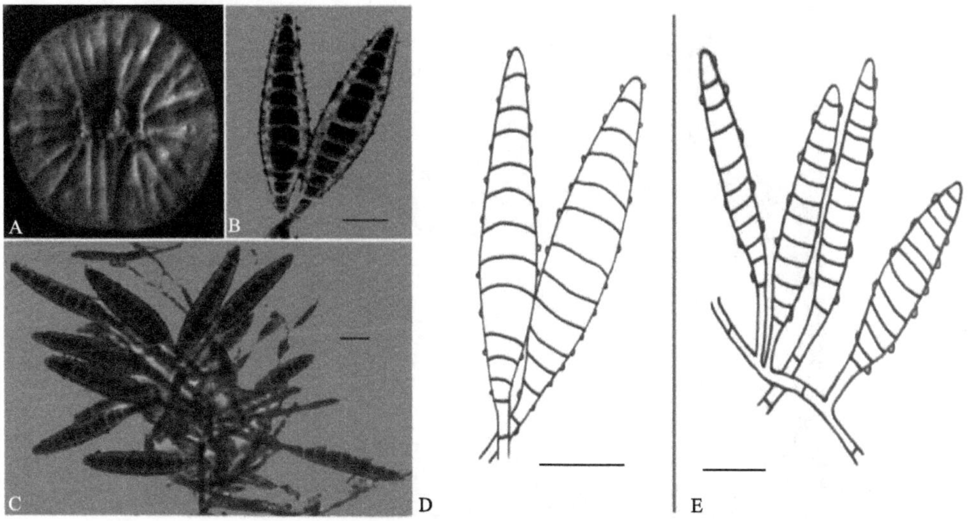

图 75 库柯帕拉癣菌 *Paraphyton cookei* (Ajello) Y. Gräser, Dukik & de Hoog（引自王端礼，2005）

A. 沙氏培养基上的菌落；B～E. 大分生孢子和产孢结构。标尺：B～E = 20 μm

奇妙帕拉癣菌 图 76，图版 II：7

Paraphyton mirabile (J.S. Choi, Y. Gräser, G. Walther, Peano, F. Symoens & de Hoog) Y. Gräser, Dukik & de Hoog, Mycopathologia, 182: 2017. Shao *et al*., Journal of Mountain Agriculture and Biology 38(2): 19, 2019.

≡ *Microsporum mirabile* J.S. Choi, Y. Gräser, G. Walther, Peano, F. Symoens & de Hoog, Medical Mycol. 50(2): 168, 2012.

在 PDA 培养基上，28℃培养 7 天，菌落直径 48 mm，中部粉状，微突起，边缘白色，中部浅棕色，边缘裂叶状，近圆形；背面浅葡萄紫色。菌丝分隔，表面光滑，无球拍状菌丝和瘤状菌丝，具节孢子状菌丝，宽 1.6~4.9 μm。大分生孢子顶生或侧生，表面光滑，分隔，1～7 隔，梭形或船形，42.4～58.8 × 7.9～14.8 μm；小分生孢子少，着生于菌丝，梨形或倒卵形，2.4～5.9 × 1.5～2 μm。厚垣孢子未见。

研究菌株：山西：临汾市，吉县，农田土，2017 年 8 月，韩燕峰，HMAS 255404，L6.7。

世界分布：中国、意大利、新西兰。

讨论：奇妙帕拉癣菌的典型特征为：无球拍状菌丝，大分生孢子以梭形孢子为主，不弯曲；小分生孢子较少，梨形或倒卵形，2.4～5.9 × 1.5～2 μm（邵秋雨等，2019）。本种形态上与小孢子菌属中的布拉尔小孢子菌相似，但后者的小分生孢子梨形或椭圆形或棒状，3.2～22.9 × 1.8～2.0 μm（Choi *et al.*，2012），此外二者的系统发育关系较远。本种为亲动物型（de Hoog *et al.*，2017）。

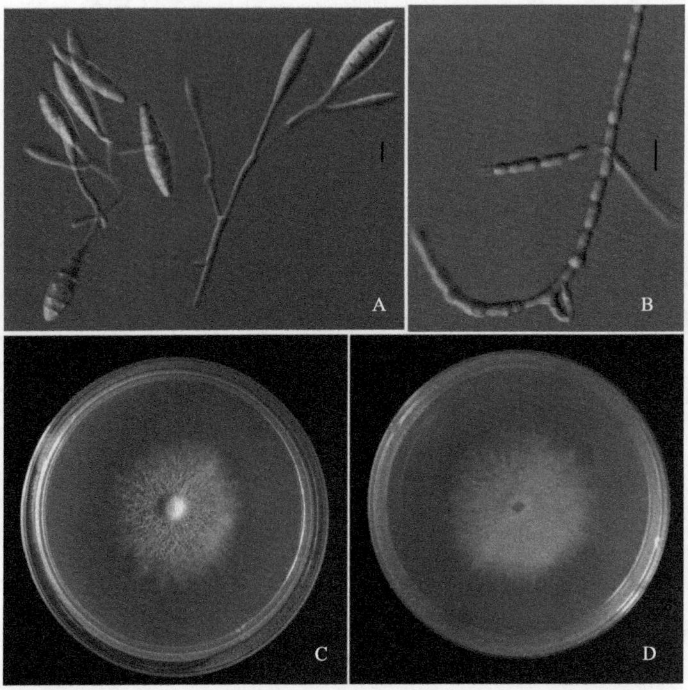

图 76　奇妙帕拉癣菌 *Paraphyton mirabile* (J.S. Choi, Y. Gräser, G. Walther, Peano, F. Symoens & de Hoog) Y. Gräser, Dukik & de Hoog（标本号：HMAS 255404）

A. 大分生孢子；B. 节孢子状菌丝；C、D. PDA 培养基上的菌落。标尺：A、B = 20 μm

瓶霉属 **Phialophora** Medlar

Mycologia 7(4): 202, 1915

菌落平展，褐色或橄榄褐色至褐色。菌丝体部分表生，部分埋生。菌丝灰色、灰褐色、褐色甚至黑色，分枝，具隔膜。子座无，刚毛和附属丝缺。分生孢子梗分化不很明显，分枝，直或弯曲，淡褐色、棕褐色至橄榄褐色，光滑。产孢细胞单瓶梗型，垂直着生于菌丝或生在菌丝分枝的顶部，单生或簇生，有限生长，烧瓶形、安瓿形或钻形；领部发育良好，浅或深，基部稍膨大，无色或淡褐色或褐色，近顶部无色。分生孢子内壁芽生，简单，直或弯曲，椭圆形或长方形，两端钝圆，无色或淡褐色或橄榄褐色，光滑，无隔膜，常聚集成黏球状。

模式种：疣状瓶霉 *Phialophora verrucosa* Medlar。

中国瓶霉属分种检索表

1. 分生孢子梗明显可见 ··· 2
1. 分生孢子梗缺乏，未见或很难区分 ··· 9
　2. 分生孢子梗无色 ··· **土壤瓶霉 *P. subterranea***
　2. 分生孢子梗有颜色 ··· 3

3. 产孢瓶梗为曲膝状 …………………………………………………………… 膝状瓶霉 ***P. geniculata***
3. 产孢瓶梗为非膝状 ………………………………………………………………………………… 4
4. 产孢瓶梗表面具疣状物 ………………………………………………… 小孢瓶霉 ***P. microspora***
4. 产孢瓶梗表面光滑 ………………………………………………………………………………… 5
5. 厚垣孢子可见，球形或卵形 ……………………………………………………… 光滑瓶霉 ***P. levis***
5. 厚垣孢子未见 ……………………………………………………………………………………… 6
6. 产孢瓶梗为细瓶状 ………………………………………………………… 仁果瓶霉 ***P. malorum***
6. 产孢瓶梗为烧瓶形或圆柱形 ……………………………………………………………………… 7
7. 菌丝具疣状物 …………………………………………………………… 海榄雌瓶霉 ***P. avicenniae***
7. 菌丝未见疣状物 …………………………………………………………………………………… 8
8. 产孢瓶梗为烧瓶形，基部稍窄 …………………………………………… 帚状瓶霉 ***P. fastigiata***
8. 产孢瓶梗近圆柱形，顶端变细 ……………………………………… 聂拉木瓶霉 ***P. nielamuensis***
9. 分生孢子无色或近无色 …………………………………………………………………………… 10
9. 分生孢子具颜色 …………………………………………………………………………………… 14
10. 产孢瓶梗呈杯状、近球形或近圆柱形 ……………………………………… 疣状瓶霉 ***P. verrucosa***
10. 产孢瓶梗为烧瓶状或圆柱形 ……………………………………………………………………… 11
11. 产孢方式中出现出芽形式 ………………………………………………………………………… 12
11. 产孢方式中未见出芽形式 ………………………………………………………………………… 13
12. 一些较大的分生孢子中形成砖格状分隔或缓慢出芽 …………………… 中华瓶霉 ***P. chinensis***
12. 分生孢子有时出现出芽形式 ………………………………………… 大孢瓶霉 ***P. macrospora***
13. 瓶梗大小可变，具花瓶形或管状领部 ………………………………… 美洲瓶霉 ***P. americana***
13. 瓶梗具漏斗状至近圆柱状领部 ……………………………………………… 膨大瓶霉 ***P. expanda***
14. 分生孢子非褐色 …………………………………………………………………………………… 15
14. 分生孢子呈浅褐色或深褐色 ……………………………………………………………………… 16
15. 分生孢子为淡黄色，并产生两种类型分生孢子，香肠形和椭圆形 ………… 布氏瓶霉 ***P. bubakii***
15. 分生孢子为灰色，球形或长柱形，有时略弯曲 ………………………… 灰质瓶霉 ***P. cinerescens***
16. 产孢瓶梗为倒棒状，至顶部渐细 ………………………………… 理查德瓶霉 ***P. richardsiae***
16. 产孢瓶梗为烧瓶形、安瓿形或圆柱形 ……………………………………………………………… 17
17. 分生孢子偏大，5.5～15 × 3～5.5 μm ……………………………… 双型瓶霉 ***P. dimorphospora***
17. 分生孢子偏小 ……………………………………………………………………………………… 18
18. 分生孢子为卵形、近球形，1.6～4 × 1.2～3 μm ……………………… 欧洲瓶霉 ***P. europaea***
18. 分生孢子主要为椭圆形 …………………………………………………………………………… 19
19. 分生孢子具隔膜，3～8 × 1.5～3.5 μm ………………………………… 西藏瓶霉 ***P. tibetensis***
19. 分生孢子不具隔膜，3～6 × 1～2 μm ……………………………………… 蜜色瓶霉 ***P. melinii***

美洲瓶霉　图 77

Phialophora americana (Nannf.) S. Hughes, Can. J. Bot. 36: 795, 1958. Li *et al*., Persoonia 38: 11, 2017.

在 OA 培养基上，30℃培养 21 天，菌落适度生长，橄榄褐色，中部颜色较浅；背面深橄榄色。

在 MEA 培养基上，菌落羊毛状，浅橄榄色；背面深橄榄色，无扩散性色素产生。菌丝不规则状，宽 1～3 μm。分生孢子梗缺乏。瓶梗为烧瓶状至圆柱状或细长柱状，深色；花瓶口或管状领部，也有直接着生于未分化的菌丝上。分生孢子无色，3.5～7.0 × 2～4 μm，近球形至宽椭圆形，偶近圆柱形，大小可变，大多数在领开口处黏附成簇，少数形成松散串状。

研究菌株：贵州：贵阳市，森林公园，土壤，2011 年 10 月 1 日，韩燕峰，E330。

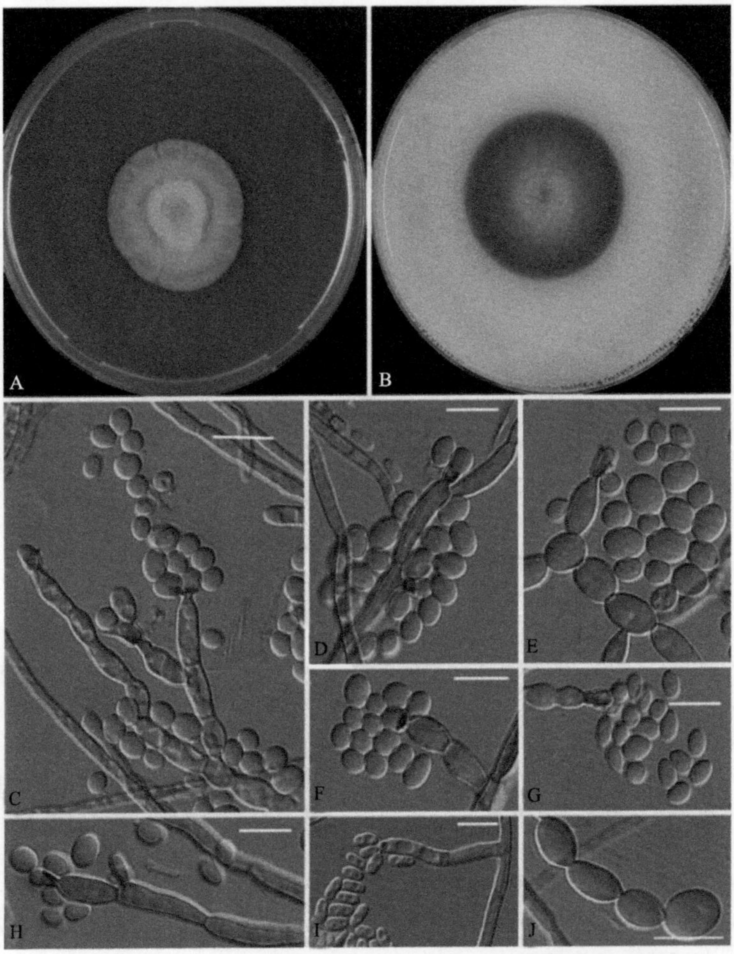

图 77　美洲瓶霉 *Phialophora americana* (Nannf.) S. Hughes（引自 Li *et al.*，2017b）

A、B. MEA、OA 培养基上的菌落；C～J. 瓶梗、孢子、近球状菌丝和砖格状细胞。标尺：C～J = 10 μm

世界分布：中国、瑞士、阿根廷。

讨论：菌株 CBS 281.35 为美洲瓶霉，分离自美国的一名患有腿部疣状皮肤病的成人患者。Schol-Schwarz（1970）首先将该分离株描述为疣状瓶霉。Yamagishi 等（1997）重新描述该菌株为美洲瓶霉。Untereiner 和 Naveau（1999）研究发现，寄生于法国杨树上的地衣的菌株 MUCL 40572 与模式标本相同，并提供带有瓶霉结构的插图，可见该菌带有深瓶形的领部。该种的生长温度：最低为 21℃，最适为 30℃，最高为 37℃。

海榄雌瓶霉　图 78

Phialophora avicenniae Yue L. Liu & Z.D. Jiang, Mycotaxon 124: 33, 2013.

菌落起初为酵母状，逐渐变为絮状，棕色。在 PDA 培养基上，27℃培养 5 天后，菌落直径可达 20～30 mm；培养 14 天后菌落颜色变为棕色或深棕色。菌丝无色至近无色，表面光滑，偶呈棕色，表面具疣，宽 1.5～3 μm。分生孢子梗无色或棕色，分枝，极少结构简单，具隔膜，表面光滑，大小可变。瓶梗通常聚集在分生孢子梗上，烧瓶形或圆柱形，无色或棕色，表面光滑，壁薄，6.5～11 × 1.8～3.5 μm；领部颜色深，漏斗状或锥状，0.5～1 × 0.3～0.5 μm。分生孢子成链，无色或棕色，单胞，表面光滑，卵形至球形，基部略微细尖，1.6～2.5 × 1.8～3.5 μm；厚垣孢子未见。

研究菌株：贵州：贵阳市，贵州森林野生动物园，土壤，2011 年 10 月 1 日，韩燕峰，E335。

世界分布：中国。

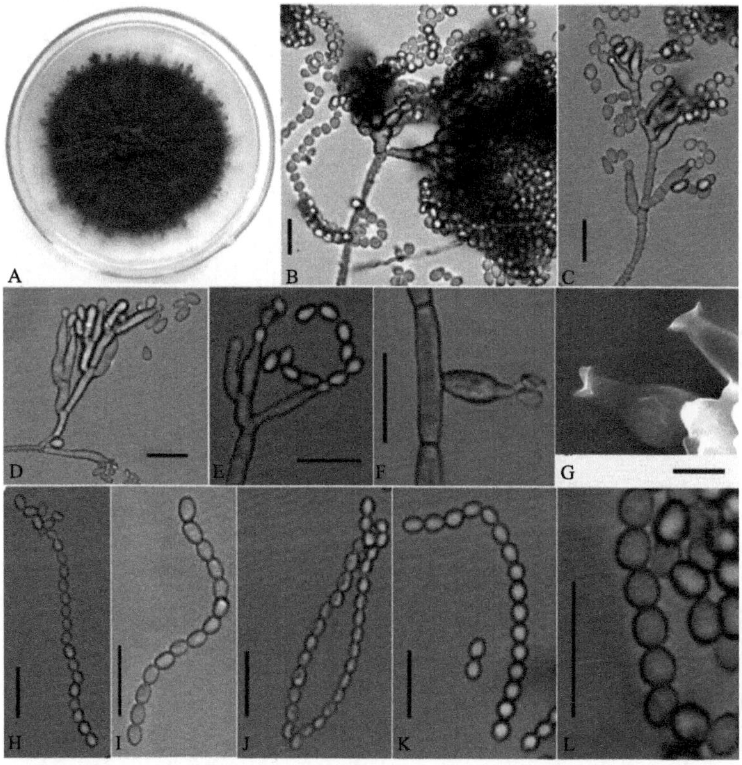

图 78　海榄雌瓶霉 *Phialophora avicenniae* Yue L. Liu & Z.D. Jiang（引自 Liu *et al.*，2013）

A. PDA 培养基上的菌落；B～G. 产孢结构；H～L. 分生孢子链。标尺：B～L = 10 μm

讨论：海榄雌瓶霉的链生孢子与棒孢瓶霉 *Phialophora clavispora* W. Gams、橄榄瓶霉和尖孢瓶霉 *P. oxyspora* W. Gams 相似，但棒孢瓶霉和橄榄瓶霉仅具有结构简单的分生孢子梗，尖孢瓶霉的分生孢子为纺锤形（Gams & Holubová-Jechová，1976）。此外，海榄雌瓶霉的营养类型为内生，而其他 3 种目前所知的均为腐生。

布氏瓶霉　图 79

Phialophora bubakii (Laxa) Schol-Schwarz, Persoonia 6(1): 66, 1970. Wang, Species diversity of dematiaceous hyphomycetes from soil in northern Qinghai Province p. 82, 2009.

在 PDA 培养基上，25℃培养 5 天，菌落平展，深灰橄榄褐色或深褐色，绒毛状，时常轮纹状。菌丝体部分表生，部分埋生。菌丝光滑，淡黄色，具隔膜，宽 1.5～3.1 μm。产孢瓶体垂直着生于菌丝上，或生在菌丝分枝的顶部，单生或簇生，有限生长，光滑，淡黄色，5～10 × 2～3 μm。产生两种形态的分生孢子：香肠形、椭圆形，淡黄色，常具 1～2 个油滴。培养时间短时，常见香肠形分生孢子，3～7 × 1～2 μm；培养时间长时，常见椭圆形孢子，4～6 × 4～5 μm。

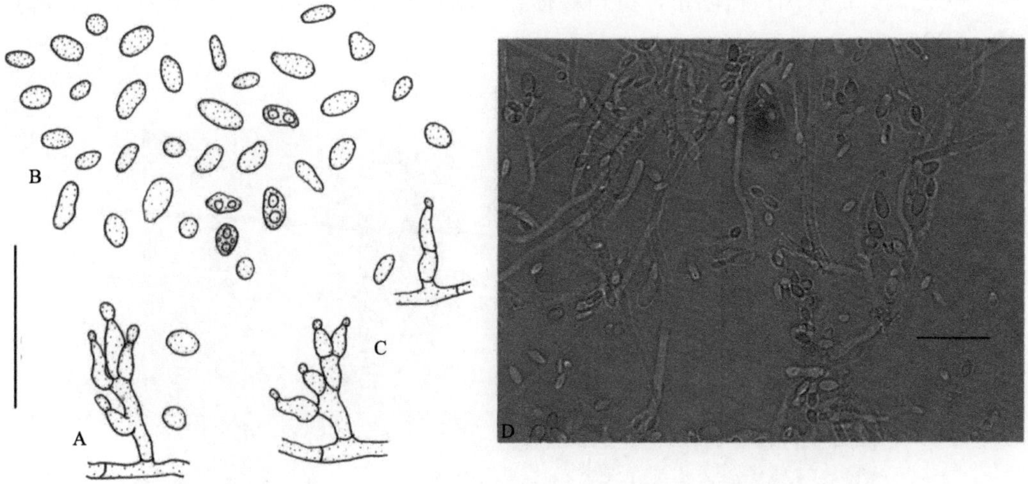

图 79　布氏瓶霉 *Phialophora bubakii* (Laxa) Schol-Schwarz（引自王洪凤，2009）

A、C、D. 产孢结构；B. 分生孢子。标尺：A～C = 25 μm；D = 20 μm

研究菌株：甘肃：兰州市，吐鲁沟土样，2005 年 8 月 10 日，韩燕峰，E340。

世界分布：中国、加拿大。

讨论：该种具有的显著特征是产生两种不同形态的分生孢子，即香肠形和椭圆形，能与其他相似种明显区分。

中华瓶霉　图 80

Phialophora chinensis Ya L. Li, de Hoog & R.Y. Li, Persoonia 38: 11, 2016.

在 OA 培养基上，30℃培养 21 天，菌落生长缓慢，深橄榄色，中部浅橄榄色；背面深橄榄色。

在 MEA 培养基上，30℃培养 21 天，菌落生长缓慢，羊毛状，具棕色的浅橄榄色，边缘光滑；背面深橄榄色，无扩散性色素产生。菌丝棕色，有规律性分隔，宽 3～4.5 μm。分生孢子梗未见。瓶梗为宽烧瓶形。分生孢子无色，表面光滑，球形至宽椭圆形，4.5 ± 0.5（3～6）× 3.5 ± 0.5（2～5.5）μm，一些较大的分生孢子可形成砖格状分隔或缓慢出芽。有性型未知。

生长温度：最低为 21℃，最适为 24℃，最高为 40℃。

研究菌株：湖北：武汉市，花坛土，2013 年 11 月 2 日，韩燕峰，E332。

世界分布：中国。

讨论：中华瓶霉的典型特征是菌落生长缓慢，菌丝棕色，瓶梗为宽烧瓶状。该种生长温度：最低为 21℃，最适为 24℃，最高为 40℃。

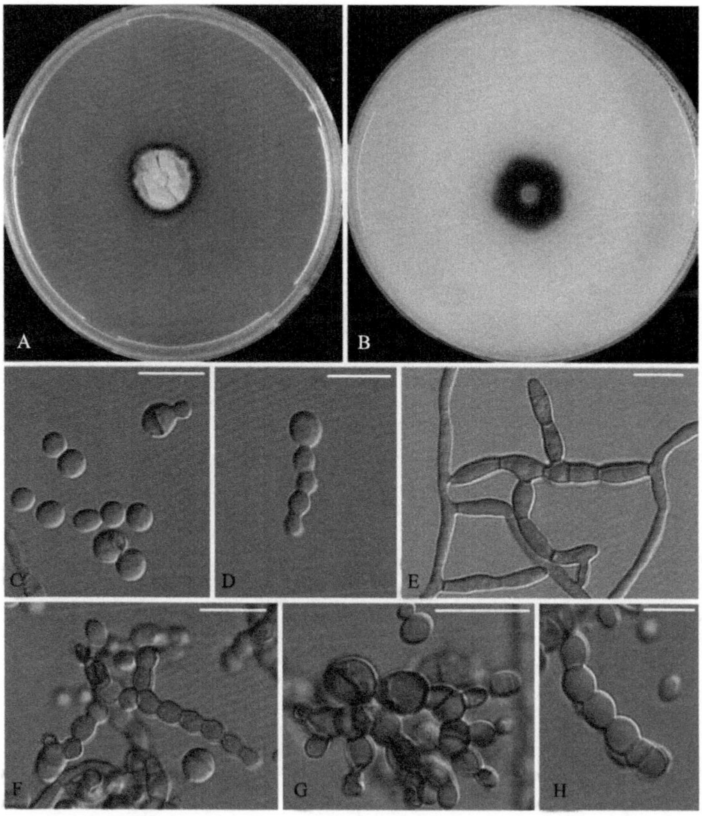

图 80　中华瓶霉 *Phialophora chinensis* Ya L. Li, de Hoog & R.Y. Li（引自 Li *et al.*，2017b）

A. MEA 培养基上的菌落；B. 在 OA 培养基上生长 21 天的菌落；C～H. 瓶梗、孢子、近球状菌丝和砖格状细胞。标尺：C～H = 10 μm

作者未观察的种

灰质瓶霉　图 81

Phialophora cinerescens (Wollenw.) J.F.H. Beyma, Antonie van Leeuwenhoek 6: 38, 1940.

Wang, Species diversity of dematiaceous hyphomycetes from soil in northern Qinghai Province p. 83, 2009.

讨论：王洪凤（2009）记载，该菌菌落灰褐色至橄榄褐色。菌丝淡褐色，常形成网结。产孢瓶体单生或簇生在菌丝的侧面，烧瓶形，浅褐色，具典型的围领，6～15 × 1.5～3 μm。分生孢子灰色，球形或长柱形，有时略弯曲，2.8～5 × 1.2～2 μm。

该种起初分离自康乃馨，其能引起康乃馨的枯萎病，发病后导管组织变色，而后上部叶片枯萎，首次在土壤中分离到。

世界分布：中国、加拿大、德国。

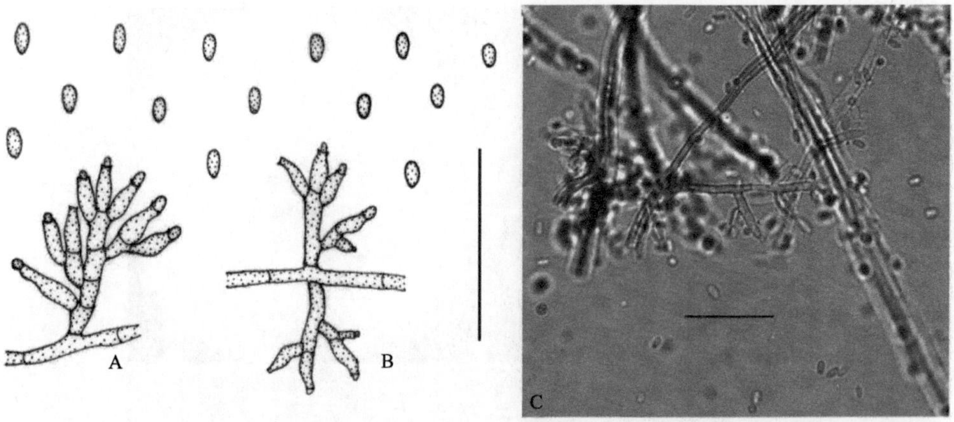

图 81　灰质瓶霉 *Phialophora cinerescens* (Wollenw.) J.F.H. Beyma（引自王洪凤，2009）

A、B. 分生孢子梗和分生孢子；C. 产孢结构。标尺：A、B = 25 μm；C = 20 μm

双型瓶霉　图 82

Phialophora dimorphospora J.J. Xu & T.Y. Zhang, Mycotaxon 132(3): 680, 2017.

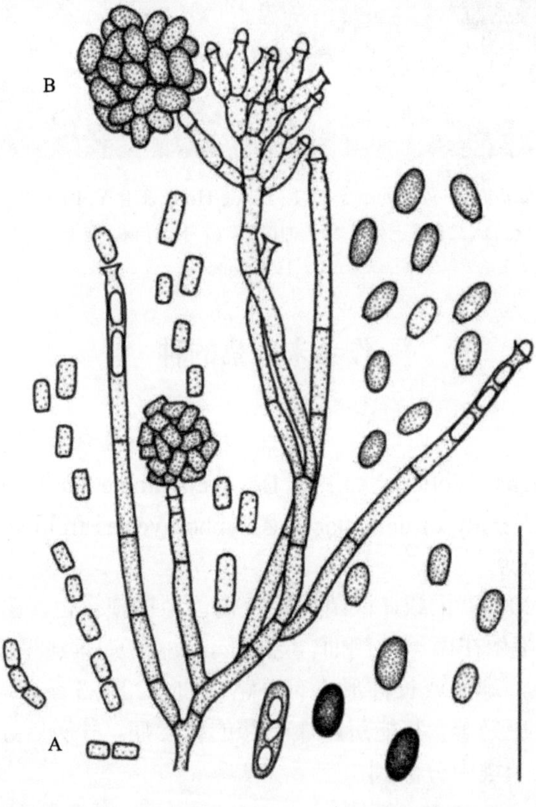

图 82　双型瓶霉 *Phialophora dimorphospora* J.J. Xu & T.Y. Zhang（引自王洪凤，2009）

A. 两端平截的分生孢子；B. 分生孢子团。标尺：25 μm

讨论：王洪凤（2009）记载，该种菌丝体多表生。通常菌丝成束。分生孢子梗分化明显，单生或者丛生，典型阔线形，淡褐色至褐色，帚状排列，3.5～15 × 1.5～4.5 μm。

瓶梗大多数烧瓶形或者圆柱形，具明显领口，顶部明显平截。分生孢子顶生，柱状或椭圆形，基部有时平截，淡褐色至深褐色，5.5～15 × 3～5.5 μm。

世界分布：中国、瑞典、挪威、加拿大、美国。

欧洲瓶霉　图 83

Phialophora europaea de Hoog, Mayser & Haase, Mycoses 43(11/12): 414, 2000. Feng, The primary study on the diversity of fungi from sediment from Yellow Sea and Yangtze Estuary p. 39, 2012.

讨论：冯春辉（2012）记载，该种在 PDA 培养基上，菌落平展，絮状，灰色或褐色。菌丝体部分表生。分生孢子梗安瓿瓶状或烧瓶状，单生或簇生，基部和瓶体间常有隔膜，8～20 × 2～4 μm。分生孢子卵形、近球形、椭圆形、球形，浅褐色，1.6～4 × 1.2～3 μm，常聚集成球。

de Hoog 等（2000b）将 4 株分离于患者的皮肤鳞屑和指甲的疣状瓶霉菌株命名为欧洲瓶霉。疣状瓶霉在形态上与欧洲瓶霉较为相似，二者的区别是，前者的围领结构更为明显。该菌与其他皮肤病的病原菌不同，欧洲瓶霉只会引起皮肤表面病变而不会感染皮下组织的真皮层结构。

世界分布：中国、荷兰、德国、法国。

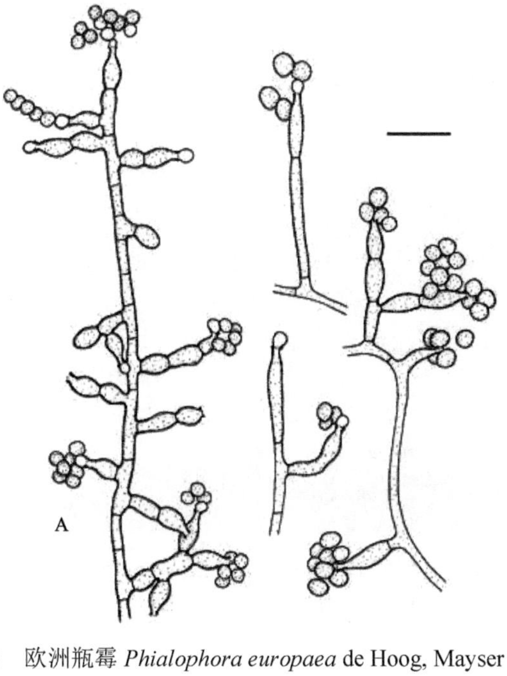

图 83　欧洲瓶霉 *Phialophora europaea* de Hoog, Mayser & Haase（引自冯春辉，2012）

A、B. 产孢结构和分生孢子。标尺：10 μm

膨大瓶霉　图 84

Phialophora expanda Ya L. Li, de Hoog & R.Y. Li, Persoonia 38: 17, 2016.

讨论：Li 等（2017b）记载，该种在 OA 培养基上，菌落生长缓慢，深橄榄色。瓶梗大多呈烧瓶状至长柱形，至顶部逐渐变窄，基部不具隔膜的短瓶梗偶出现；领部颜色深于瓶梗基部，漏斗状至近圆柱状，通常带有大或小、明显着色、纤细且宽口的顶端。分生孢子椭圆形，2～5 × 1.5～3.5 μm，偶现出芽，聚集在顶部，偶成短链。生长温度：最低为 21℃，最适为 30℃，最高达 40℃。

基于系统发育分析，膨大瓶霉与椭圆瓶霉具有较近的关系。但膨大瓶霉的领状基部膨大可与椭圆瓶霉相区别。

世界分布：中国。

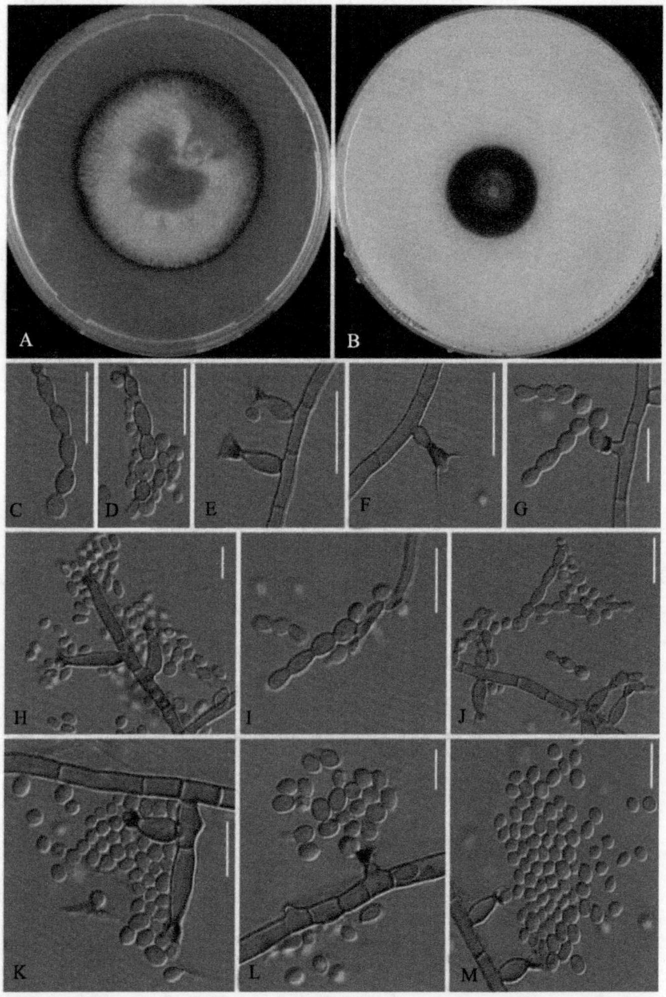

图 84　膨大瓶霉 *Phialophora expanda* Ya L. Li, de Hoog & R.Y. Li（引自 Li *et al.*，2017b）

A. MEA 培养基上的菌落；B. OA 培养基上的菌落；C～M. 瓶梗、孢子、近球状菌丝和砖格状细胞。标尺：C～M = 10 μm

帚状瓶霉　图 85

Phialophora fastigiata (Lagerb. & Melin) Conant, Mycologia 29(5): 598, 1937. Wang, Species diversity of dematiaceous hyphomycetes from soil in northern Qinghai Province p. 84, 2009.

讨论：王洪凤（2009）记载，该种在 PDA 培养基上，菌落生长缓慢，平展，橄榄褐色。菌丝体表生或埋生，常形成分枝并形成网结。分生孢子梗分枝，直或弯曲，淡褐色、中度褐色至橄榄褐色。产孢瓶体常单侧生于菌丝上，或着生于菌丝分枝的末端。产孢瓶体烧瓶状，基部稍窄，逐渐膨大，有 1 瓶颈。瓶梗孢子阔卵形或椭圆形，3～7 × 1.5～2.5 μm，常聚集形成黏球状，强光透视下常有 2 个小油滴。

该种的形态与疣状瓶霉相似，但后者产孢瓶体形成暗色、显著加厚的领部；与理查德瓶霉的不同在于后者产生两种类型的产孢瓶体和分生孢子，分生孢子带有明显黑褐色，展开领部的产孢瓶体产生近球形的分生孢子，直径 2～4 μm；淡褐色领部不明显的产孢瓶体产生椭圆形的分生孢子，2～6.5 × 0.8～3.2 μm。本种是该属中作为土壤真菌分布比较广泛的种之一。最初发现于木头，引起黑污病，也发现于腐烂的木头上、喷防腐剂的木头、储藏的木材上。本种也存在于土壤、空气、江河沉积物、污染水源中。还能从小麦、蚕豆、大戟科植物、紫菀、小麦根围、松科植物和草种子上分离到（Domsch *et al.*，1980；Ellis，1971；刘云龙等，2005）。

世界分布：中国、俄罗斯、瑞典、乌干达、加拿大、美国和澳大利亚。

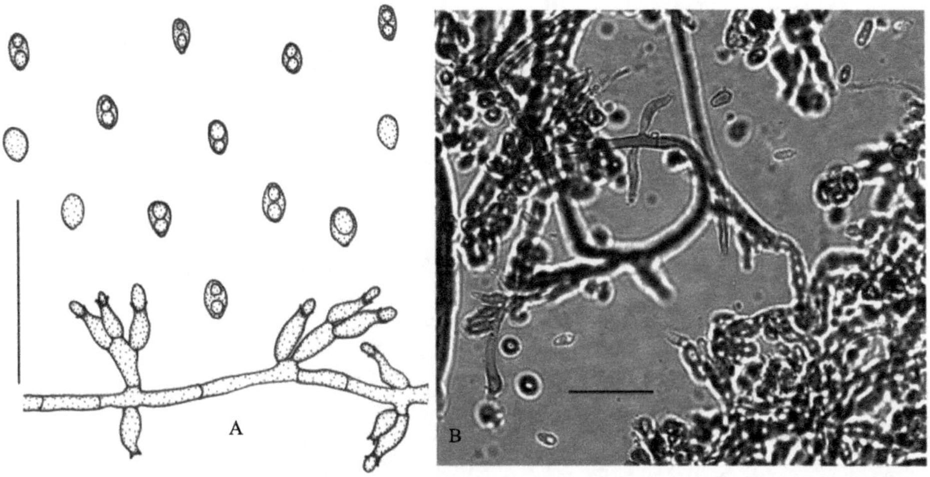

图 85 帚状瓶霉 *Phialophora fastigiata* (Lagerb. & Melin) Conant（引自王洪凤，2009）

A、B. 分生孢子梗和分生孢子。标尺：A = 25 μm；B = 20 μm

膝状瓶霉 图 86

Phialophora geniculata Emden, Acta Bot. Neerl. 24(2): 194, 1975. Zhang, Taxonomic studies of dematiaceous hyphomycetes from soil of middle-southern subtropical and tropical zones in China p. 123, 2007.

讨论：张悦丽（2007）记载，该菌在 PDA 培养基上，菌落平展，深黑色，生长缓慢。菌丝体埋生。分生孢子梗简单，淡褐色至褐色。产孢细胞瓶梗状，合轴式延伸，曲膝状，随着产孢不规则加厚。分生孢子常聚集成小滴，柱状，两端钝圆，淡褐色，4～5 × 2～2.5 μm。

该种与理查德瓶霉的产孢细胞都是合轴式延伸，但该种具柱状的孢子，无气生菌丝体及生长慢等特征，可以与后者相区别。

世界分布：中国、印度尼西亚。

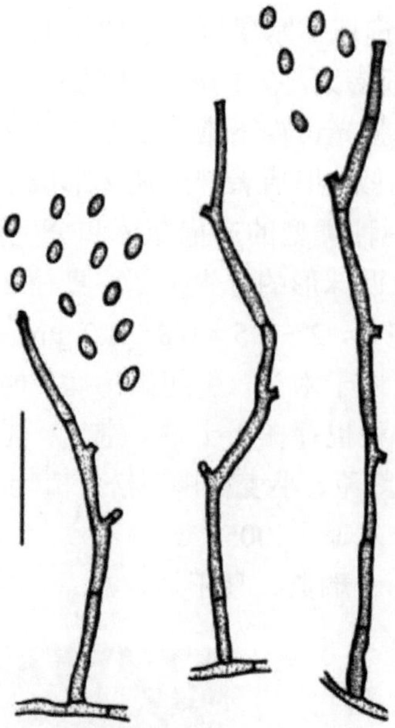

图 86　膝状瓶霉 *Phialophora geniculata* Emden（引自张悦丽，2007）

分生孢子梗和分生孢子。标尺：25 μm

光滑瓶霉　图 87

Phialophora levis Y.L. Jiang & Yong Wang bis, Mycosystema 29(6): 784, 2010.

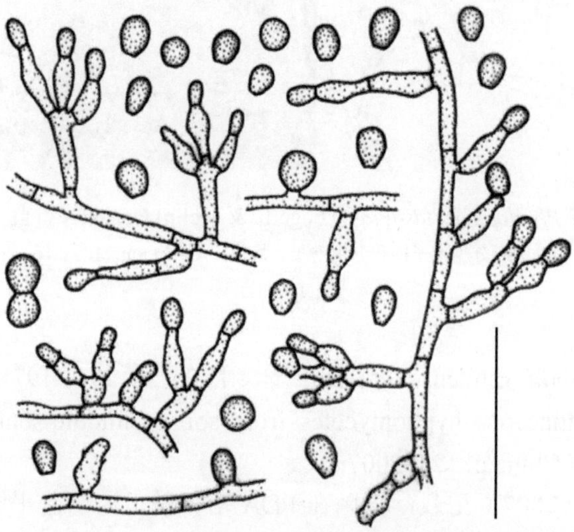

图 87　光滑瓶霉 *Phialophora levis* Y.L. Jiang & Yong Wang bis（引自 Jiang & Wang，2010）

分生孢子梗和分生孢子。标尺：25 μm

讨论：Jiang 和 Wang（2010）记载，该种在 PDA 培养基上，菌落天鹅绒状，灰黑色。菌丝体部分埋生。分生孢子梗为微孢梗束型、半微孢梗束型、菌丝型，分枝。产孢细胞单瓶梗型，长瓶颈状，浅棕色，3～11 × 2～4 μm；领部明显，颜色略深。分生孢子倒卵形，基部截形，有时呈球形或近球形，浅棕色，3～6 × 2.5～5 μm；厚垣孢子球形或卵形。

布氏瓶霉与光滑瓶霉在瓶梗和孢子的大小上较相似，但布氏瓶霉瓶梗的形状是可变的，且分生孢子为双型，香肠状或卵圆形（Schol-Schwarz，1970）。

世界分布：中国。

大孢瓶霉 图 88

Phialophora macrospora M. Moore & F.P. Almeida, Ann. Mo. Bot. Gdn 23: 545, 1937. Li *et al*., Persoonia 38: 17, 2017.

讨论：Li 等（2017b）记载，该种在 OA 培养基上，菌落生长快，深橄榄色。菌丝浅棕色，弯曲。瓶梗直接着生于菌丝，烧瓶状；领部小且短，花瓶口状至漏斗状。分生孢子出现出芽形式，3～5.5 × 1.5～2.5 μm。生长温度：最低为 21℃，最适为 30℃，最高为 40℃。

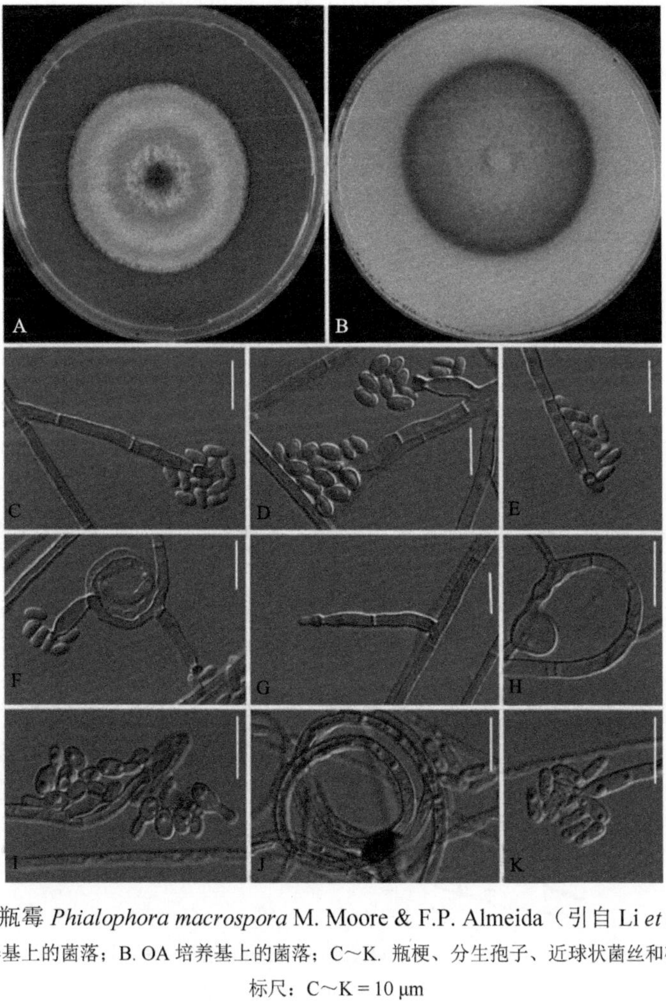

图 88　大孢瓶霉 *Phialophora macrospora* M. Moore & F.P. Almeida（引自 Li *et al.*，2017b）

A. MEA 培养基上的菌落；B. OA 培养基上的菌落；C～K. 瓶梗、分生孢子、近球状菌丝和砖格状细胞。

标尺：C～K = 10 μm

该种以其分生孢子大小能与该属内的近似种台湾瓶霉 *Phialophora taiwanensis* Matsush.相区别，后者的分生孢子分别为 7～11 × 2.5～3.5 μm，明显大于该种，且其分生孢子近圆柱形或桶状。该种首次报道分离自印度和美国纽约的土壤中，Matsushima

（1983）从日本三处土壤中分离到该菌。

世界分布：中国、日本、印度、美国、墨西哥。

仁果瓶霉　图 89

Phialophora malorum (Kidd & Beaumont) McColloch, Phytopathology 32: 1094, 1942. Zhang, Soil dematiaceous Hyphomycetes from cold temperate zone-temperate zone of Northeast China p. 125, 2006.

讨论：张伟（2006）记载，在 PDA 培养基上，菌落生长缓慢，深橄榄褐色。菌丝体表生或埋生。分生孢子梗分化不很明显，褐色至橄榄褐色，宽 2～3 μm。产孢瓶体常单生或簇生，细瓶状，6～12.5 × 1.5～2.5 μm。产孢瓶体包括领部长 7.5～15 μm，烧瓶状，基部稍窄，膨大 2～3.5 μm，瓶颈细长可达 6 μm，领部最外围宽 1.2～2.5 μm。瓶梗孢子椭圆形至圆柱状，4～7 × 1.5～3 μm，长约是宽的 3 倍。

此种比较常见，分布广，基质也比较广，可从木头、土壤、化学材料、污水、钟乳石、空气中分离到，在叙尔特赛岛的火山岛的沙滩、北极冻原、森林、西伯利亚一带的大草原土壤、小麦根围、防腐处理的木材、腐烂的苹果木材、腐烂的苹果等也有被分离到的报道。该种生长最适 pH 4.4～6，能够氧化丹宁和五倍子酸，降解棉花纤维和淀粉，能够在五倍子酸和腐殖酸环境下生长。37℃条件下不生长（Domsch *et al.*，1980）。

本种与帚状瓶霉形态相似，但本种瓶梗孢子长宽比明显较后者大，且本种的瓶颈较后者长，易于区分。Domsch 等（1980）描述本种瓶梗孢子基部稍尖，且常产生厚垣孢子，但该特征不稳定，可能与不同的地理来源和培养条件有关。

世界分布：中国、美国、澳大利亚、巴布亚新几内亚等。

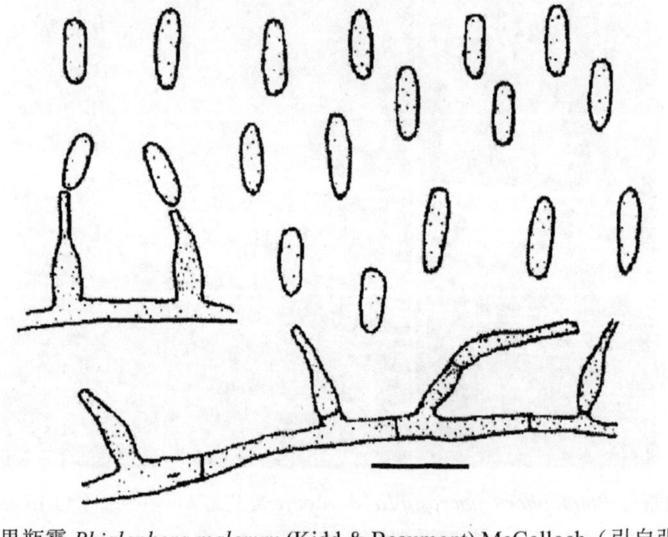

图 89 仁果瓶霉 *Phialophora malorum* (Kidd & Beaumont) McColloch（引自张伟，2006）

分生孢子梗和分生孢子。标尺：10 μm

蜜色瓶霉　图 90

Phialophora melinii (Nannf.) Conant, Mycologia 29(5): 598, 1937. Wang, Species diversity of dematiaceous hyphomycetes from soil in northern Qinghai Province p. 86, 2009.

讨论：王洪凤（2009）记载，该种在PDA培养基上，菌落直径15～20 mm，平展，灰色或者橄榄褐色，绒毛状，菌丝体多表生，少部分埋生。菌丝多个隔膜，褐色或黑褐色，光滑或粗糙，20～75 μm，常形成菌丝束。产孢瓶体分化明显，直接着生于菌丝或者菌丝束上，单生或者丛生，典型阔线形，淡褐色至褐色，围领发育完整，7～20 × 1.5～3 μm。分生孢子椭圆球形至圆柱形，淡褐色，3～6 × 1～2 μm。

本种多分离自软质或硬质木材或木浆中（Ellis，1976）。本种的主要特征是菌丝能形成菌丝束，分生孢子比较小，围领明显。

世界分布：中国、瑞典、美国

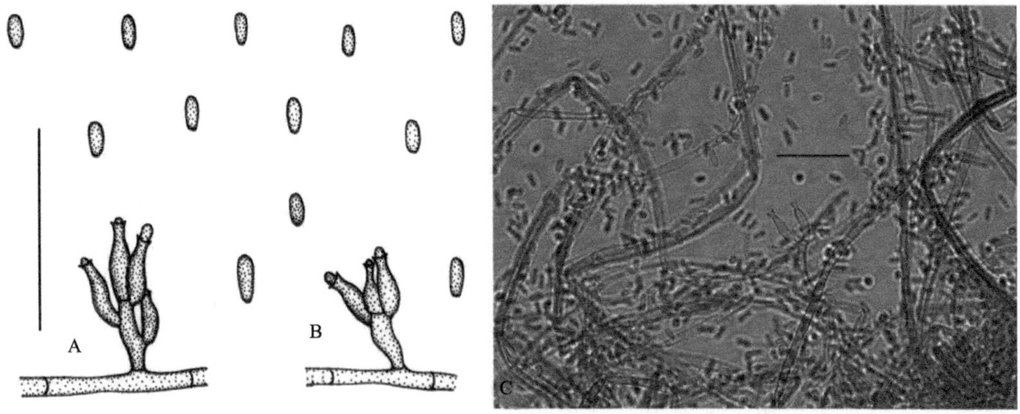

图 90　蜜色瓶霉 *Phialophora melinii* (Nannf.) Conant（引自王洪凤，2009）

A～C. 分生孢子梗和分生孢子。标尺：A、B = 25 μm；C = 20 μm

小孢瓶霉　图 91

Phialophora microspora Y.M. Wu & T.Y. Zhang, Mycotaxon 115(1): 253, 2011.

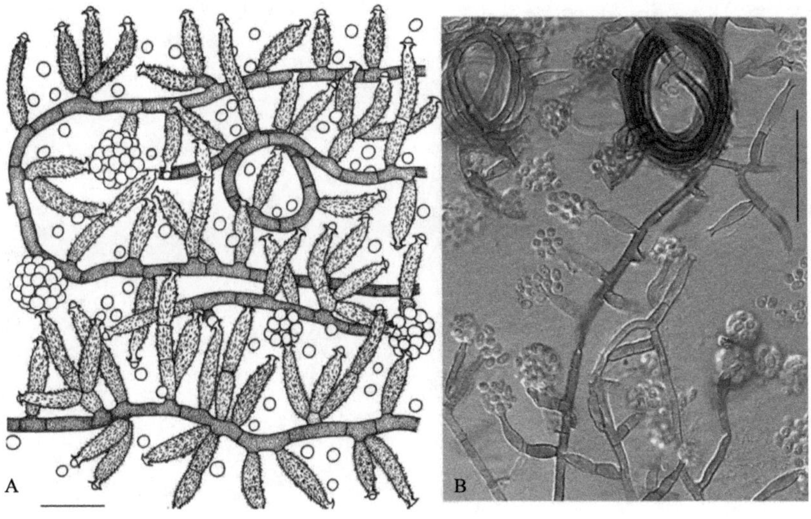

图 91　小孢瓶霉 *Phialophora microspora* Y.M. Wu & T.Y. Zhang（引自王洪凤，2009）

A、B. 分生孢子梗和分生孢子。标尺：A、B = 25 μm

讨论：王洪凤（2009）记载，该种在 PDA 培养基上，菌落绒状，灰橄榄色。菌丝体大多数表生，部分埋生。分生孢子梗半孢梗束型、菌丝型，垂直着生于菌丝，结构简单或偶有分枝，浅棕色，具疣，30～80 × 4～7 μm。产孢细胞单瓶梗型，直接聚集在菌丝末端或分散着生，近圆柱形，至顶端渐狭，浅棕色，具疣，20～30 × 4～7 μm，有时膨大；领部明显，漏斗至花瓶状，颜色略深。分生孢子单胞，球形或近球形，无色，表面光滑，2～3 × 1.5～2.5 μm，聚集形成黏液团。

小孢瓶霉与理查德瓶霉（Conant，1937）和深海瓶霉（Millar，1990）在分生孢子的形状和颜色上相似，但后两个种的产孢细胞表面光滑，小孢瓶霉的分生孢子小于深海瓶霉（2.5～4 × 2～2.5 μm）和理查德瓶霉（2～4 μm）的分生孢子。

世界分布：中国。

聂拉木瓶霉 图 92

Phialophora nielamuensis Y.M. Wu & T.Y. Zhang, Mycotaxon 115(1): 251, 2011.

讨论：Wu 和 Zhang（2011）记载，该种在 PDA 培养基上，菌落天鹅绒状，灰橄榄色。菌丝体大多表生，部分埋生。分生孢子梗微孢梗束型至半微孢梗束型、菌丝型，垂直着生于菌丝上，浅棕色，20～110 × 1.5～3 μm。产孢细胞单瓶梗型，近圆柱形，顶部变细，浅棕色至近无色，15～60 × 1.5～3 μm，有时膨大；领部明显，呈漏斗状至花瓶口状，颜色微深。分生孢子单胞，梨形、椭圆形或近球形，基部具脐，浅橄榄色，3.5～5 × 2～2.5 μm，聚集形成黏孢团。

聂拉木瓶霉与日本瓶霉（Iwatsu & Udagawa，1985）在分生孢子的形态结构上相似。但前者分生孢子较小（3.5～5 × 2～2.5 μm）；后者的分生孢子较大（4.5～7 × 2～2.8 μm），且颜色深褐色。

世界分布：中国。

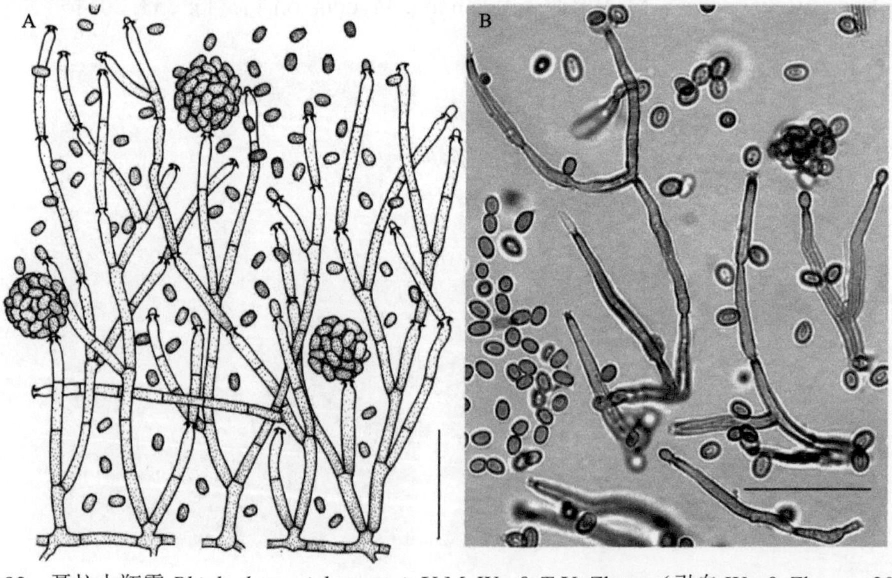

图 92 聂拉木瓶霉 *Phialophora nielamuensis* Y.M. Wu & T.Y. Zhang（引自 Wu & Zhang，2011）

A、B. 分生孢子梗和分生孢子。标尺：A、B = 25 μm

理查德瓶霉　图 93

Phialophora richardsiae (Nannf.) Conant, Mycologia 29(5): 598, 1937. Xu, Taxonomic studies on soil dematiaceous Hyphomycetes from South-eastern Zones and Yunnan Province of China p. 181, 2006.

讨论：许俊杰（2006）记载，该种在 PDA 培养基上，菌落平展，羊毛状或者有时粉状，棕色或者橄榄褐色。产孢瓶体倒棒状，至顶部渐细，基部或者中部略膨大，5～28 × 1.5～4 μm。产孢瓶体有两种类型：一类是黑褐色的具有发达明显的领部结构，领部深褐色，明显向外展开，领口宽 2.5～5 μm，深 0.5～1.5 μm；另一类是淡褐色领部不明显。分生孢子球形或近球形，有时具油滴，淡褐色中等褐色，光滑，2～4 μm，或椭圆形，2～6.5 × 0.8～3.2 μm。

该种是由 Conant（1937）将理查德淡领瓶霉 *Cadophora richardsiae* Nannf.组合而来，世界性分布，其基质有烂木头、地面木头、纸浆、污水、土壤、塑料、棕榈橄榄油和松木等（Chabert & Nicot，1966；Ellis，1976）。此外，从人体的前列腺和皮下胆囊的肉瘤中也曾分离到过（Schwartz & Emmons，1968）。

世界分布：中国、法国、尼日利亚、美国、加拿大。

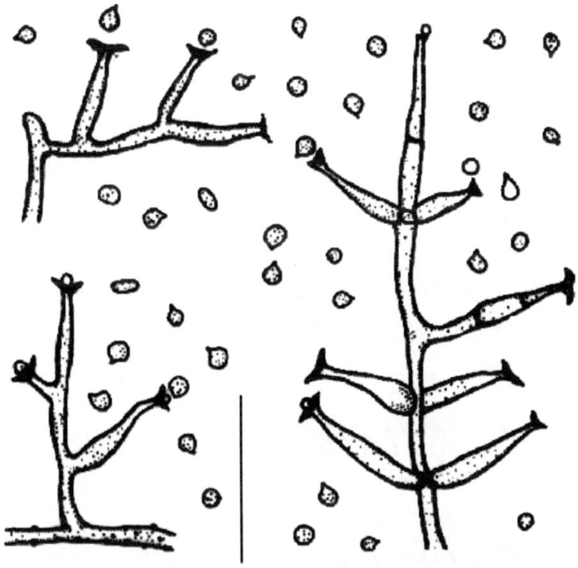

图 93　理查德瓶霉 *Phialophora richardsiae* (Nannf.) Conant（引自许俊杰，2006）
分生孢子梗和分生孢子。标尺：25 μm

土壤瓶霉　图 94

Phialophora subterranea Y.L. Jiang & Yong Wang bis, Mycosystema 29(6): 784, 2010.

讨论：Jiang 和 Wang（2010）记载，该种在 PDA 培养基上，菌落扩散，天鹅绒状，灰色。菌丝体部分埋生，部分表生。分生孢子梗微孢梗束型或半微孢梗束型、菌丝型，无色或浅棕色。产孢细胞单瓶梗型，直接着生于菌丝的侧面或末端或分生孢子梗的分叉处，安瓿形或长瓶颈状，浅棕色，4～9.5 × 2.5～4 μm；领部明显，颜色略深。分生孢子椭圆形或卵形，基部截形，壁薄，浅棕色，光滑，3～5 × 2～4 μm。

该种的分生孢子大小能与相似种深海瓶霉和疣状瓶霉区分。

世界分布：中国。

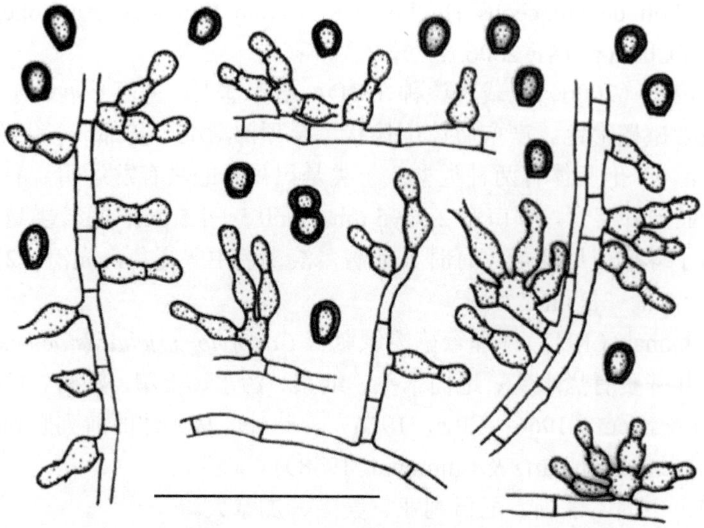

图 94　土壤瓶霉 *Phialophora subterranea* Y.L. Jiang & Yong Wang bis（引自 Jiang & Wang，2010）

分生孢子梗和分生孢子。标尺：25 μm

西藏瓶霉　图 95

Phialophora tibetensis Y.H. Geng & T.Y. Zhang, Mycosystema 28(5): 662, 2009.

讨论：Geng 和 Zhang（2009）记载，该种在 PDA 培养基上，菌落生长迅速，平展，褐色或黑褐色，绒毛状。菌丝体多表生，少部分埋生。瓶生式产孢。产孢瓶体分化明显，直接着生于菌丝上，单生或者丛生，直立，典型阔线形，淡褐色，长达 32 μm，常帚状排列。分生孢子淡褐色至中度褐色，多长椭圆形、球形、梨形、卵形，具隔膜，3～8 × 1.5～3.5 μm。

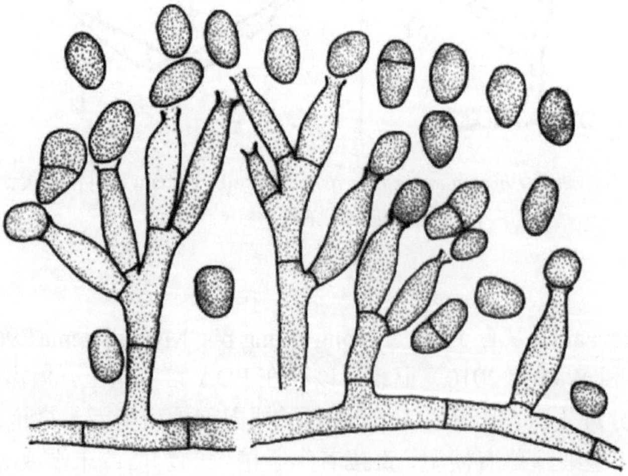

图 95　西藏瓶霉 *Phialophora tibetensis* Y.H. Geng & T.Y. Zhang（引自 Geng & Zhang，2009）

分生孢子梗和分生孢子。标尺：25 μm

西藏瓶霉与疣状瓶霉相似，最主要的区别在于疣状瓶霉的分生孢子梗大多数分枝，西藏瓶霉分生孢子通常为单胞。

世界分布：中国。

疣状瓶霉 图 96

Phialophora verrucosa Medlar, Mycologia 7(4): 203, 1915. Xu, Taxonomic studies on soil dematiaceous Hyphomycetes from South-eastern Zones and Yunnan Province of China p. 178, 2006.

讨论：Li 等（2017b）记载，该种在 PDA 培养基上，菌落生长缓慢，平展，橄榄褐色或中等褐色，绒毛状或粉状。菌丝体部分表生，部分埋生。产孢瓶体常单侧生于菌丝上或着生于菌丝分枝的末端，分枝常由球形或椭圆形的链生的细胞组成。产孢瓶体明显具有领口，领部长 5～21 μm，常杯状、近球形或近圆柱形，基部宽 2.5～4 μm，逐渐缢缩，形成花瓶口形且暗色的领部，领部最外围宽 2.5～5.5 μm，深 1～4.5 μm。产孢瓶体光滑，常淡褐色，领部明显暗色，显著加厚。分生孢子阔卵形或椭圆形，2～5.5 × 2～4.5 μm，常聚集形成黏球状孢子堆或孢子团。

该种与理查德瓶霉近似，但后者产生球形或近球形的分生孢子；产孢梗（产孢瓶体）领口呈钝角三角形。该种作为该属的模式种，最初由 Medlar 分离自人体感病皮肤上。该种可分离自植物材料、腐烂的木头、松树树皮、土壤和人体上（Medlar，1915；Schol-Schwarz，1970；Liu & Zhang，2004；Iwatsu *et al.*，1981）。

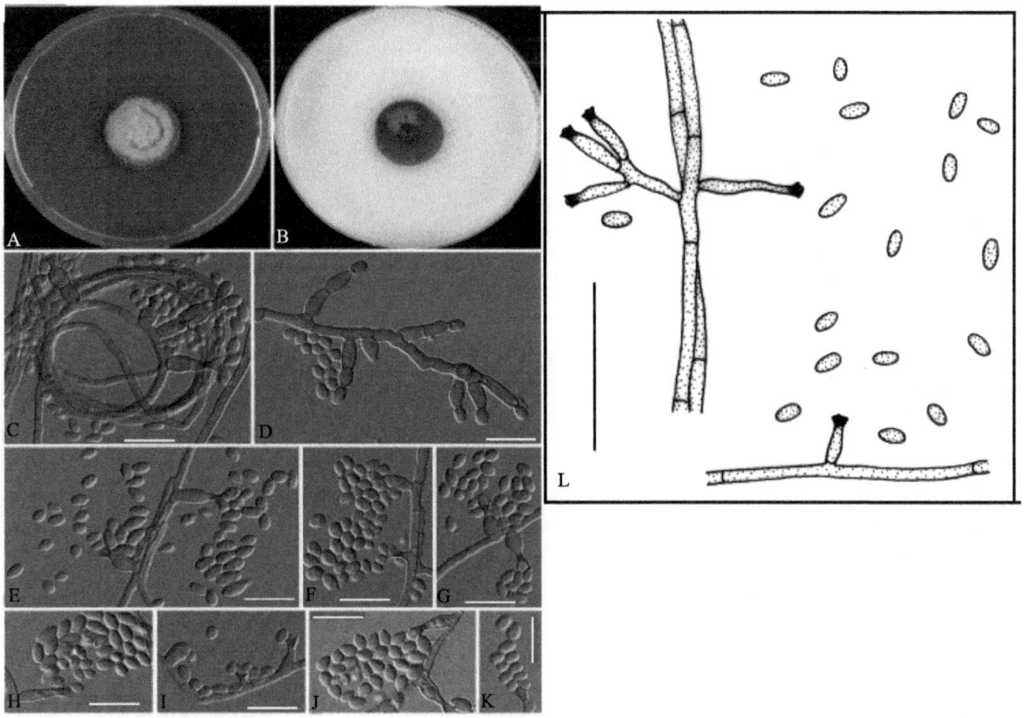

图 96 疣状瓶霉 *Phialophora verrucosa* Medlar（左图引自许俊杰，2006；右图引自 Li *et al.*，2017b）

A、B. PDA 培养基上的菌落；C～L. 产孢结构和分生孢子。标尺：C～K = 10 μm；L = 25 μm

世界分布：中国、日本、美国。

假裸囊菌属 **Pseudogymnoascus** Raillo

Centbl. Bakt. ParasitKde, Abt. II 78: 520, 1929

子囊膜开始由以分枝状着生于营养菌丝的卷曲产囊体组成，且不具有可识别的雄器，在菌丝丛中产生由子囊菌丝形成的松散的丝状物。子囊果球形，散生或聚集着生，呈黄色、红色或深棕色。包被为薄的、扭曲的、交错分布的菌丝组成的松散丝状物形成的网状结构，菌丝表面光滑或由颗粒状硬壳覆盖，且随着时间变成薄壁且具有较深的颜色。附属物不显著或简单，通常较短，壁薄，不分枝，具小刺或疣状物。子囊含有 8 个子囊孢子，球形至卵形，通常具有由产囊丝钩形成的圆柱状结构。子囊孢子单胞，椭圆形至梭形，表面光滑，无色、淡黄色或淡粉色。无性型为地丝霉型。

模式种：玫红假裸囊菌 *Pseudogymnoascus roseus* Raillo。

中国假裸囊菌分种检索表

1. 产生节孢子……2
1. 不产生节孢子……12
  2. 在 PDA 培养基上产生大量肉桂色色素，不产扩散性色素……**构树假裸囊菌 *P. papyriferae***
  2. 在 PDA 培养基上不产生色素和扩散性色素……3
3. 在 PDA 培养基上的菌落为粉色色调……4
3. 在 PDA 培养基上的菌落为白色色调……10
  4. 间生孢子 1～6 个呈长链状……**链状假裸囊菌 *P. catenatus***
  4. 间生孢子单生，或短链状……5
5. 主轴顶生分生孢子较大，6.5～9 × 2.5～4.5 μm……**云南假裸囊菌 *P. yunnanensis***
5. 主轴顶生分生孢子与其他分生孢子大小无明显差异……6
  6. 间生孢子为桶状……7
  6. 间生孢子为倒卵形或近球形……8
7. 成熟分生孢子和节孢子近粗糙或疣状……**疣状假裸囊菌 *P. verrucosus***
7. 成熟分生孢子和节孢子光滑……**葡状假裸囊菌 *P. botryoides***
  8. 分生孢子椭圆形至倒梨形，6～11 × 1.8～3.5（～4.5）μm……**毡状假裸囊菌 *P. pannorum***
  8. 分生孢子较小……9
9. 人工培养基上无分泌物……**校园假裸囊菌 *P. campensis***
9. 人工培养基上分泌物明显……**贵州假裸囊菌 *P. guizhouensis***
  10. PDA 培养基上菌落表面可见明显渗出物和可溶性色素……**陕西假裸囊菌 *P. shaanxiensis***
  10. 在 PDA 培养基上菌落表面无明显渗出物和可溶性色素……11
11. 分生孢子梨形，偶尔倒卵形至近球形，3～5 × 2.5～3.5 μm……**宗琦假裸囊菌 *P. zongqii***
11. 分生孢子倒卵球形，2.5～5.5 × 2.5～4 μm……**福建假裸囊菌 *P. fujianensis***
  12. 主轴顶生分生孢子较大，6～10 × 3～3.5 μm……**香樟假裸囊菌 *P. camphorae***
  12. 主轴顶生分生孢子与其他分生孢子大小无明显差异……13
13. 间生孢子光滑……**中国假裸囊菌 *P. sinensis***
13. 间生孢子光滑或粗糙……**浙江假裸囊菌 *P. zhejiangensis***

葡状假裸囊菌 图 97

Pseudogymnoascus botryoides Zhi Y. Zhang, Y.F. Han & Z.Q. Liang, MycoKeys 98: 189, 2023.

在 PDA 培养基上，25℃培养 14 天，菌落直径 29～30 mm，圆形，边缘气生菌丝稀疏，中心至边缘淡粉色至白色，平展，致密，无色素和渗出物；背面棕色。在 MEA 培养基上，25℃培养 14 天，菌落直径 28～29 mm，白色，平展，致密，近圆形，边缘规则，分泌大量淡红色色素，无渗出物；背面棕色至白色。在 OA 培养基上，25℃培养 14 天，菌落直径 25～28 mm，中心至边缘卡其色至白色，絮状，颗粒状，近圆形，边缘微波纹状，分泌大量白色扩散性渗出物，无色素；背面中心至边缘卡其色至白色。菌丝分枝，分隔，无色，光滑，宽 0.5～2.5 μm。可育菌丝上着生分生孢子和/或间生孢子，无柄。分生孢子和间生孢子着生于一起形成类葡萄簇状结构。分生孢子梗大量，致密，交织成网状，弯曲，无色，表面粗糙，常着生 2～8 轮锐角分枝。分生孢子光滑，梨形、倒卵形，细长，具 1 宽的截形基痕，2～4.5 × 1.5～2.5 μm；间生孢子桶状、肾形，2.5～

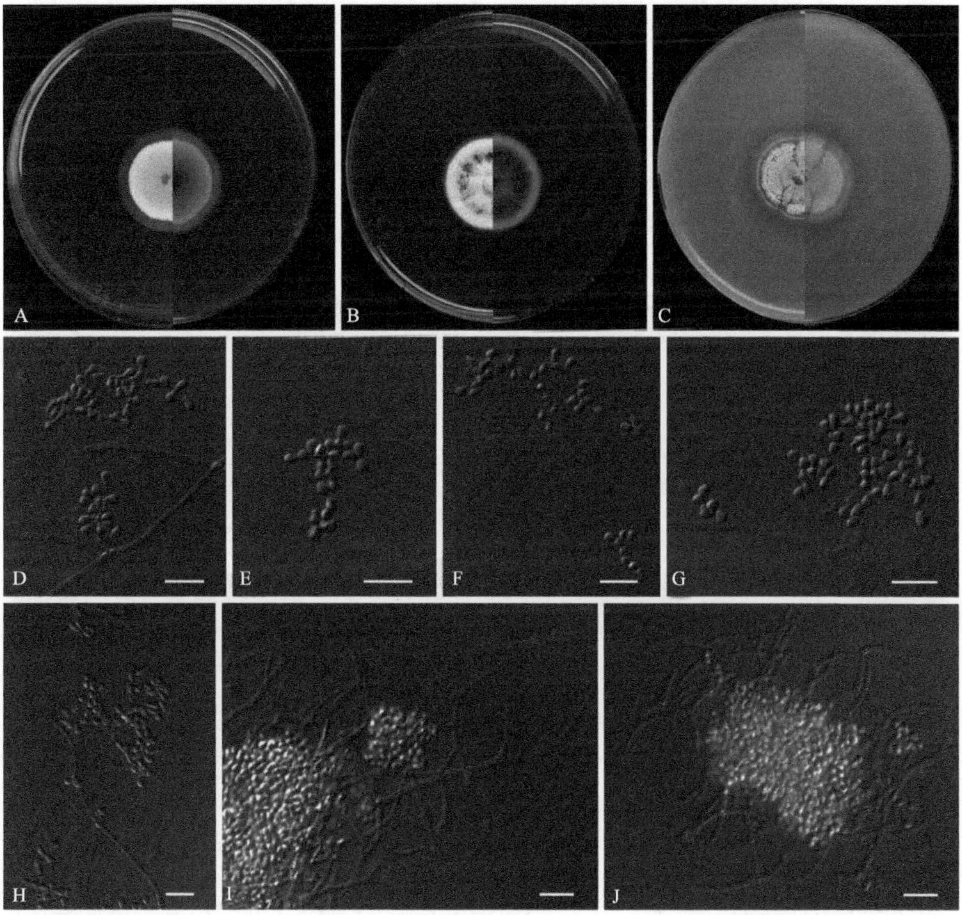

图 97 葡状假裸囊菌 *Pseudogymnoascus botryoides* Zhi Y. Zhang, Y.F. Han & Z.Q. Liang（标本号：HMAS 351904）

A～C. PDA、MEA 和 OA 培养基上的菌落；D～H. 分生孢子梗和分生孢子；I、J. 簇状分生孢子。标尺：D～J =10 μm

5 × 1.5～2.5 μm，由连接细胞分隔，着生无柄分生孢子。节孢子稀少，光滑，柱状，偶尔稍弯曲，2.5～5 × 1～2 μm。

研究菌株：广西：北海市，绿化地土，2016 年 7 月 10 日，张芝元，HMAS 351904（主模式），CGMCC 3.20875 = GZUIFR 22.024；广东：湛江市，广东医科大学附属医院，绿化地土，2019 年 8 月 25 日，张芝元，GZUIFR 22.045，GZUIFR 22.046。

世界分布：中国。

讨论：形态上，葡状假裸囊菌因其可育菌丝上着生分生孢子和/或间生孢子，无柄；分生孢子和间生孢子着生于一起形成类葡萄簇状结构；分生孢子梗大量，致密，交织成网状，弯曲，无色，表面粗糙，常着生 2～8 轮锐角分枝而明显异于该属的其他物种。

校园假裸囊菌 图 98

Pseudogymnoascus campensis Zhi Y. Zhang & Y.F. Han, MycoKeys 95: 55, 2023.

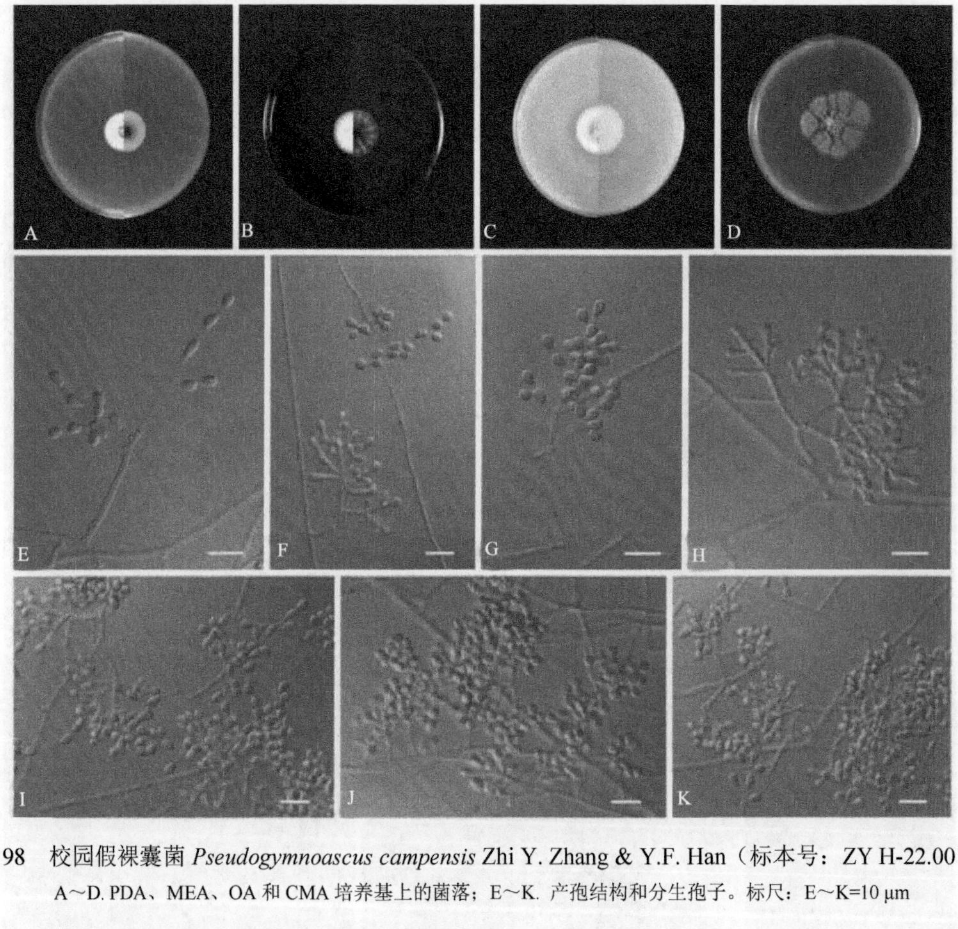

图 98 校园假裸囊菌 *Pseudogymnoascus campensis* Zhi Y. Zhang & Y.F. Han（标本号：ZY H-22.001）

A～D. PDA、MEA、OA 和 CMA 培养基上的菌落；E～K. 产孢结构和分生孢子。标尺：E～K=10 μm

在 PDA 培养基上，25℃培养 14 天，菌落直径 20～21 mm，白色至浅绿色，绒毛状，边缘近圆，整齐；背面从中心至边缘为酒红色至白色。气生菌丝无色，光滑，有隔膜，宽 1.5～2.5 μm。有时侧生菌丝顶端产生桶状、肾形或梨形的粉孢子链。分生孢子梗大量产生，单生，直立，与主干菌丝成锐角，无色，光滑，常常从成锐角的柄处产生

2～3 个垂直分枝的轮生体。分生孢子（粉孢子）梨形至倒卵圆形，3～5 × 2～2.5 μm，具有明显的基痕，光滑或粗糙；间生孢子桶状、肾形、梨形至长不规则形，3.5～5.5 × 2～3 μm，光滑或粗糙。

研究菌株：贵州：贵州民族大学，陆地土，2022 年 7 月，张芝元，ZY H-22.001（主模式）。

世界分布：中国。

讨论：该种在系统发育上与陕西假裸囊菌、澳洲假裸囊菌和灰假裸囊菌亲缘关系相近，但该种因在人工培养基上无分泌物而能与其近缘种相区别。

香樟假裸囊菌　图 99

Pseudogymnoascus camphorae Zhi Y. Zhang, Y.F. Han & Z.Q. Liang, MycoKeys 98: 191, 2023.

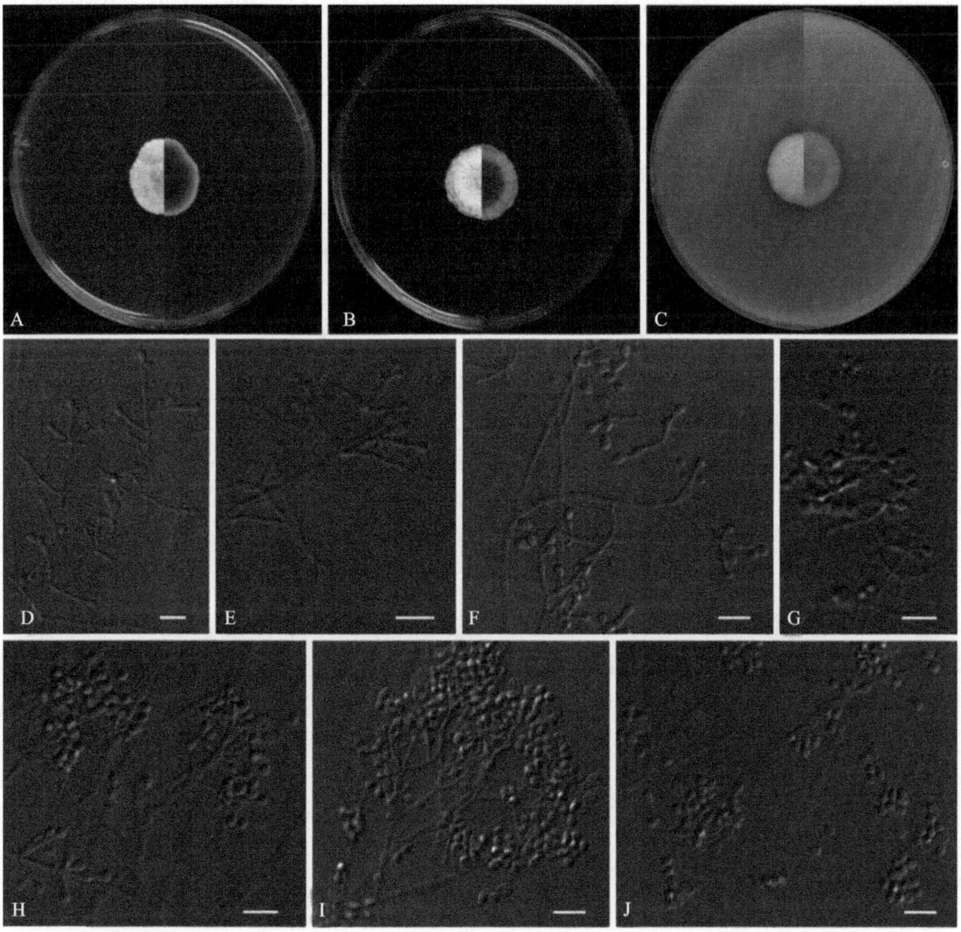

图 99　香樟假裸囊菌 *Pseudogymnoascus camphorae* Zhi Y. Zhang, Y.F. Han & Z.Q. Liang（标本号：HMAS 351901）

A～C. PDA、MEA 和 OA 培养基上的菌落；D～I. 分生孢子梗和分生孢子；J. 分生孢子。标尺：D～J = 10 μm

在 PDA 培养基上，25℃培养 14 天，菌落直径 20～22 mm，中心至边缘灰白色至白色，稍隆起，絮状，边缘不规则，无渗出物和扩散性色素；背面中心至边缘棕色至白色。在 MEA 培养基上，25℃培养 14 天，菌落直径 20～21 mm，中心至边缘淡黄色至白色，絮状，近圆形，边缘波纹状，无渗出物和扩散性色素；背面中心至边缘棕色至淡黄色。在 OA 培养基上，25℃培养 14 天，菌落直径 19～20 mm，灰色，稍隆起，绒状，近圆形，边缘规则，无渗出物，青灰色扩散性色素大量；背面浅酒红色。菌丝分枝，分隔，无色，光滑，宽 1.0～3.0 μm。分生孢子梗大量，单生，偶尔来自菌丝分化形成，无色，光滑，自直立菌丝产生，常着生 2～4 轮与柄呈锐角的分枝。分生孢子梨形、倒卵形，具 1 宽的截形基痕，4～5.5 × 3～3.5 μm。主轴顶生分生孢子，倒卵形、棒状至不规则，单生或 2 个成链，6～10 × 3～3.5 μm，具 1 宽的截形基痕；间生孢子近球形、桶状、倒卵形，3～4 × 2～3 μm，偶尔由连接细胞隔开。

研究菌株：福建：厦门市，五缘湾湿地公园，绿化地土，2019 年 8 月 19 日，张芝元，GZUIFR 22. 22.049。贵州：贵阳市，贵州大学，南校区，香樟树根部附生土，2018 年 8 月 8 日，张芝元，HMAS 351901（主模式），CGMCC 3.20876 = GZUIFR 22.021。

世界分布：中国。

讨论：形态上，香樟假裸囊菌与浙江假裸囊菌和链状假裸囊菌较为相似。香樟假裸囊菌不产生节孢子而有别于链状假裸囊菌。香樟假裸囊菌有别于浙江假裸囊菌之处为：分生孢子的形状和大小（分别为：梨形、倒卵形，4～5.5 × 3～3.5 μm vs. 倒卵形至球形，2.5～4.5 × 2.5～4 μm），顶生分生孢子（分别为：倒卵形、棒状至不规则，单生或 2 个成链，6～10 × 3～3.5 μm vs. 棒状、长倒卵形，5～9 × 2.5～4 μm）。

链状假裸囊菌　图 100

Pseudogymnoascus catenatus Zhi Y. Zhang, Y.F. Han & Z.Q. Liang, Microbiology Spectrum 9: e00867-21, 2021b.

在 PDA 培养基上，25℃培养 14 天，菌落直径 6～10 mm，绒状，短而蓬松，边缘不规则，浅灰色至白色，无色素和渗出物；背面棕色。37℃不生长。菌丝分隔，无色，光滑，分枝，宽 1～3 μm；无球拍状菌丝。分生孢子梗丰富，分枝，呈锐角，通常 1～2 个轮生分生孢子梗，每轮有 1～4 个分枝，二级和三级分枝仍可再分枝。分生孢子丰富，着生于轮生分枝的末端或横向着生于短突或短侧枝；无色至半透明，光滑或粗糙；倒卵形、近球形，3～6 × 3～4 μm。间生孢子着生于轮状菌丝，单生，或 1～6 个成链，壁光滑，或粗糙，倒卵状、近球形、纺锤形、腰鼓状，两端平，3.5～6.5 × 3.0～4.5 μm；或柱状，桶状，两端截断，5.5～6.5 × 2.5～3.5 μm。节孢子无色，筒状、倒卵形，3.0～6.0 × 2.0～3.5 μm。

研究菌株：福建：武夷山市，列宁公园，绿化地土，2019 年 8 月 18 日，张芝元，HMAS 350322（主模式），CGMCC 3.20472 = GZUIFR 21.815；浙江：宁波市，月亮湖，绿化地土，2019 年 8 月 16 日，张芝元，GZUIFR 21.813，GZUIFR 21.814。

世界分布：中国。

讨论：形态上，链状假裸囊菌与疣状假裸囊菌相似，均产生节孢子。两物种的主要区别在于：链状假裸囊菌的分生孢子倒卵形、近球形，3～6 × 3～4 μm，而疣状假裸囊菌的分生孢子近球形至宽梨形，2.5～3 × 1～2.5 μm 。系统发育上，链状假裸囊菌的 4

个分离物聚为一单独的分支，并与其他物种相分离。

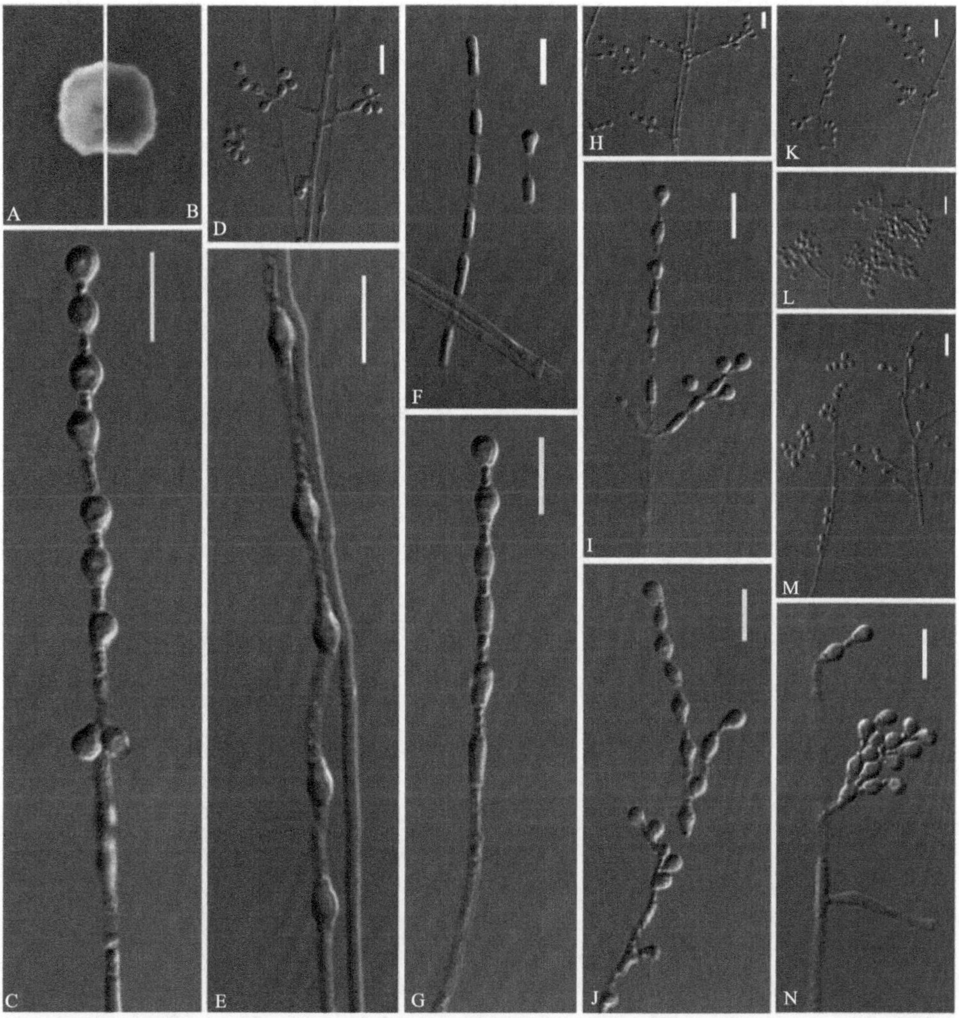

图 100　链状假裸囊菌 *Pseudogymnoascus catenatus* Zhi Y. Zhang, Y.F. Han & Z.Q. Liang（标本号：HMAS 350322）

A、B. PDA 培养基上的菌落；C、E、G、I、J. 间生孢子；D、H、K～N. 分生孢子梗和分生孢子；F. 节孢子。标尺：C～N = 10 μm

福建假裸囊菌　图 101

Pseudogymnoascus fujianensis Zhi Y. Zhang, Y.F. Han & Z.Q. Liang, Microbiology Spectrum 9: e00867-21, 2021b.

在 PDA 培养基上，25℃培养 14 天，菌落直径 19～20 mm，平展，絮状，角变，边缘清晰，白色至粉红色，无色素和渗出物；背面为褐色。37℃不生长。菌丝分隔，无色，分枝，光滑，宽 0.5～3.5 μm；无球拍状菌丝。分生孢子梗丰富，分枝呈锐角，不规则。分生孢子丰富，顶生或侧生，无柄或着生于菌丝、短突起或侧枝上；单生，或融合，或 2 个成链；无色，壁光滑；倒卵球形，2.5～5.5 × 2.5～4 μm。间生孢子丰富，单生或簇生，粗糙，倒卵形、腰鼓形，2.5～5 × 2.5～3.5 μm。

研究菌株：福建：武夷山市，列宁公园，绿化地土，2019 年 8 月 18 日，张芝元，HMAS 350324（主模式），CGMCC 3.20474 = GZUIFR 21.819；福建：武夷山市，武夷大学，绿化地土，2019 年 8 月 18 日，张芝元，GZUIFR 21.821，GZUIFR 21.822。

世界分布：中国。

讨论：形态和系统发育分析均支持福建假裸囊菌的成立。福建假裸囊菌在系统发育上与疣状假裸囊菌、玫红假裸囊菌和毁害假裸囊菌密切相关（Zhang *et al.*，2021b）。该种与其他物种的区别在于其产生间生孢子，无节孢子。

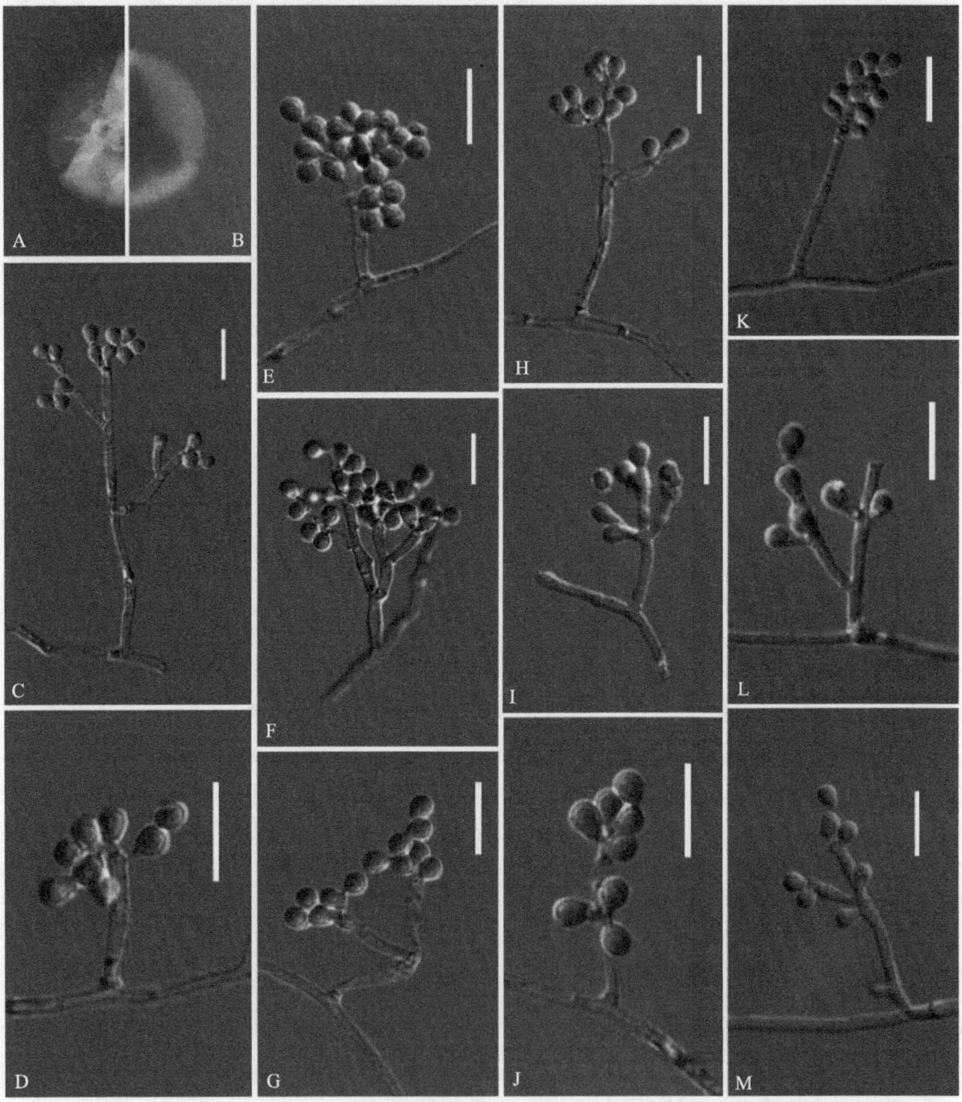

图 101　福建假裸囊菌 *Pseudogymnoascus fujianensis* Zhi Y. Zhang, Y.F. Han & Z.Q. Liang（标本号：HMAS 350324）

A、B. PDA 培养基上的菌落；C～M. 分生孢子梗、分生孢子、间生孢子。标尺：C～M =10 μm

贵州假裸囊菌　图 102

Pseudogymnoascus guizhouensis Zhi Y. Zhang, Y.F. Han & Z.Q. Liang, Frontiers in Microbiology 11: 576, 2020.

在 PDA 培养基上，25℃培养 14 天，菌落直径 20～23 mm，隆起，粉状，边缘整齐，局部内陷，分界明显，中心浅紫色，边缘白色，分泌物清晰；背面棕色。气生菌丝大量，壁薄且光滑，有隔膜，宽 1.5～3.0 μm；无球拍状菌丝。分生孢子通常着生于轮生分枝上，顶生、与间生孢子串生、侧生，单生于菌丝上，壁光滑或刺状，倒卵形、梨形、近球形或棍棒状，3.0～5.5 × 3.0～3.5 μm。间生孢子生于菌丝或轮状菌丝的外侧分枝上，间生，1～2 个成链，壁光滑或刺状，楔形、桶状，3.5～6.0 × 3.0～3.5 μm。有性型未见。

研究菌株：贵州：贵阳市，香樟树附生土，2019 年 6 月 6 日，张芝元，干模 GZAC 376（主模式），GZUIFR 376.1。该模式标本及其模式菌株 GZUIFR 376.1 均保存于贵州大学真菌资源研究所。

世界分布：中国。

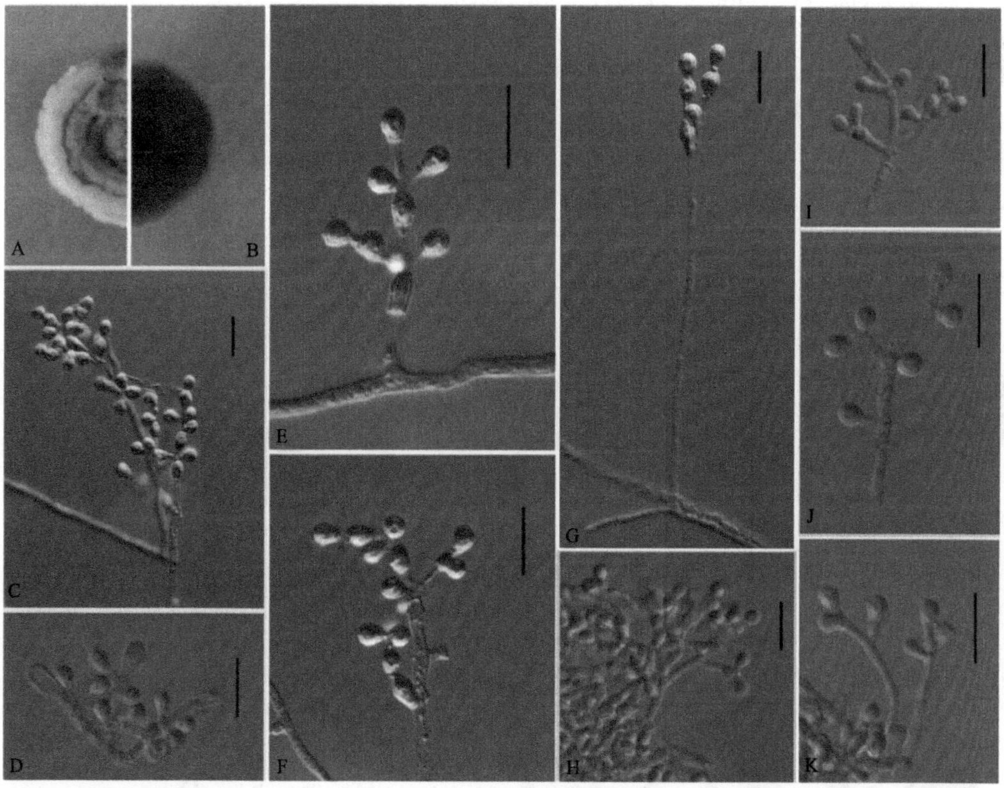

图 102　贵州假裸囊菌 *Pseudogymnoascus guizhouensis* Zhi Y. Zhang, Y.F. Han & Z.Q. Liang（标本号：GZAC 376）

A、B. PDA 培养基上的菌落；C～K. 分生孢子梗和孢子。标尺：C～K=10 μm

讨论：形态上，贵州假裸囊菌与林氏假裸囊菌、特纳假裸囊菌（均从沉积物中分离获得）和毁害假裸囊菌（从小棕蝠 *Myotis lucifugus* 的翅膀中分离获得）相似，均产生倒卵形和近球形的分生孢子，主要鉴别特征为：林氏假裸囊菌和特纳假裸囊菌的间生孢子球状，两端平（Crous *et al.*，2019）；毁害假裸囊菌（5.0～12.0 × 2.0～3.5 μm）比本菌的孢子大（3.0～5.5 × 3.0～3.5 μm）（Gargas *et al.*，2009）。系统发育上，贵州假裸囊菌的 4 个分离株（GZUIFR 376.1、GZUIFR 376.2、GZUIFR 376.3 和 GZUIFR 376.4）以极高的支持率聚为一支，并与该属的其余物种相分离（Zhang *et al.*，2020）。

毡状假裸囊菌 图 103

Pseudogymnoascus pannorum (Link) Minnis & D.L. Lindner, Fungal Biology 117(9): 646, 2013.

≡ *Sporotrichum pannorum* Link, Sp. Pl., Edn 4 6(1): 13, 1824.

= *Chrysosporium pannorum* (Link) S. Hughes, Can. J. Bot. 36: 749, 1958.

= *Chrysosporium verrucosum* Tubaki, Neue Denkschr. Allg. Schweiz. Ges. Gesammten Naturwiss. 14: 6, 1961.

= *Geomyces pannorum* (Link) Sigler & J.W. Carmich. [as '*pannorus*'], Mycotaxon 4(2): 377, 1976. Liang *et al*., Guizhou Agricultural Sciences 41(8): 136, 2013.

在查氏培养基上，25℃培养 14 天，菌落直径 25～30 mm，平展，朱红色，后期土灰色，粉状，具放射状纵沟；背面淡棕红色。菌丝无色，分隔，宽 1.2～3 μm；具球拍状菌丝。分生孢子侧生于菌丝上或着生于短突起上，无色，光滑或具细刺，椭圆至倒梨形，6～11 × 1.8～3.5（～4.5）μm；具少量间生孢子，桶状，4～8 × 2～3 μm；基痕宽 1～3 μm。

研究菌株：贵州：贵阳市，花溪区，孟关，农田土，2010 年 8 月 2 日，韩燕峰，GZM1008。

世界分布：中国、德国、南极。

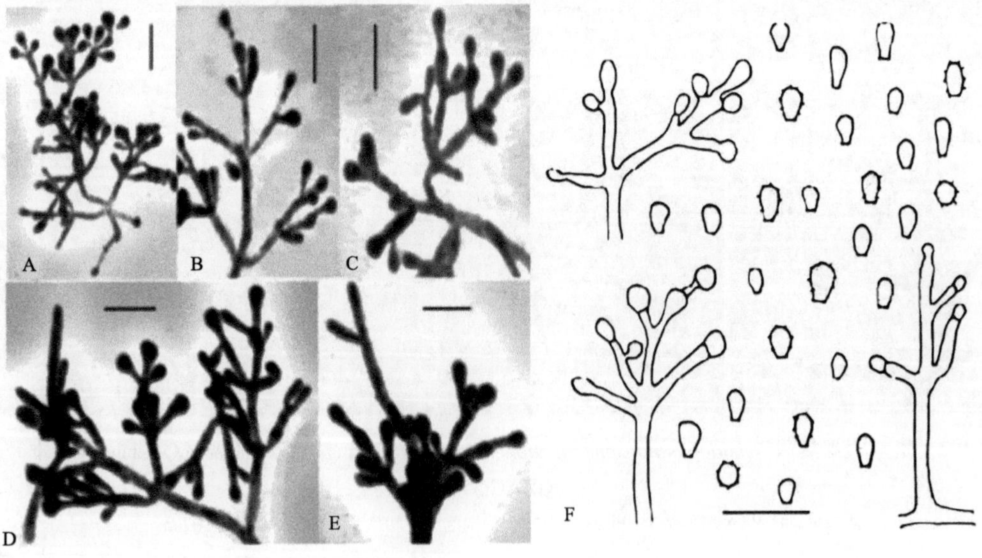

图 103 毡状假裸囊菌 *Pseudogymnoascus pannorum* (Link) Minnis & D.L. Lindner（标本号：GZM1008）

A～F. 分生孢子梗、分生孢子和间生孢子。标尺：A～F=10 μm（F 引自 Song *et al.*，2008）

讨论：该种分布广泛，习生于土壤、冷藏肉、贸易蘑菇、腐烂冷杉树液、啮齿动物的肺和脾、人及动物的皮肤和毛发等多种基质上（Carmichael，1962；胡以仁，1986；王娜，2012；慕东艳，2012；Song *et al.*，2008；宋伟，2008；黄悦华，2009；潘好芹，2009），为实验室的污染物之一。该种多数菌株可分解玻璃质且适宜低温生存，但不能分解高压处理的毛发，也不能在 37℃时生存（黄悦华，2009），该种真菌在高海拔地区分布广泛，比较适应低温环境（潘好芹，2009）。形态上，毡状假裸囊菌与福建假裸囊菌因分生孢子丰富，主轴顶生分生孢子与其他分生孢子大小无明显差异，无节孢子而较为相似；但该种的分生孢子较大，可与后两个物种相区别。

构树假裸囊菌　图 104

Pseudogymnoascus papyriferae Zhi Y. Zhang, Y.F. Han & Z.Q. Liang, MycoKeys 98: 193, 2023.

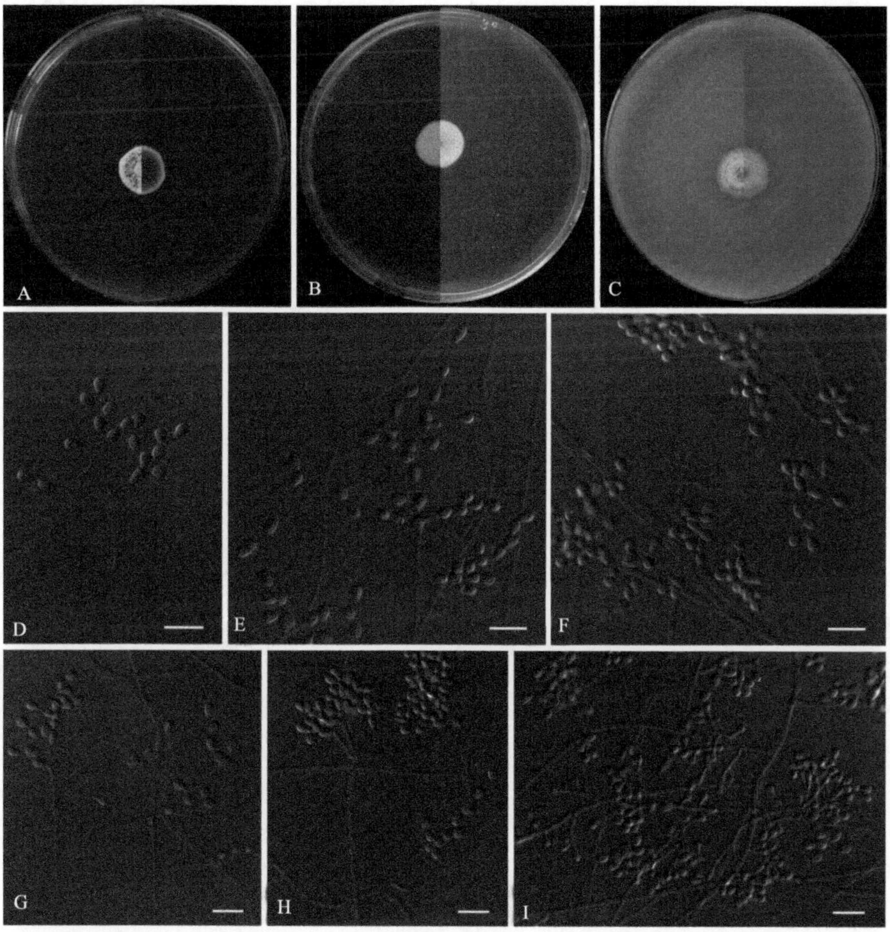

图 104　构树假裸囊菌 *Pseudogymnoascus papyriferae* Zhi Y. Zhang, Y.F. Han & Z.Q. Liang（标本号：HMAS 351878）

A～C. PDA、MEA 和 OA 培养基上的菌落；D～I. 分生孢子梗、分生孢子和间生孢子。标尺：D～I = 10 μm

在 PDA 培养基上，25℃培养 14 天，菌落直径 13～15 mm，白色，稍隆起，棉状、絮状，近圆形，边缘轻微波纹状，大量渗出物呈大的肉桂色滴状，无扩散性色素；背面棕色。在 MEA 培养基上，25℃培养 14 天，菌落直径 13～14 mm，白色菌丝缠结成束，在中心微隆起，近圆形，无渗出物与扩散性色素；背面边缘至中心白色至淡黄色。在 OA 培养基上，25℃培养 14 天，菌落直径 14～15 mm，白色，粉状，中间致密，边缘稀疏，中部稍隆起，近圆形，边缘轻微波纹状，无渗出物，产生白色扩散性色素；背面边缘至中心白色至棕色。菌丝分枝，分隔，无色，光滑，宽 1～2.5 μm。偶尔侧生菌丝末端着生两端钝的桶状或梭形节孢子链，偶尔着生分生孢子，无柄或短柄。分生孢子梗大量，单生，直立，与主轴呈锐角，无色，光滑，常具 2～4 轮锐角分枝。分生孢子倒卵形、梨形至近球形，具 1 宽的截形基痕，3.5～6 × 2.5～4 μm，经连接细胞隔开。间生孢子腰鼓状、桶状、梨形至细长，一端或两端具宽的截痕，3.5～5.5 × 2.5～3.5 μm。节孢子稀少，柱状至中部稍膨胀，2.5～4.5 × 2～2.5 μm，链状或经连接细胞隔开，偶尔着生无柄分生孢子。有性型未见。

研究菌株：陕西：汉中市，构树附生土，2018 年 9 月 11 日，张芝元，HMAS 351878（主模式），CGMCC 3.20877 = GZUIFR 22.020。

世界分布：中国。

讨论：形态上，构树假裸囊菌与陕西假裸囊菌、澳洲假裸囊菌和灰假裸囊菌较为相似（Zhang *et al.*，2020；Villanueva *et al.*，2021）。构树假裸囊菌因产生节孢子而区别于陕西假裸囊菌。构树假裸囊菌与澳洲假裸囊菌的主要区别在于其产生腰鼓状、桶状、梨形至细长的间生孢子，以及较少的节孢子。此外，构树假裸囊菌与灰假裸囊菌的主要区别为：前者的间生孢子为腰鼓状、桶状、梨形至细长，3.5～5.5 × 2.5～3.5 μm，而后者的间生孢子为近球形至细长或桶状，3.5～9.6 × 1.7～3.9 μm。

陕西假裸囊菌 图 105

Pseudogymnoascus shaanxiensis Zhi Y. Zhang, Y.F. Han & Z.Q. Liang, Frontiers in Microbiology 11: 580, 2020.

在 PDA 培养基上，25℃培养 14 天，菌落直径 21～23 mm，绒状至絮状，边缘规则，白色，产生可扩散的黄色色素和清晰的渗出物；背面棕色。菌丝分隔，无色，光滑，宽 1.5～2.5 μm；无球拍状菌丝。分生孢子丰富，光滑，梨形，有时近球形，3.5～5.0 × 2.5～3.0 μm；间生孢子近球形、梨形，或不规则，光滑，3.5～4.0 × 3.0～4.0 μm；顶生和侧生分生孢子着生于菌丝、短柄或侧枝上，单生，形成轮生和对生的分枝，与主轴呈锐角。有性型未见。

研究菌株：陕西：西安市，新城区，构树附生土，2018 年 9 月 10 日，张芝元，HMAS 255395（模式标本），GZUIFR 173.1；陕西：汉中市，南郑区，棕榈树附生土，2018 年 9 月 11 日，张芝元，GZUIFR HZ5.7；湖北：宜昌市，滨江公园，绿地土，2018 年 9 月 18 日，张芝元，GZUIFR CY 1.8。

世界分布：中国、美国。

讨论：形态上，陕西假裸囊菌与附属假裸囊菌、疣状假裸囊菌相似，其分生孢子梨形，有时近球形，主要区别在于后两者的间生孢子近球形至桶状（Rice & Currah，2006）。系统发育上，该种的菌株 GZUIFR 173.1、GZUIFR HZ5.7 和 GZUIFR CY 1.8 聚集到一个亚分支，并与其他分支相分离。

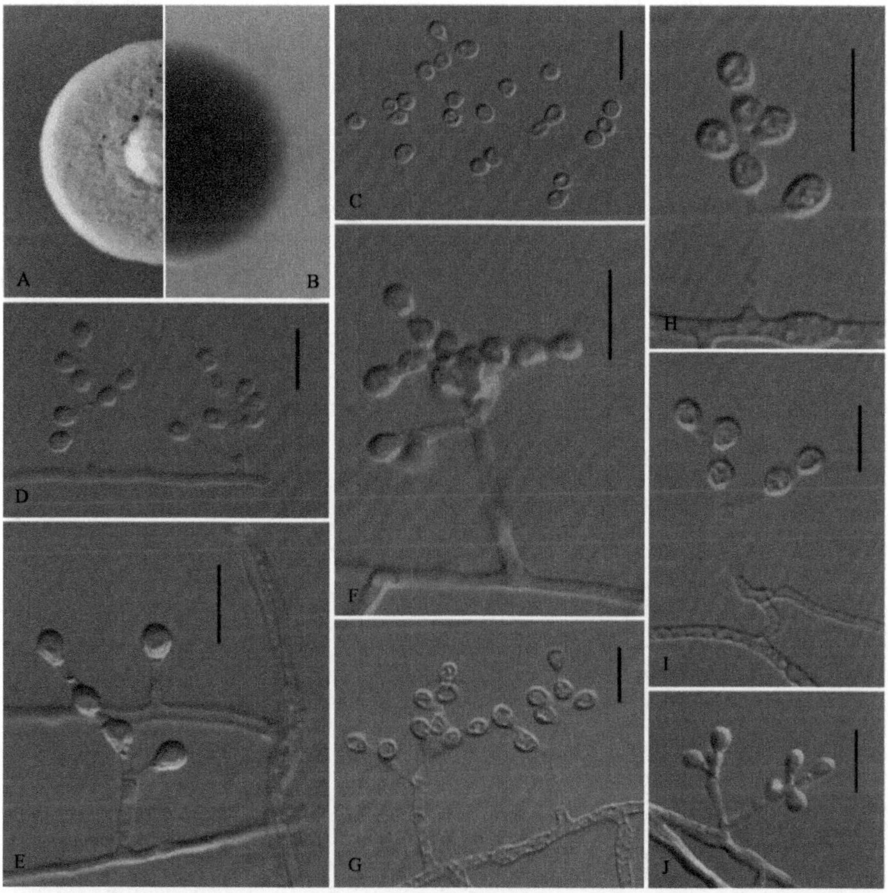

图 105　陕西假裸囊菌 *Pseudogymnoascus shaanxiensis* Zhi Y. Zhang, Y.F. Han & Z.Q. Liang（标本号：HMAS 255395）

A、B. PDA 培养基上的菌落；C. 分生孢子；D～J. 分生孢子梗和孢子。标尺：C～J = 10 μm

中国假裸囊菌　图 106

Pseudogymnoascus sinensis Zhi Y. Zhang, Y.F. Han & Z.Q. Liang, Frontiers in Microbiology 11: 582, 2020.

在 PDA 培养基上，25℃培养 14 天，菌落直径 20～21 mm，粉状至絮状，边缘清晰，中心浅粉色，边缘紫铜色；背面褐色。气生菌丝丰富，壁薄且光滑，有隔膜，宽 1～2 μm；无球拍状菌丝。分生孢子丰富，顶生和侧生于菌丝、短柄或侧枝上，有时形成轮生和对生分枝，近先端与主轴呈锐角，单生，倒卵球形，3～5 × 2.5～3 μm。间生孢子着生于菌丝或轮生的外侧分枝上，壁光滑，腰鼓状、倒卵状、梨形或不规则，3.0～4.5 × 2.5～5 μm。

研究菌株：贵州：贵阳市，贵州医科大学附属医院，绿化地土，2016 年 9 月 10 日，张芝元，HMAS 255394（模式标本），CGMCC 3.18493 = GZUIFR K278.1。

世界分布：中国。

讨论：形态上，中国假裸囊菌与林氏假裸囊菌和特纳假裸囊菌相似，均产生倒卵形的分生孢子，主要区别在于林氏假裸囊菌和特纳假裸囊菌的间生孢子为球形，两端平（Crous *et al.*，2019）。系统发育上，分离株 CGMCC 3.18493 和 CGMCC 3.18494 以较高的支持率聚为一支，并形成独立的分支，且与其他物种相分离。

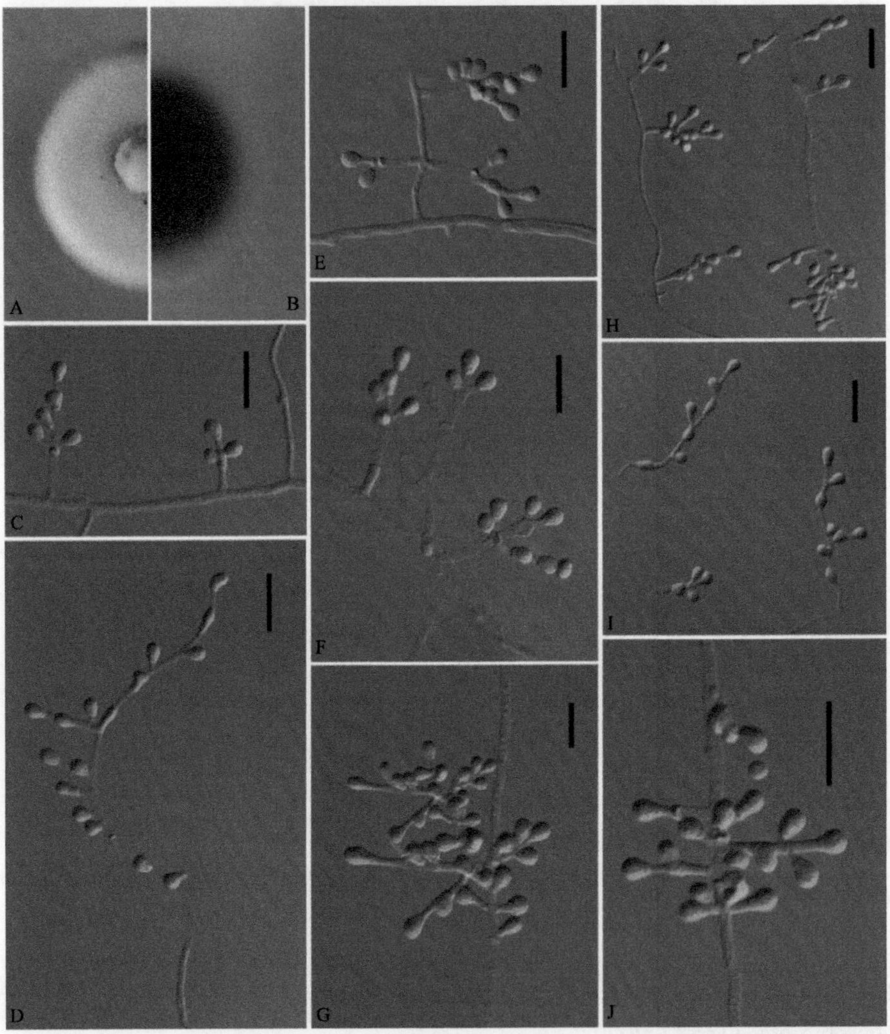

图 106　中国假裸囊菌 *Pseudogymnoascus sinensis* Zhi Y. Zhang, Y.F. Han & Z.Q. Liang（标本号：HMAS 255394）

A、B. PDA 培养基上的菌落；C～J. 分生孢子梗和孢子。标尺：C～J = 10 μm

疣状假裸囊菌　图 107

Pseudogymnoascus verrucosus A.V. Rice & Currah, Mycologia 98: 311. 2006.

在 PDA 培养基上，25℃培养 14 天，菌落直径 19～23 mm，中心至边缘浅粉色至白色，平展，绒状，近圆形，边缘规则，无渗出物和扩散性色素；背面浅粉色至红棕色。菌丝分枝，分隔，无色，表面光滑，宽 0.5～1.5 μm。分生孢子白色，成团时浅桃红色，顶生，近球形至宽梨形，具 1 突出的截形基痕，壁厚，无色至近无色，成熟时近粗糙或疣状，2.5～3 × 1～2.5 μm。节孢子近球形至椭圆形、桶状，末端截形，壁厚，无色至近无色，成熟时近粗糙或疣状，2.5～4 × 3～2.5 μm。分生孢子梗直立，无色，壁薄，表面光滑，树状具轮生状分枝，5～22 × 1～2.5 μm。

研究菌株：贵州：贵阳市，贵州大学，北校区，绿化地土，2019 年 8 月 28 日，张芝元，GZUIFR 21.932，GZUIFR 21.933。

世界分布：中国、加拿大。

讨论：Rice 和 Currah（2006）首次报道了疣状假裸囊菌，他们对该种的有性型与无性型均进行了描述，但作者仅观察到了该种的无性型。基于无性型，疣状假裸囊菌和葡状假裸囊菌较为相似，它们在 PDA 培养基上均不产生色素和扩散性色素，且均具节孢子；但疣状假裸囊菌成熟的分生孢子和节孢子近粗糙或疣状等特征明显不同于遵义假裸囊菌和葡状假裸囊菌。

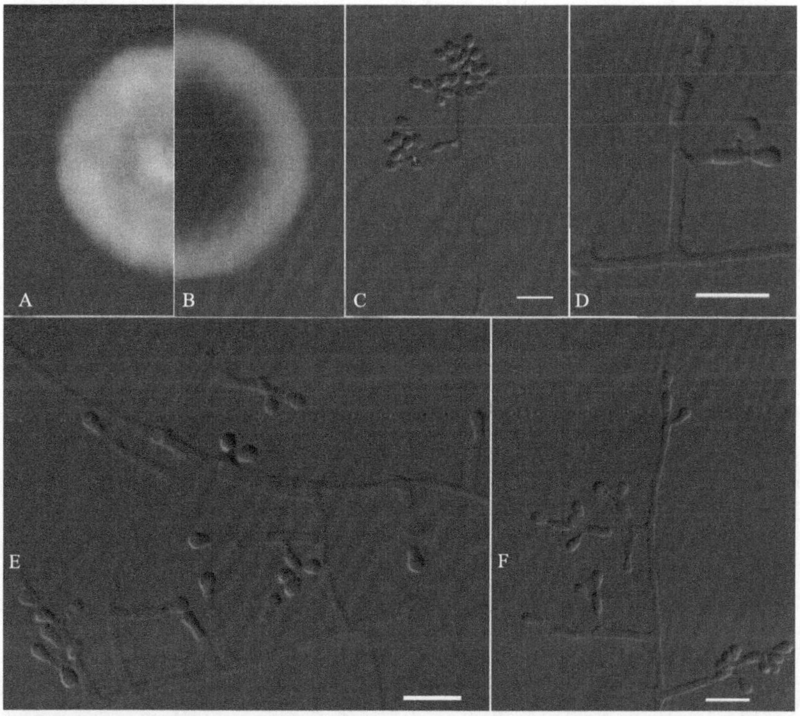

图 107　疣状假裸囊菌 *Pseudogymnoascus verrucosus* A.V. Rice & Currah（标本号：GZUIFR 21.932）

A、B. PDA 培养基上的菌落；C～F. 分生孢子梗和分生孢子。标尺：C～F = 10 μm

云南假裸囊菌　图 108

Pseudogymnoascus yunnanensis Zhi Y. Zhang, Y.F. Han & Z.Q. Liang, Microbiology Spectrum 9: e00867-21, 2021b.

在 PDA 培养基上，25℃培养 14 天，菌落直径 19～20 mm，绒状至粉状，边缘规则，局部内陷，粉红色，边缘白色，无色素和渗出物；背面棕色。37℃时不生长。菌丝分隔，无色，分枝，光滑，宽 1～3 μm；无球拍状菌丝。分生孢子梗丰富，呈锐角分枝，通常 2～3 个轮生，每轮有 1～4 个分枝，二级和三级分枝仍能再分枝。分生孢子丰富，着生于轮生分枝末端，或横向着生于短的突起或短的侧枝上；无色至半透明，壁光滑或刺状；倒卵形、近球形到球形、梨形，2.5～4.5 × 2.5～3.5 μm；顶生孢子，棍棒状、纺锤形，基部有疤痕，6.5～9 × 2.5～4.5 μm；间生孢子着生于菌丝或轮生菌丝的外部分枝，单生，或两个成链，壁光滑，或粗糙，肾形、纺锤形，两端平截，2.5～5.5 × 2.5～4 μm。

研究菌株：云南：大理市，大理白族自治州人民医院，绿化地土，2019 年 9 月 3 日，张芝元，HMAS 350320（主模式），CGMCC 3.20475 = GZUIFR 21.807；云南：大理市，大理大学，绿化地土壤，2019 年 9 月 2 日，张芝元，GZUIFR 21.809。

世界分布：中国。

讨论：形态上，云南假裸囊菌与林氏假裸囊菌、特纳假裸囊菌和贵州假裸囊菌相似，均产生倒卵形的分生孢子。然而，云南假裸囊菌与林氏假裸囊菌、特纳假裸囊菌的主要区别为：其产生棍棒状、纺锤形与基部有疤痕的顶生分生孢子，且未观察到有性型态。云南假裸囊菌与贵州假裸囊菌的主要区别为：其产生肾形、纺锤形的间生孢子，且间生孢子的两端平截。系统发育上，云南假裸囊菌的 3 个分离株以极高的支持率聚为一亚支，与贵州假裸囊菌形成姐妹支，但它们很容易被区分开来。

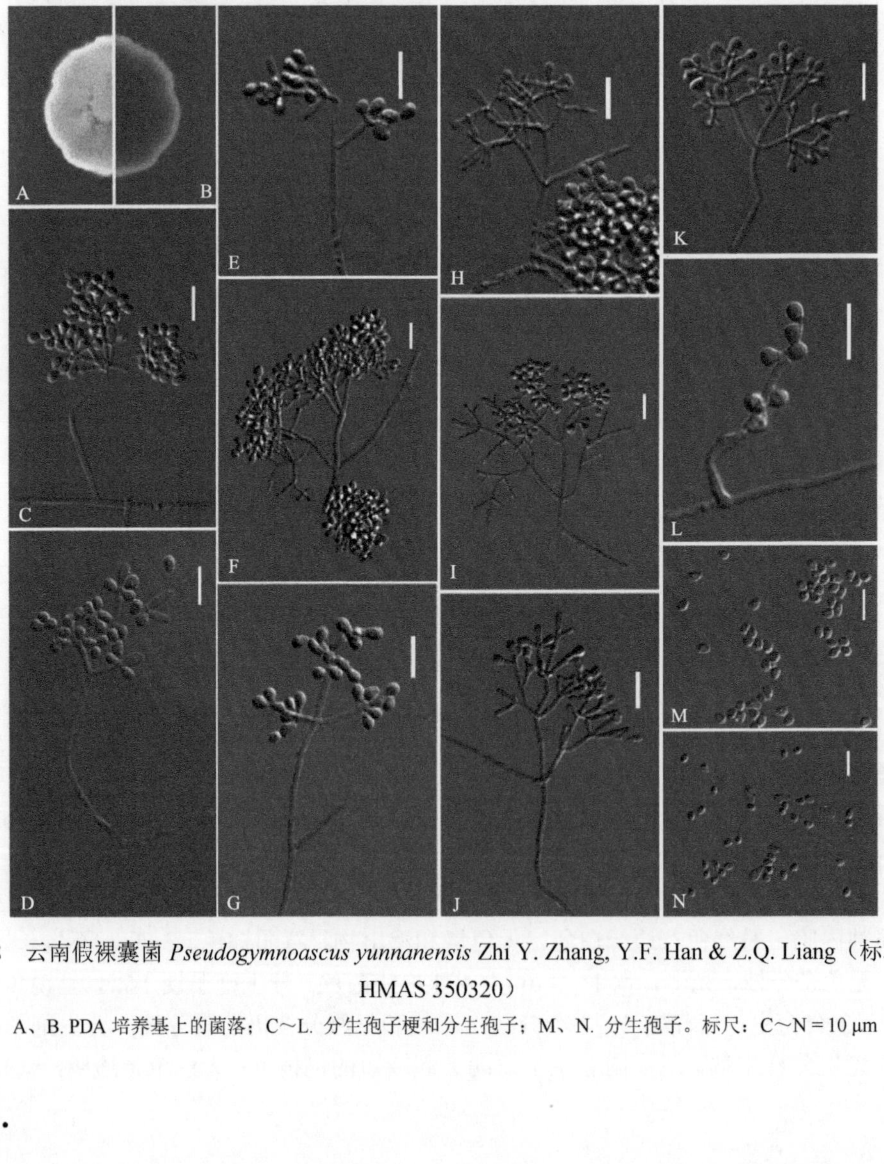

图 108　云南假裸囊菌 *Pseudogymnoascus yunnanensis* Zhi Y. Zhang, Y.F. Han & Z.Q. Liang（标本号：HMAS 350320）

A、B. PDA 培养基上的菌落；C～L. 分生孢子梗和分生孢子；M、N. 分生孢子。标尺：C～N = 10 μm

浙江假裸囊菌　图 109

Pseudogymnoascus zhejiangensis Zhi Y. Zhang, Y.F. Han & Z.Q. Liang, Microbiology Spectrum 9: e00867-21, 2021b.

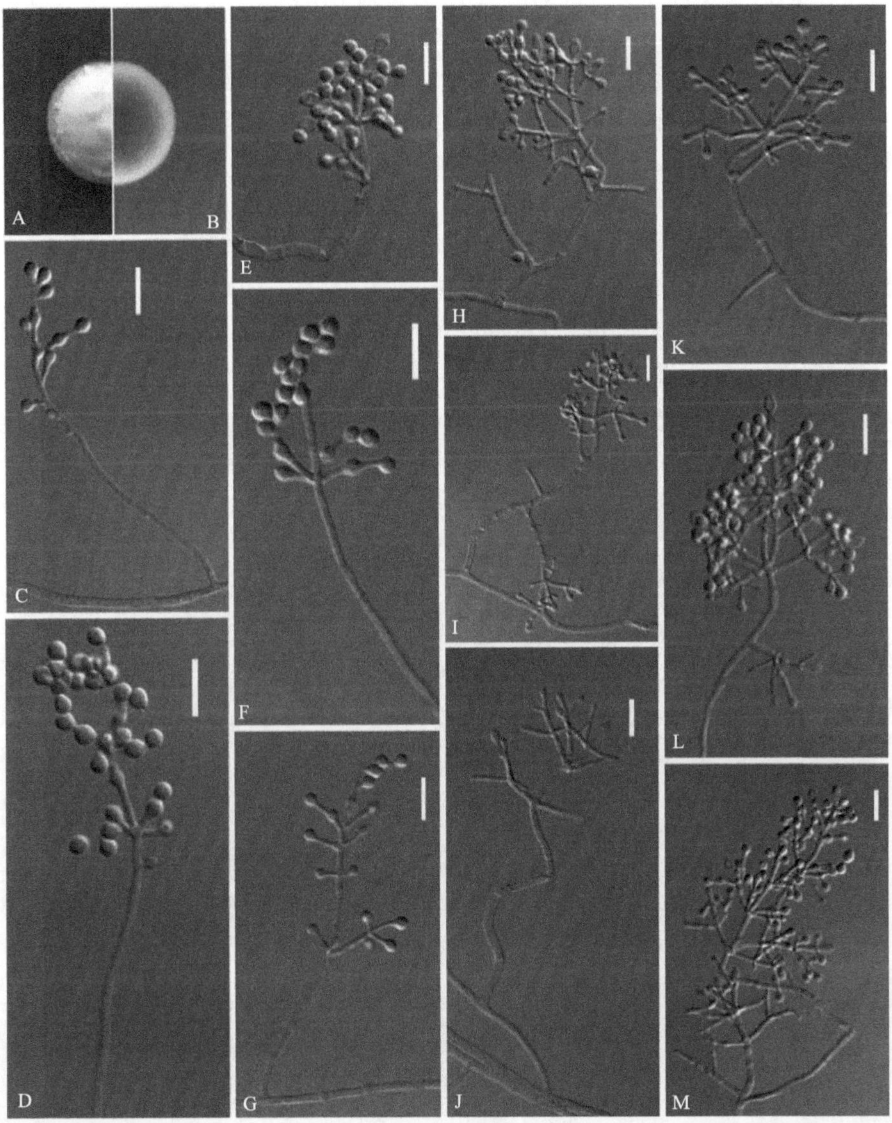

图 109　浙江假裸囊菌 *Pseudogymnoascus zhejiangensis* Zhi Y. Zhang, Y.F. Han & Z.Q. Liang（标本号：HMAS 350321）

A、B. PDA 培养基上的菌落；C～M. 分生孢子梗、分生孢子和间生孢子。标尺：C～M =10 μm

在 PDA 培养基上，25℃培养 14 天，菌落直径 20 mm，绒状至絮状，边缘完整，白色，无色素和渗出物；背面粉红色，边缘白色。37℃时不生长。菌丝分隔，无色，分枝，光滑，宽 1～3 μm；无球拍状菌丝。分生孢子梗丰富，呈锐角分枝，通常 1～4 轮，每轮有 1～4 个分枝，二级和三级分枝仍可再分枝。分生孢子丰富，着生于轮生分枝末端或横向着生于短突或短侧枝上，无色至半透明，光滑或粗糙，倒卵形至球形，2.5～

4.5 × 2.5～4 μm；棒状、长倒卵形，5～9 × 2.5～4 μm。间生孢子着生于轮状菌丝或菌丝上，单生，壁光滑，或粗糙，倒卵形、近球形至球形，3.5～4.5 × 3～4 μm。

研究菌株：浙江：宁波市，月亮湖，绿化地土，2019 年 8 月 16 日，张芝元，HMAS 350321（主模式），CGMCC 3.20476 = GZUIFR 21.810。

世界分布：中国。

讨论：形态上，浙江假裸囊菌与林氏假裸囊菌、特纳假裸囊菌和云南假裸囊菌相似，均产生倒卵形的分生孢子，不同的是林氏假裸囊菌与特纳假裸囊菌的间生孢子球形，两端平截；云南假裸囊菌的间生孢子肾形、纺锤形，两端平截。系统发育上，浙江假裸囊菌的 3 个分离株聚为一亚分支，并与别的支系相分离。

宗琦假裸囊菌　图 110，图版 II：8

Pseudogymnoascus zongqii Zhi Y. Zhang, Y.F. Han & Z.Q. Liang, MycoKeys 98: 195, 2023.

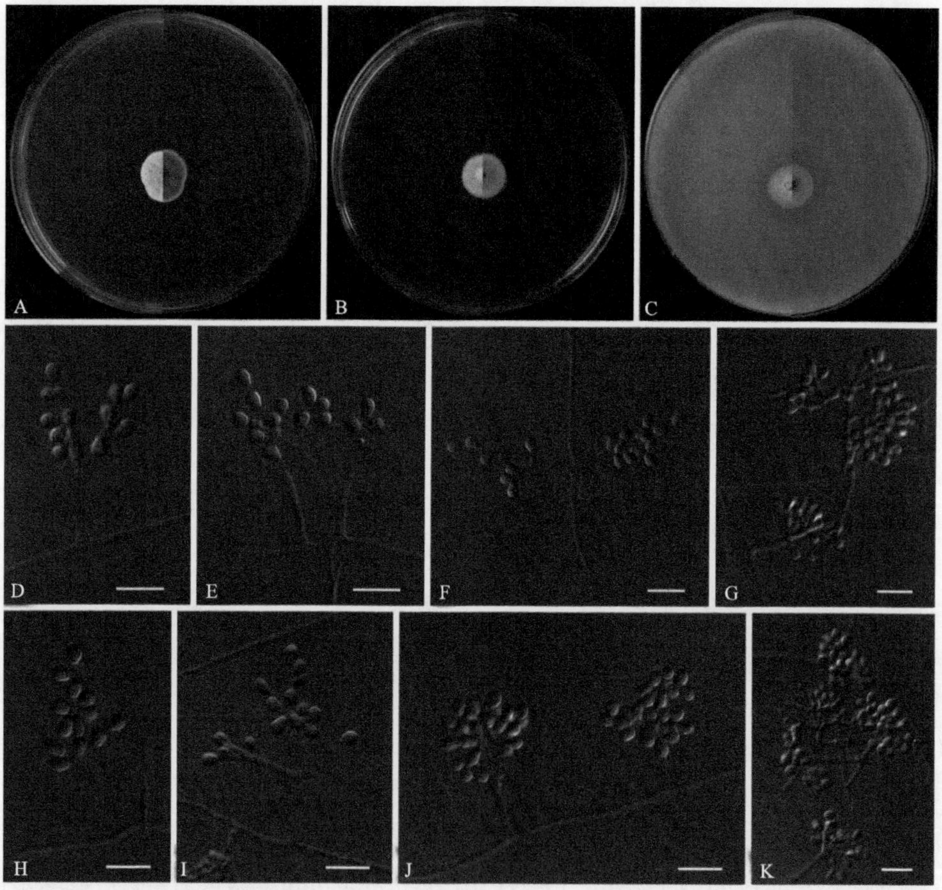

图 110　宗琦假裸囊菌 *Pseudogymnoascus zongqii* Zhi Y. Zhang, Y.F. Han & Z.Q. Liang（标本号：HMAS 351905）

A～C. PDA、MEA 和 OA 培养基上的菌落；D～K. 分生孢子梗、分生孢子和间生孢子。标尺：D～K =10 μm

在 PDA 培养基上，25℃培养 14 天，菌落直径 13～15 mm，中心至边缘粉色至白色，绒毛状、絮状，近圆形，边缘轻微凹陷，无渗出物，扩散性色素色浅不明显；背面棕色。在 MEA 培养基上，25℃培养 14 天，菌落直径 12～13 mm，边缘至中心白色至淡黄色，菌丝扭结成束，中部稍隆起，近圆形，无渗出物和扩散性色素；背面中心至边缘棕色至淡黄色。在 OA 培养基上，25℃培养 14 天，菌落直径 12 mm，灰色，絮状，中心致密，边缘稀疏，近圆形，边缘规则；背面边缘至中心白色至棕色。菌丝分枝，分隔，无色，光滑，宽 1～3 μm。分生孢子梗大量，单生，有时自菌丝细微分化，无色，光滑，自直立菌丝产生，常具 2～5 轮锐角分枝。分生孢子和间生孢子丰富，无色，光滑，或粗糙。分生孢子梨形，偶尔倒卵形至近球形，具宽的截形基痕，3～5 × 2.5～3.5 μm；间生孢子梨形至倒卵形，3.5～5 × 2.5～4.5 μm，经连接细胞而隔开，偶尔着生无柄分生孢子。

研究菌株：四川：成都市，花坛土，2016 年 8 月 8 日，张芝元，HMAS 351905（主模式），CGMCC 3.20878 = GZUIFR 22.025；贵州：遵义市，遵义医科大学附属医院，绿化带土，2016 年 9 月 11 日，张芝元，GZUIFR 22.042，GZUIFR 22.043。

世界分布：中国。

讨论：形态上，宗琦假裸囊菌与中国假裸囊菌较为相似。但宗琦假裸囊菌在 PDA 培养基上粉色至白色，背面棕色等特征可与中国假裸囊菌相区别（Zhang *et al.*，2020）。

赛多孢属 **Scedosporium** Sacc. ex Castell. & Chalm.

Manual of Tropical Medicine (London): 1122, 1919

菌落粉状或绒毛状，颜色多样；菌丝有隔，光滑，分支；分生孢子梗有或无，分生孢子单生；具有大、小分生孢子；大分生孢子仅出现在尖端赛多孢；小分生孢子形状多样。有性型状态下出现深色、非开口、球形子囊，薄且表面具有表皮样纹路，含有 8 个子囊孢子，子囊很快就消失。子囊孢子单胞，浅黄色或浅红棕色，广梭形到椭圆形，两端通常各有 1 个芽孔。

模式种：尖端赛多孢 *Scedosporium apiospermum* Sacc. ex Castell. & Chalm.。

中国赛多孢属分种检索表

1. 分生孢子呈卵形、倒卵形或圆柱形……………………………………………………2
1. 分生孢子呈椭圆形，产孢结构分赛多型和黏束孢型，分生孢子多以单个存在，分生孢子梗细长……7
 2. 产孢细胞基部膨大……………………………………………………………………3
 2. 产孢细胞基部柱状……………………………………………………………………4
3. 分生孢子呈倒卵圆形，少数近球形……………………………………**三亚赛多孢 *S. sanyaense***
3. 分生孢子呈卵形，可见厚壁的球形孢子……………………………**多育赛多孢 *S. prolificans***
 4. 产孢数量少，分生孢子呈倒卵形 ……………………………………**少孢赛多孢 *S. rarisporum***
 4. 产孢量较多……………………………………………………………………………5
5. 产孢细胞圆柱状和锥形，分生孢子常形成黏性分生孢子团…………………**分生赛多孢 *S. dehoogii***
5. 产孢细胞柱状，分生孢子单生，不形成孢子团……………………………………………6
 6. PDA 培养基上菌落浅黄色至白色，产孢细胞不膨大……………………**海口赛多孢 *S. haikouense***
 6. PDA 培养基上菌落为橘黄色，产孢细胞柱状或轻微膨大烧瓶状……**橘黄赛多孢 *S. aurantiacum***
7. 分生孢子单生……………………………………………………………**尖端赛多孢 *S. apiospermum***

7. 分生孢子单生或 2～3 个簇生……………………………………………………………………………… 8
8. PDA 培养基上菌落背面白色，分生孢子卵圆至近球形………………… 多孢赛多孢 ***S. multisporum***
8. PDA 培养基上菌落背面青灰色，分生孢子卵圆或椭圆形……………… 海南赛多孢 ***S. hainanense***

橘黄赛多孢 图 111

Scedosporium aurantiacum Gilgado, Cano, Gené & Guarro, J. Clin. Microbiol. 43(10): 4938, 2005. Wang *et al*., Microbiology China 41(10): 2092, 2016.

在查氏培养基上，25℃培养 7 天，菌落直径 30 mm 左右，绒状，较稀疏，灰白色至淡棕色，隆起达 5 mm；背面淡棕色至灰白色。在 PDA 培养基上，25℃培养 7 天，菌落直径 20～25 mm，细绒状，向上隆起，高 3～6 mm，中间褐色，边缘灰白色，中央突出，边缘不规则；背面红棕色，周围可见扩散的黄色素。菌丝具隔膜，无色，光滑，直径 1.6～5.4 μm。分生孢子梗单生或分枝；单生的分生孢子梗常退化为产孢细胞，常由 2～3 个产孢细胞组成轮生体。产孢细胞顶生或侧生，无色至近无色，薄壁，柱状或稍膨大的烧瓶状，3.5～45 × 1.5～3 μm。分生孢子顶生、侧生，单生，无色至近无色，薄壁，光滑，椭圆形、倒卵圆形或近柱状，4.3～14 × 2.5～5.5 μm。

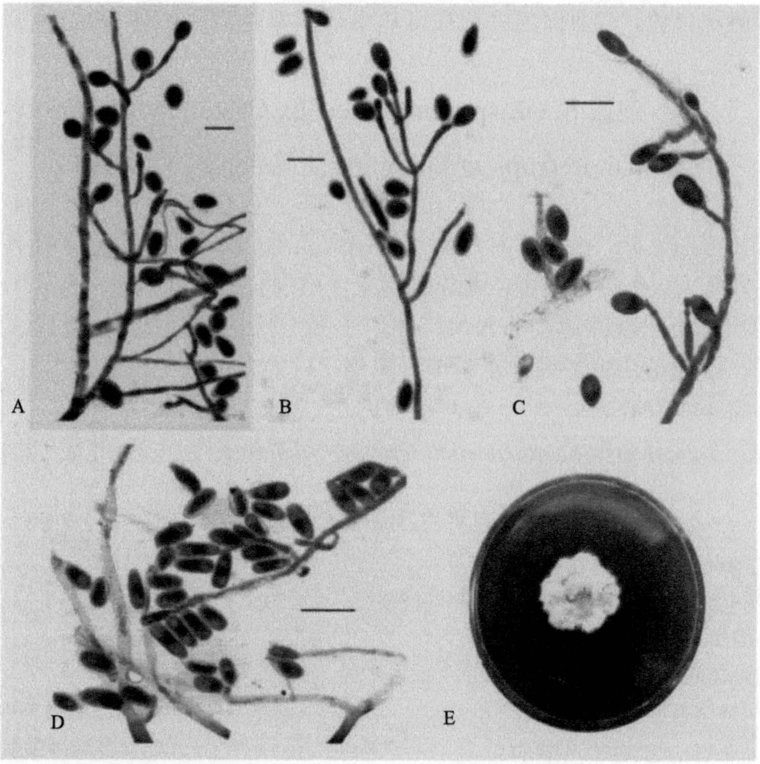

图 111　橘黄赛多孢 *Scedosporium aurantiacum* Gilgado, Cano, Gené & Guarro（标本号：EM12901）
A～D. 产孢结构；E. PDA 培养基上的菌落。标尺：A～D = 10 μm

研究菌株：贵州：遵义市，遵义县，龙坪镇，中兴养殖场，含猪粪土样，2011 年 7 月，王玉荣，EM12901。

世界分布：中国、日本、美国、阿根廷。

讨论：橘黄赛多孢产孢细胞柱状，与海口赛多孢相似，但海口赛多孢的产孢细胞不膨大，而本种的产孢细胞轻微膨大成烧瓶状（王玉荣等，2014），二者可明显区分。

海口赛多孢 图 112

Scedosporium haikouense Zhi Y. Zhang, Y.F. Han & Z.Q. Liang, Microbiology Spectrum 9(2): e00867-21, 2021.

在 PDA 培养基上，25℃培养 5 天，菌落直径 54～56 mm，绒毛状，浅黄色至白色，边缘灰色，中部可见轮纹，边缘浅波浪状；背面奶酪色至黑色。在 PDA 培养基上，37℃培养 5 天，菌落直径 68～70 mm。菌丝具隔膜，无色，光滑，直径 0.5～5.5 μm。分生孢子梗单生，常退化为产孢细胞，顶生或侧生于菌丝上，无色，光滑，柱状，5～26 × 1～2 μm。分生孢子着生于菌丝、短突起或侧分枝上，单生或 2～3 个簇生，浅褐色至褐色，卵圆形、椭圆形、近柱状且两端钝圆，5～9 × 3～4.5 μm。

研究菌株：海南：海口市，海南大学，校园绿地土，2019 年 8 月 28 日，张芝元，GZUIFR 21.833 = CGMCC 3.20468。

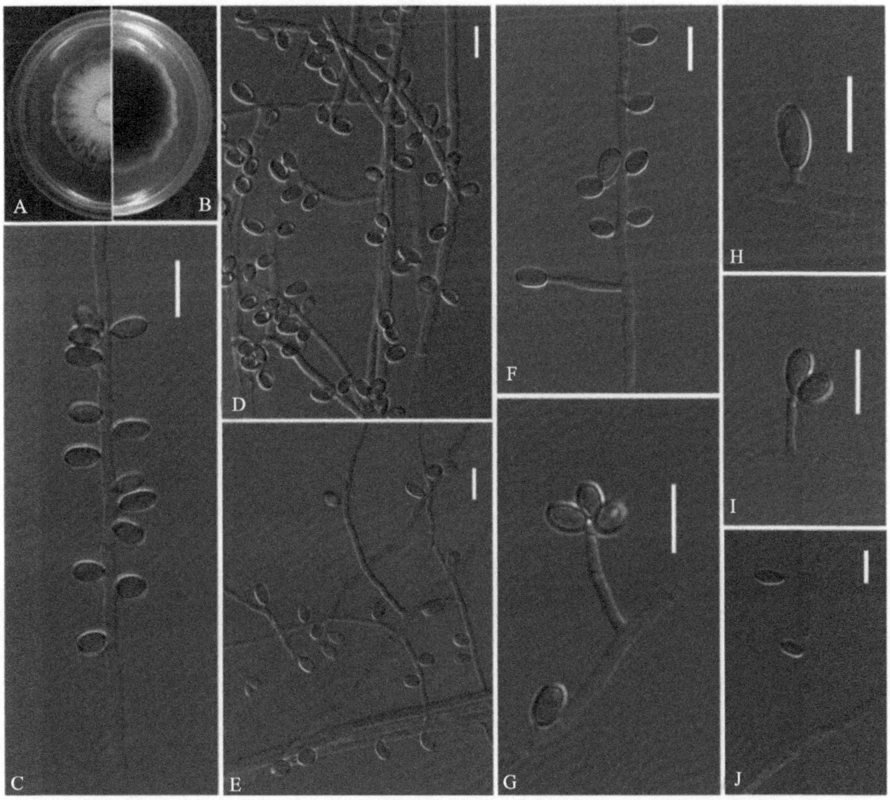

图 112 海口赛多孢 *Scedosporium haikouense* Zhi Y. Zhang, Y.F. Han & Z.Q. Liang（标本号：CGMCC 3.20468）

A、B. PDA 培养基上的菌落；C～J. 分生孢子和产孢细胞。标尺：C～J = 10 μm

世界分布：中国。

讨论：基于系统发育分析，海口赛多孢与少孢赛多孢、蜡状孢赛多孢和橘黄赛多孢相似。但海口赛多孢可通过大量卵圆形、椭圆形和近柱状孢子与少孢赛多孢相区别；通过产孢细胞单生或 2～3 个簇生与蜡状孢赛多孢相区别；通过不产生色素和渗出物及孢梗束与橘黄赛多孢相区别（Zhang *et al.*，2021b）。

海南赛多孢　图 113

Scedosporium hainanense Zhi Y. Zhang, Y.F. Han & Z.Q. Liang, Microbiology Spectrum 9(2): e00867-21, 2021.

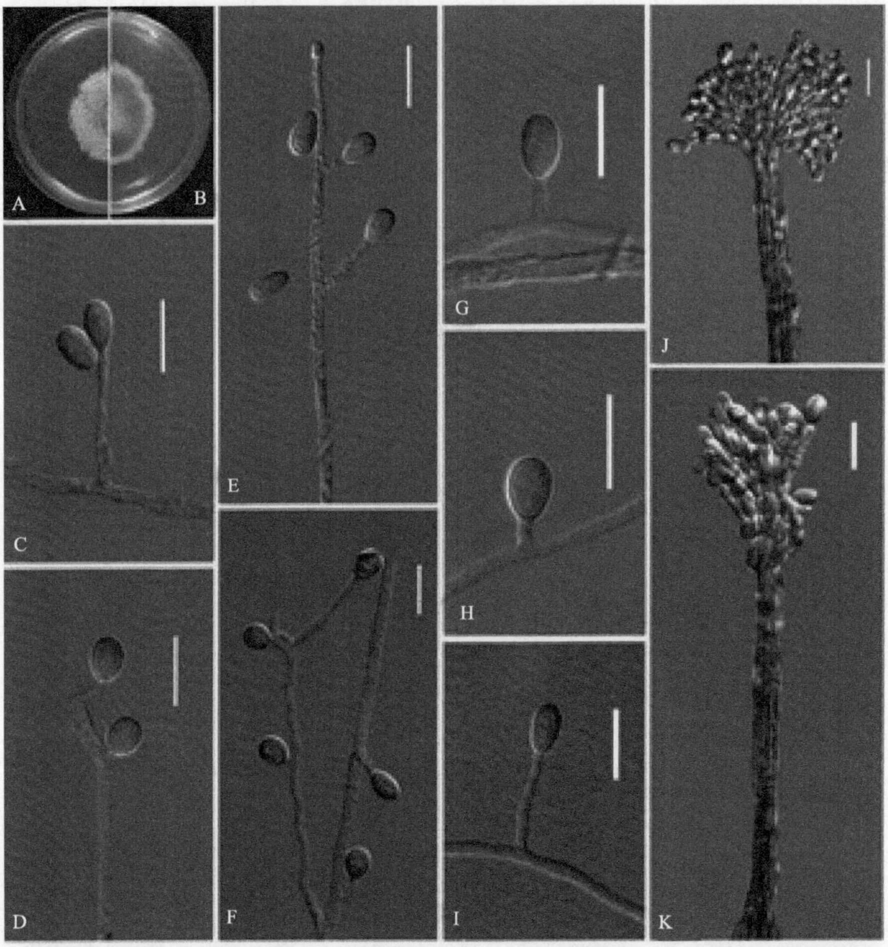

图 113　海南赛多孢 *Scedosporium hainanense* Zhi Y. Zhang, Y.F. Han & Z.Q. Liang（标本号：CGMCC 3.20469）

A、B. PDA 培养基上的菌落；C～I. 分生孢子和产孢细胞；J、K. 子座状载孢体。标尺：C～K = 10 μm

在 PDA 培养基上，25℃培养 5 天，菌落直径 38～43 mm，绵毛状或绒毛状，浅灰色，边缘不规则；背面青灰色，边缘白色。在 PDA 培养基上，37℃培养 5 天，菌落直径 64～66 mm。菌丝具隔膜，无色，光滑，直径 0.5～4.5 μm。分生孢子梗单生，常退化成为单个产孢细胞，或由 2～3 个产孢细胞组成轮生体，顶生或侧生于菌丝或短柄上。产孢细胞无色，薄壁光滑，侧生或顶生，柱状或基部较宽，随着培养时间的延长，顶部可见环纹，2.5～33 × 1～2.5 μm。分生孢子着生于菌丝、短突起或侧分枝上，单胞，单生，无色，卵圆形、5.0～8 × 2.5～6 μm，椭圆形、5.5～7 × 5～5.5 μm；载孢体子座状，直立，梗柱状，无色，光滑，其上着生的分生孢子柱状或棒状，两端平截，4.5～8.5 × 2.5～3.5 μm。

研究菌株：海南：三亚市，海南热带海洋学院，绿地土，2019 年 8 月 26 日，张芝元，GZUIFR 21.829 = CGMCC 3.20469；海南：儋州市，儋州大学，校园绿地土，2019 年 8 月 26 日，张芝元，GZUIFR 21.827。

世界分布：中国。

讨论：基于系统发育分析，海南赛多孢与尖端赛多孢相近，但海南赛多孢分生孢子的形状、具有明显的子座状载孢体等特征可与尖端赛多孢区分（Zhang *et al.*，2021b）。

多孢赛多孢 图 114

Scedosporium multisporum Zhi Y. Zhang, Y.F. Han & Z.Q. Liang, Microbiology Spectrum 9(2): e00867-21, 2021.

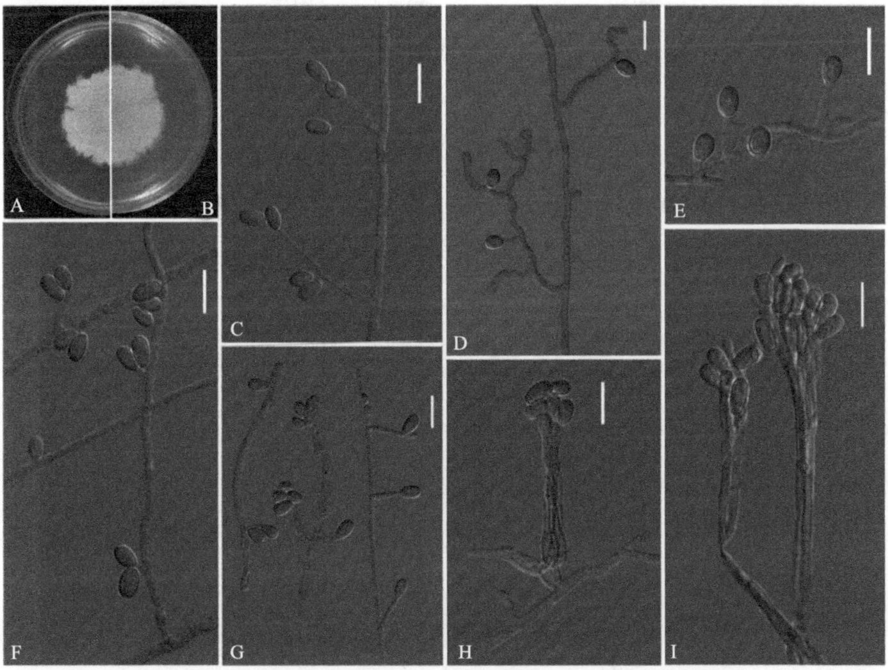

图 114 多孢赛多孢 *Scedosporium multisporum* Zhi Y. Zhang, Y.F. Han & Z.Q. Liang（标本号：CGMCC 3.20470）

A、B. PDA 培养基上的菌落；C～G. 分生孢子和产孢细胞；H、I. 子座状载孢体。标尺：C～I = 10 μm

在 PDA 培养基上，25℃培养 5 天，菌落直径 45～50 mm，绵毛状，中部粉状，白色；背面白色，中部淡黄色。在 PDA 培养基上，37℃培养 5 天，菌落直径 70～73 mm。菌丝具隔膜，分枝，无色，光滑，直径 1～4 μm。分生孢子梗单生，常退化为单个产孢细胞，或由 2～3 个产孢细胞组成轮生体，顶生或侧生于菌丝或短柄上；产孢细胞无色，薄壁光滑，侧生或顶生，柱状或基部较宽，随着培养时间的延长，顶部可见环纹，0.5～16 × 1.0～3.5 μm。分生孢子着生于菌丝、短突起或侧分枝上，单胞，单生或 2～3 个簇生，无色，卵圆形至近球形，3～7.5 × 3～5 μm。载孢体子座状，直立，梗柱状，无色，光滑，其上着生的分生孢子柱状或卵圆形，基部平截，两端平截，5～10 × 2～4 μm。

研究菌株：湖南：怀化市，怀化学院，绿地土，2019 年 8 月 12 日，张芝元，GZUIFR 21.830 = CGMCC 3.20470，GZUIFR 21.831。

世界分布：中国。

讨论：基于系统发育分析，多孢赛多孢与尖端赛多孢复合种具有较近的关系。多孢赛多孢与其他赛多孢可通过单生或 2～3 个簇生的分生孢子，产孢结构呈孢梗束型相区别（Zhang *et al.*，2021b）。

少孢赛多孢 图 115

Scedosporium rarisporum Y.F. Han, H. Zheng, Y. Luo, Y.R. Wang & Z.Q. Liang, Mycosystema 36(2): 149, 2017.

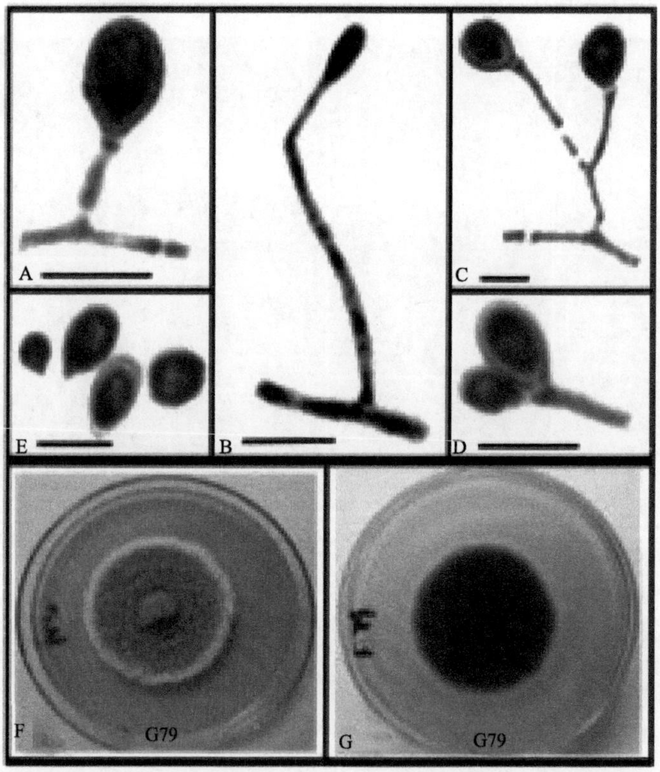

图 115 少孢赛多孢 *Scedosporium rarisporum* Y.F. Han, H. Zheng, Y. Luo, Y.R. Wang & Z.Q. Liang（标本号：GZUIFR G79）

A～D. 产孢结构；E. 分生孢子；F、G. PDA 培养基上的菌落。标尺：A～E = 10 μm

在 PDA 培养基上，25℃培养 7 天，菌落直径 52 mm，细绒状，中间灰褐色至灰色，边缘白色，中央隆起，菌落边缘规则；背面黑色。气生菌丝具隔膜，无色，光滑，宽 0.8～2 μm。分生孢子生于菌丝或分枝上，顶生或侧生，非常稀少，大多单生，少见双生，分生孢子无色至近无色，薄壁，光滑，倒卵形，5.4～10.8 × 3.2～5.4 μm。无间生孢子和厚垣孢子。

研究菌株：广西：贵港市，马草江狐尾椰根系土，2015 年 3 月 25 日，罗韵，GZUIFR G79。

世界分布：中国。

讨论：在基于 ITS 序列分析的系统发育树中，该种与沙生赛多孢和橘黄赛多孢聚在一起，但由于沙生赛多孢是基于分子数据建立的系统发育种，并未提供任何有关形态学特征，故无法通过形态特征比较出两者之间的差异；而与橘黄赛多孢可通过产生较少的分生孢子这一特征明显区分。该种的典型特征是：分生孢子生于菌丝或分枝上，顶生或侧生，非常稀少，孢子无色至近无色，倒卵形；无间生孢子和厚垣孢子。

三亚赛多孢 图 116

Scedosporium sanyaense Y.F. Han, H. Zheng, Y. Luo, Y.R. Wang & Z.Q. Liang, Mycosystema 36(2): 150, 2017.

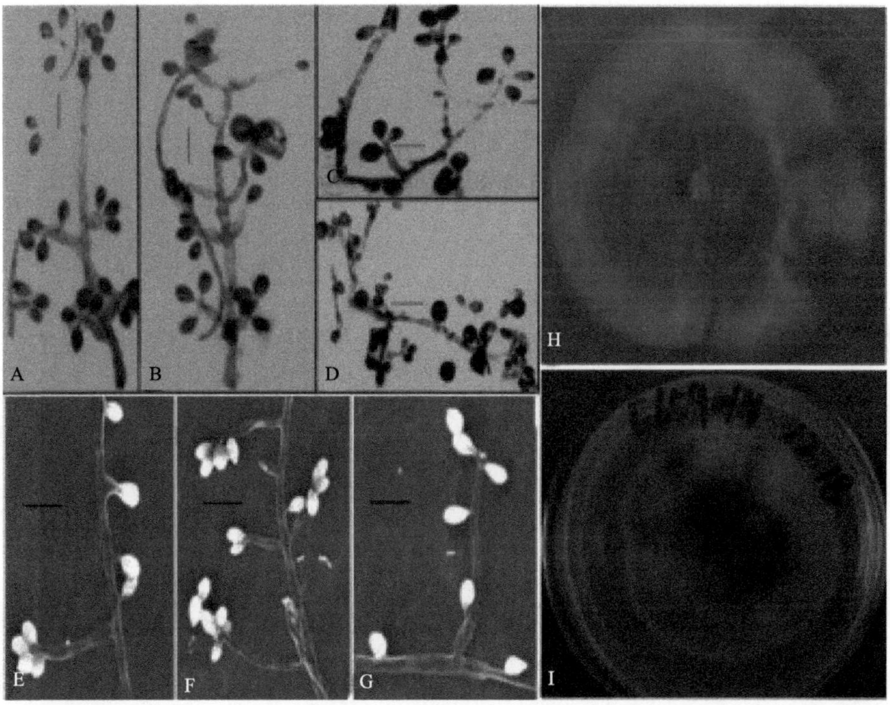

图 116 三亚赛多孢 *Scedosporium sanyaense* Y.F. Han, H. Zheng, Y. Luo, Y.R. Wang & Z.Q. Liang（标本号：GZUIFR EM65901）

A～D. 光学显微镜下的产孢结构；E～G. 电镜下的产孢结构；H、I. PDA 培养基上的菌落。标尺：A～G = 10 μm

在 PDA 培养基上，25℃培养 14 天，菌落直径达 65 mm，细绒状，深褐色、浅褐色至灰白色，边缘不规则，中央稍隆起；背面深褐色、浅褐色到浅黄色，具不明显的放射状沟纹。在查氏培养基上，26℃培养 14 天，菌落直径达 38～50 mm，绒状，深褐色，气生菌丝十分稀疏，边缘不规则；背面深褐色。菌丝无色或近无色，表面光滑，有隔膜，宽 1～4 μm。分生孢子着生于菌丝上或支撑细胞上，顶生或侧生，单生或 1～4 个着生于膨大或不膨大的支撑细胞上，光滑，无色至近无色；倒卵圆形、4～7.5 × 3～4 μm，少数近球形、直径 3～5 μm，大多数单胞，偶见双胞；无间生孢子；厚垣孢子椭圆形，5～7.5 × 4～6.5 μm，或近球形，直径 5～7.5 μm。

研究菌株：海南：三亚市，槟榔村，田埂沟渠污泥，2012 年 12 月，王玉荣，GZUIFR EM65901。

世界分布：中国。

讨论：基于 ITS 序列构建的系统发育树显示，三亚赛多孢与多育赛多孢聚为一亚分支，但两者在形态特征上存在较大差别，多育赛多孢的黑色酵母样菌落，分生孢子卵形，2～5 × 3～12 μm，可与三亚赛多孢的深褐色到浅褐色再到灰白色菌落，分生孢子倒卵圆形，4～7.5 × 3～4 μm 相区别。该菌的典型特征为分生孢子着生于菌丝上或支撑细胞上，顶生或侧生，无色至近无色，倒卵圆形，少数近球形，大多数单胞，偶见双胞；无间生孢子；厚垣孢子椭圆形或近球形（韩燕峰等，2017b）。

作者未观察的种

尖端赛多孢　图 117

Scedosporium apiospermum Sacc. ex Castell. & Chalm., Manual of Tropical Medicine (London): 1122, 1919. Li, Diversity of marine-derived fungi from seaweeds in intertidal zone of Qingdao and sediments of Bohai Sea p. 34, 2013.

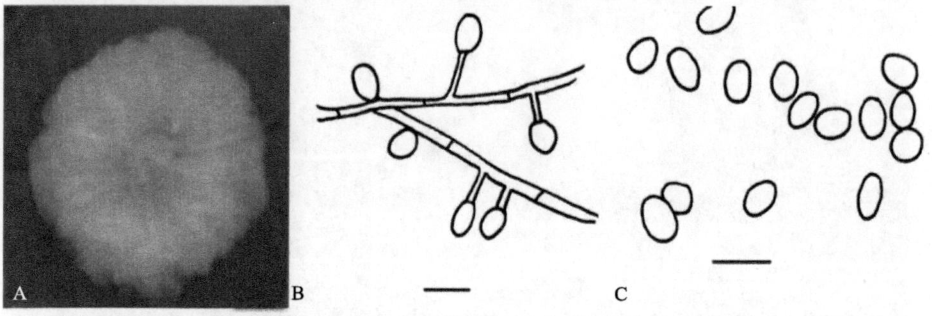

图 117　尖端赛多孢 *Scedosporium apiospermum* Sacc. ex Castell. & Chalm.（引自王澎等，2007）

A. SDA 培养基上的菌落；B. 产孢细胞；C. 分生孢子。标尺：B、C = 10 μm

讨论：李长林（2013）记载，该种在沙氏（SDA）培养基上，菌落生长迅速，深褐色，呈同心圆向四周扩散，可见晶莹透彻的黏液珠。菌丝具隔膜，较粗。分生孢子梗侧生或顶生，分生孢子梗端着生单个分生孢子。分生孢子椭圆形，环痕产孢，有时可产生多个孢子。尖端赛多孢的典型特征为分生孢子呈椭圆形，产孢结构分赛多型和黏束孢型，分生孢子多以单个存在，分生孢子梗细长，分生孢子单生。

世界分布：中国、印度、荷兰、法国。

分生赛多孢　图 118

Scedosporium dehoogii Gilgado, Cano, Gené & Guarro, J. Clin. Microbiol. 46(2): 768, 2008. Feng, The primary study on the diversity of fungi from sediment from Yellow Sea and Yangtze Estuary p. 39, 2012.

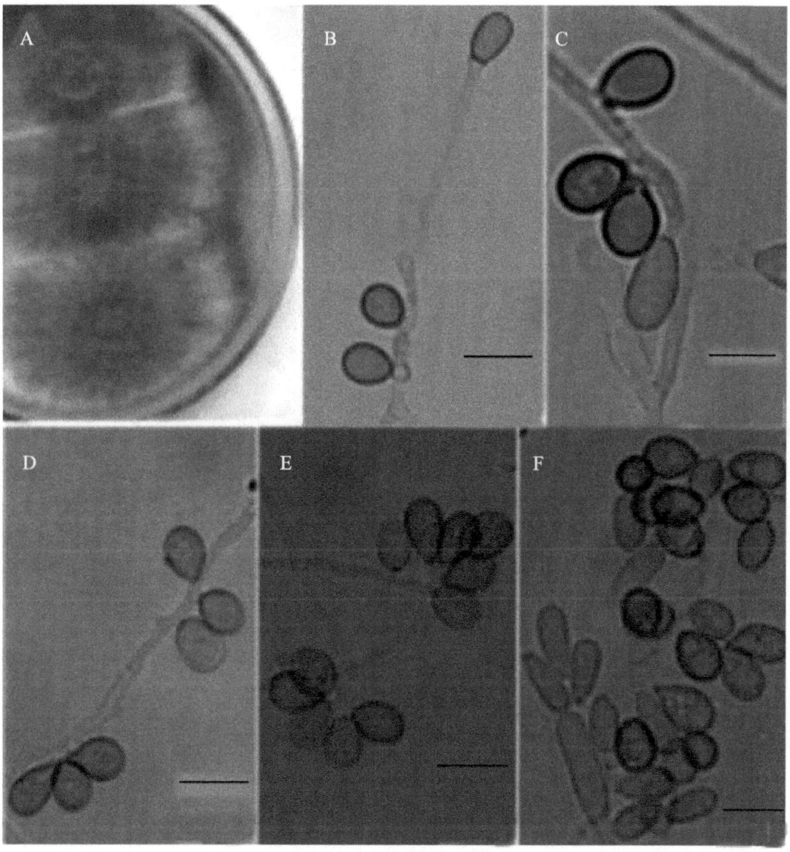

图 118　分生赛多孢 *Scedosporium dehoogii* Gilgado, Cano, Gené & Guarro（引自冯春辉，2012）

A. PDA 培养基上的菌落；B. 无柄的分生孢子；C. 锥瓶状的产孢细胞；D、E. 合生的分生孢子；F. 分生孢子。标尺：B～F = 10 μm

讨论：Gilgado 等（2008）记载，该种在 PDA 培养基上，菌落絮状，灰白色或浅褐色。菌丝体表生或埋生。分生孢子梗圆柱形或锥形，6～50 × 1～1.5 μm。分生孢子淡褐色，倒卵形或椭圆形，5～14 × 3～6 μm。该种产孢细胞圆柱状和锥形，明显区别于属内其他种（Frank *et al.*，2010），曾从土壤和海洋中分离到（Hu *et al.*，2016；冯春辉，2012）。

世界分布：中国、西班牙、荷兰、比利时。

多育赛多孢　图 119

Scedosporium prolificans (Hennebert & B.G. Desai) E. Guého & de Hoog, J. Mycol. Médic. 1(1): 8, 1991. Wang *et al.*, Chinese Journal of Mycology 2(4): 211, 2007.

≡ *Lomentospora prolificans* Hennebert & B.G. Desai, Mycotaxon 1(1): 47, 1974.

讨论：王澎等（2007）记载，该种在 SDA 培养基上，分生孢子以环痕产孢方式产生环痕孢子。环痕孢子与菌丝相连的基部膨大。孢子呈合轴状形成小堆，卵形，3～12 × 2～5 μm，并可见厚壁的球形孢子。多育赛多孢与近似种尖端赛多孢的主要区别为：①尖端赛多孢分赛多型和黏束孢型，尖端赛多孢的有性期是波氏假阿利什菌

Pseudallescheria boydii (Shear) McGinnis, A.A. Padhye & Ajello，而多育赛多孢没有有性阶段。②多育赛多孢的环痕孢子成小堆，而尖端赛多孢的环痕孢子多以单个存在；尖端赛多孢的分生孢子梗细长，而多育赛多孢分生孢子梗基部膨大。③尖端赛多孢菌落是白色至灰色羊毛样，而多育赛多孢菌落可在黑色酵母样菌落与白色短绒样丝状菌落之间转变。

世界分布：中国、德国、西班牙、澳大利亚。

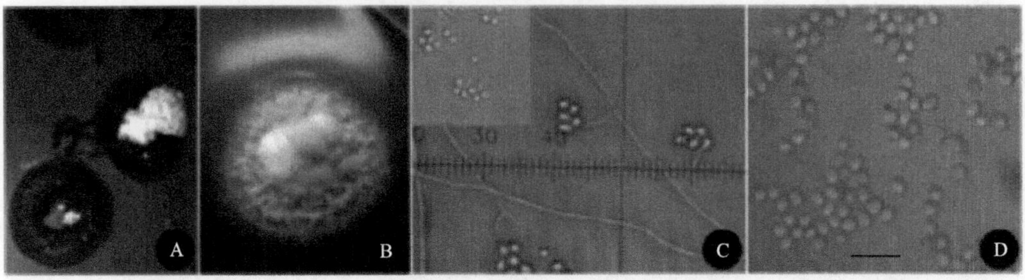

图 119　多育赛多孢 *Scedosporium prolificans* (Hennebert & B.G. Desai) E. Guého & de Hoog（引自王澎等，2007）

A. SDA 培养基上的黑色酵母样菌落；B. 白色短绒状气生菌丝；C. 环痕孢子与菌丝相连的基部膨大；D. 血琼脂上 37℃培养 7 天出现球形厚壁孢子。标尺：D = 10 μm

帚霉属 **Scopulariopsis** Bainier

Bull. Soc. Mycol. Fr. 23(2): 98, 1907

菌落扩散快，绒状，绳索状或似颗粒状，由白色和灰白色变成几种色调的浅黄色、褐色或暗褐色，但从未出现过绿色或黑色。子囊壳，埋生或表生，生长缓慢，散生；球形至近球形，梨形，光滑，具孔，乳头状或具有圆柱形的颈部；包被黑色，由薄壁、略平的角形组织构成。子囊单囊壁，含 8 个分生孢子，近球形、不规则卵圆形或椭圆形，存在时间短。子囊孢子单胞，不对称，短，宽肾形或新月形，幼嫩时似糊精，有或无不明显的芽孔。产孢细胞环痕式，着生于分枝的帚状分生孢子梗上，偶单独着生于营养菌丝或 2～3 个聚生在短柄上，圆柱形，基部通常膨大，后有圆柱形环痕，末端具平且宽的产孢开口，无色，表面光滑或粗糙，分生孢子单胞，无色、淡褐色至褐色，球形至卵圆形，顶部圆形或锥形，具有明显的突起，基部平截，表面光滑或粗糙，壁薄，形成长的向基式链。

模式种：短帚霉 *Scopulariopsis brevicaulis* (Sacc.) Bainier。

中国帚霉属分种检索表

1. 分生孢子球形、近球形或宽椭圆形 …… 2
1. 分生孢子椭圆形、卵形或倒卵形 …… 11
　2. 分生孢子表面光滑 …… 3
　2. 分生孢子多具疣或小刺 …… 9
3. 分生孢子直径< 5 μm …… 4
3. 分生孢子直径> 5 μm …… 7

4. 菌丝和分生孢子梗具疣，浅褐色或褐色，分生孢子直径 3～5 μm ········ **具瘤帚霉 *S. verrucifera***

4. 菌丝和分生孢子梗表面光滑，透明或浅褐色 ··· 5

5. 分生孢子梗简单，单胞或未见 ··· 6

5. 分生孢子梗分隔 ·· **长梗帚霉 *S. longipes***

6. 环痕梗具有两种类型 ·· **番红花帚霉 *S. croci***

6. 环痕梗仅具有一种类型 ··· **云南帚霉 *S. yunnanensis***

7. 分生孢子梗简单，单胞或未见 ··· 8

7. 分生孢子梗分隔 ··· **粗糙帚霉 *S. crassa***

8. 分生孢子透明，顶生厚垣孢子偶现 ·· **短帚霉 *S. brevicaulis***

8. 分生孢子橄榄色或棕色，未见厚垣孢子 ······································· **暗色帚霉 *S. fusca***

9. 环痕梗具有两种类型 ·· **细小帚霉 *S. parvula***

9. 环痕梗仅具有一种类型 ··· 10

10. 分生孢子近球形、卵圆形，浅棕色，光滑，成熟后表面粗糙，基部平截，5～8 × 4～7.5 μm ·· **车叶草帚霉 *S. asperula***

10. 分生孢子浅黄褐色至黄褐色，球形，光滑或粗糙，表面具细密刺状突起，3～5 μm ·· **雪白帚霉 *S. nivea***

11. 在 PDA 培养基上，菌落白色至灰白色 ··· 12

11. 在 PDA 培养基上，菌落灰色、褐色、黑色、浅灰棕色 ·· 14

12. 分生孢子直径< 10 μm ·· 13

12. 分生孢子直径> 10 μm ·· **白帚霉 *S. candida***

13. 菌落灰白色；分生孢子梗顶部帚状分枝；分生孢子浅褐色，顶端钝圆，基部平截，椭圆形，5～7 × 3～4 μm ··· **青霉状帚霉 *S. penicillioides***

13. 菌落白色至浅褐色；分生孢子梗顶部无帚状分枝；分生孢子圆形或卵圆形，基底部有切迹，5～8 × 5～7 μm，表面粗糙 ·· **黄帚霉 *S. flava***

14. 分生孢子卵形 ··· 15

14. 分生孢子倒卵形、近椭圆形、短棒状 ··· 17

15. 分生孢子梗较短，长度< 10 μm ·· 16

15. 分生孢子梗较长，长度> 10 μm ·· **马杜拉帚霉 *S. maduramycosis***

16. 分生孢子梗（产孢细胞）5～10 × 2.5～3.5 μm，分生孢子 3～4 × 2.5～3.5 μm ·· **纸帚霉 *S. chartarum***

16. 分生孢子梗短小，4.5～6.5 × 2～3.5 μm，分生孢子 4～5.5 × 3.5～4.5 μm ·· **黑帚霉 *S. carbonaria***

17. 菌落橄榄黑色，分生孢子倒卵形至短棒状，环痕梗（瓶梗）3.5～20 × 1.5～3 μm ·· **草生帚霉 *S. hibernica***

17. 菌落灰色至褐色，分生孢子倒卵形至近椭圆形，环痕梗 4～10 × 2.5～3.5 μm ·· **布鲁姆帚霉 *S. brumptii***

短帚霉　图 120

Scopulariopsis brevicaulis (Sacc.) Bainier, Bull. Soc. Mycol. Fr. 23(2): 99, 1907. Wang *et al*., Journal of Mountain Agriculture and Biology 32(4): 296, 2013.

在查氏培养基上，25℃培养 14 天，菌落直径 67～72 mm，白色，粉状，平展，稀薄，边缘较规则，表面可见不规则放射状波纹；背面奶酪色。菌丝具隔膜，无色，表面光滑，宽 1.1～3.2 μm；无球拍状菌丝。分生孢子梗由简单到复杂变化，单生或 2～3 个为一组，11～38 × 2～5 μm，或者有分枝，9～15 × 2～5 μm。环痕产孢细胞柱状，长 6.5～

23.8 μm，基部宽 2～4 μm；顶端环痕处宽 1.6～3.2 μm。分生孢子成链，近无色，光滑，大多数近球形，6～9 μm；少数椭圆形，7～9 × 8～10 μm；极少数卵圆形；无间生孢子。具顶生厚垣孢子，单生，偶尔在分生孢子链上形成，光滑，近球形至椭圆形，8～9 × 9～12 μm。

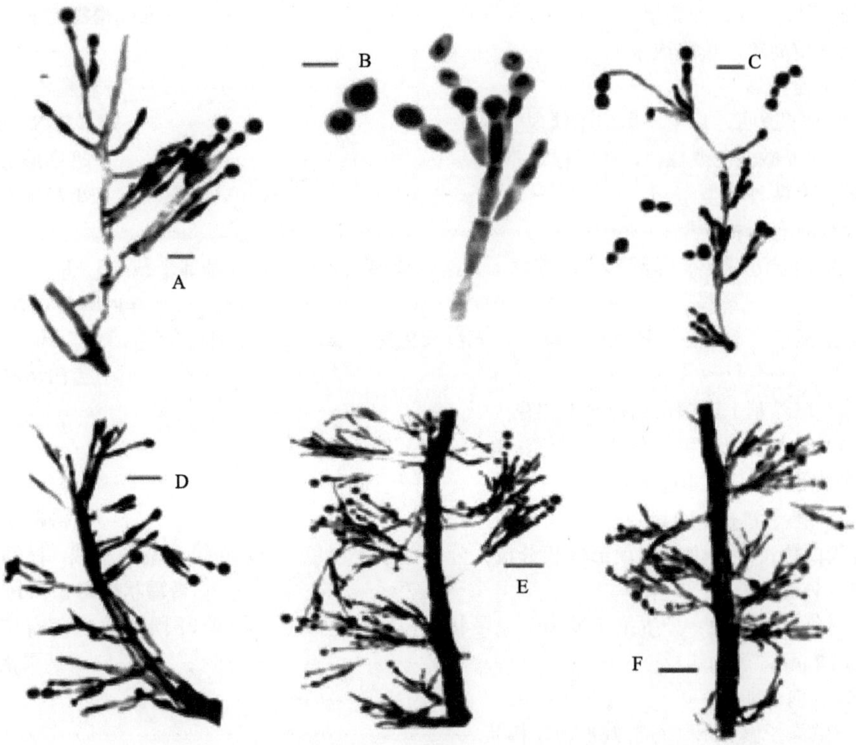

图 120　短帚霉 *Scopulariopsis brevicaulis* (Sacc.) Bainier（标本号：Chengdu A441）

A、D、E. 产孢结构；B. 顶生厚垣孢子；C. 分生孢子；F. 成束的菌丝。标尺：A～F = 10 μm

在查氏培养基上，40℃培养 14 天，菌落直径 40～41 mm，最初白色，然后迅速变成中间榛色，边缘白色，绒状到粉状，边缘不规则；背面浅黄色。菌丝具隔膜，无色，表面光滑，宽 2～4 μm；无球拍状菌丝。分生孢子梗由简单到复杂变化，单生或 2～3 个为一组，11～41 × 2～4 μm；或有分枝，8～14 × 2～4 μm。环痕产孢细胞柱状，长 8～29 μm，基部宽 2.5～4 μm，顶端环痕处宽 2.5～4 μm。分生孢子无色，成链，球形到近球形，4～7 × 5～8 μm，光滑或粗糙。单生的顶生厚垣孢子偶尔在分生孢子链上形成，光滑或粗糙，近球形至椭圆形，6～8 × 8～10 μm。

研究菌株：四川：成都市，室内一分解成粉末的鸡毛掸，2012 年 5 月，梁宗琦，Chengdu A441。

世界分布：中国、印度、西班牙、意大利。

讨论：对比形态特征，短帚霉与云南帚霉（姜成林等，1985）较为相似，二者的分生孢子均为球形至近球形；但本种的分生孢子比云南帚霉的要大（6～9 μm 或 4～7 × 5～8 μm vs. 2.8～3.3 × 2～2.5 μm）（王玉荣等，2013），二者能明显区分。

草生帚霉　图 121

Scopulariopsis hibernica A. Mangan, Trans. Br. Mycol. Soc. 48(4): 617, 1965. Wang *et al*., Journal of Mountain Agriculture and Biology 33(3): 52, 2014.

在 PDA 培养基上，25℃培养 7 天，菌落直径 12～14 mm，平展，中部稍有隆起，近粉状，橄榄黑色，易发生角变现象；背面边缘一轮白色，其余黑色。菌丝无色或近无色，无色或淡褐色，光滑，1.5～3 μm。瓶梗 3.5～20 × 1.5～3 μm，常单生于气生菌丝上，或在梗基上青霉状排列，环痕区长 1～2 μm。梗基显著膨大，6～8.5 × 2～5 μm。分生孢子幼时光滑，随着培养时间的延长会产生疣突，倒卵形至短棒状，4.5～6 × 2.5～4 μm，链生。

研究菌株：陕西：合阳县，葱根际土，2013 年 12 月，韩燕峰，E71702M。

世界分布：中国、爱尔兰。

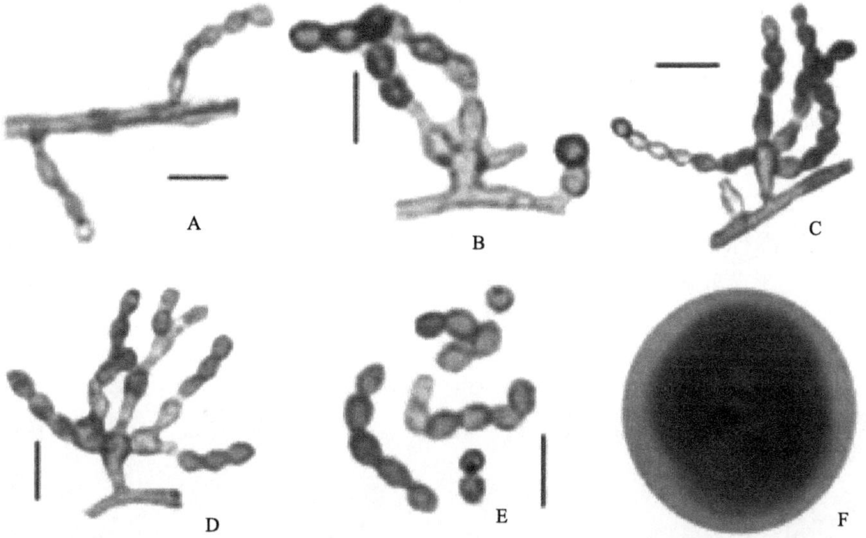

图 121　草生帚霉 *Scopulariopsis hibernica* A. Mangan（标本号：E71702M）

A～D. 产孢结构；E. 分生孢子；F. PDA 培养基上的菌落。标尺：A～E = 10 μm

讨论：该种与车叶草帚霉、番红花帚霉形态相似，但车叶草帚霉的环痕梗不膨大，番红花帚霉的环痕梗中部膨大或基部稍膨大，而该种的环痕梗基部膨大（王垚等，2014）。这一特征可与相似种明显区分。

作者未观察的种

车叶草帚霉　图 122

Scopulariopsis asperula (Sacc.) S. Hughes, Can. J. Bot. 36: 803, 1958. Zhang, The taxonomy of *Cephalotrichum* and *Scopulariopsis* p. 58, 2020.

≡ *Torula asperula* Sacc., Michelia 2(8): 560, 1882.

= *Scopulariopsis repens* Bainier, Bull. Soc. Mycol. Fr. 23: 125, 1907.

≡ *Penicillium repens* (Bainier) Biourge, La Cellule 33: 225, 1923.

= *Monilia arnoldii* L. Mangin & Pat., Bull. Soc. Mycol. Fr. 24: 164, 1908.

≡ *Scopulariopsis arnoldii* (L. Mangin & Pat.) Vuill., Bull. Soc. Mycol. Fr. 27: 148, 1911.

= *Acaulium nigrum* Sopp, Skrifter udgivne af Videnskabs-Selskabet i Christiania. Mathematisk-Naturvidenskabelig Klasse 11: 47, 1912.

≡ *Penicillium nigrum* (Sopp) Biourge, La Cellule 33: 104, 1923.

≡ *Microascus niger* (Sopp) Curzi, Bolletino della Stazione di Patologia Vegetale Roma 11: 8, 1931.

= *Scopulariopsis ivorensis* H. Boucher, Bulletin de la Société de Pathologie Exotique 11: 312, 1918.

= *Torula bestae* Pollacci, Riv. Biol. 4: 317, 1922.

≡ *Phaeoscopulariopsis bestae* (Pollacci) M. Ota, Jap. J. Derm. Urol. 28: 405, 1928.

≡ *Scopulariopsis bestae* (Pollacci) Nann., Trattato di Micopatologia Umana (Firenze) 4: 254, 1934.

= *Scopulariopsis fusca* Zach, Öst. Bot. Z. 83: 174, 1934.

= *Scopulariopsis roseola* N. Inagaki, Transactions of the Mycological Society of Japan 4: 1, 1962.

讨论：张欣（2020）记载，该菌在 OA 培养基上，菌落生长较快，菌落平展，初粉状，黄褐色或深棕色。菌丝表生或埋生。分生孢子梗无或分化不明显。产孢细胞（环痕梗）主要顶生于菌丝上，浅棕色，圆柱状，8～25 × 2.5～4.5 μm。分生孢子近球形、卵圆形，浅棕色，光滑，成熟后表面粗糙，壁厚，基部平截，5～8 × 4～7.5 μm，单生或短链生。

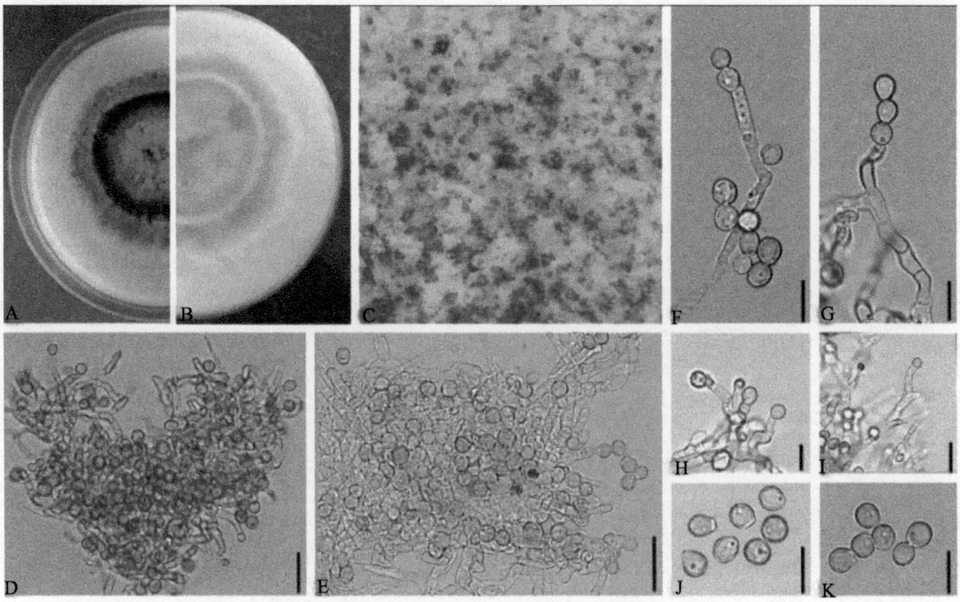

图 122　车叶草帚霉 *Scopulariopsis asperula* (Sacc.) S. Hughes（引自张欣，2020）

A、B. OA 培养基上的菌落；C. 菌丝；D～I. 产孢细胞；J、K. 分生孢子。标尺：D～K = 10 μm

车叶草帚霉多见于土壤、纸张和奶酪等基质中，也有分离自空气的记录。形态上，该种与长梗帚霉和番红花帚霉相似，但长梗帚霉分生孢子梗具隔膜，为多胞，环痕梗单生或 2～3 个簇生于产孢梗顶端；番红花帚霉的环痕梗膨大；本种的分生孢子梗为单胞，简单，环痕梗不膨大，能与其相似种区分。

世界分布：中国、意大利、法国、澳大利亚。

布鲁姆帚霉　图 123

Scopulariopsis brumptii Salv.-Duval, Thèse Fac. Pharm. Paris 23: 58, 1935. Geng, A survey on soil dematiaceous hyphomycetes at species level from Tibetan Plateau p. 88, 2008.

讨论：耿月华（2008）记载，该菌在 PDA 培养基上，菌落生长缓慢，灰色至褐色。环痕梗时常单生，从菌丝侧面生出，形成菌索，有时 2～3 个簇生，安瓿瓶状或烧瓶状，淡灰色或橄榄褐色，4～10 × 2.5～3.5 μm。分生孢子淡褐色，聚集时黑色或暗褐色，倒卵形至近椭圆形，基部平截，光滑或有时产生疣突，4～6 × 3.5～4.5 μm。

该种与其他种的区别在于其分生孢子明显小于其他种，光滑或疣突。该种经常分离自空气、土壤、水稻、蓖麻及麦秆。

世界分布：中国、印度、巴基斯坦、美国。

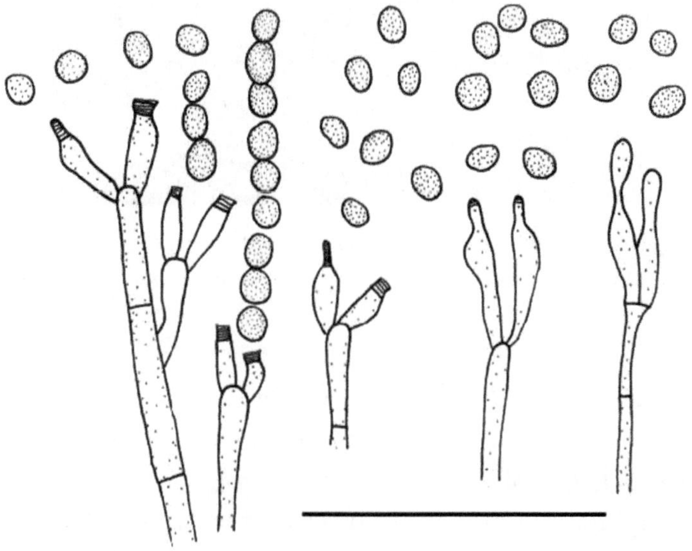

图 123　布鲁姆帚霉 *Scopulariopsis brumptii* Salv.-Duval（引自耿月华，2008）
分生孢子梗和分生孢子。标尺：25 μm

白帚霉　图 124

Scopulariopsis candida Vuill., Bull. Soc. Mycol. Fr. 27(2): 143, 1911. Yu *et al.*, Journal of Clinical Dermatology 9: 565, 2008.

讨论：余进等（2008）记载，该菌在 PDA 培养基上，菌落中等扩展，白色，有皱褶。产孢细胞位于气生菌丝或分生孢子梗上，圆柱状。分生孢子无色，呈链状，卵圆形或圆柱形，11～22 × 2～4 μm。

白帚霉、青霉状帚霉和马杜拉帚霉都能产生椭圆形、卵形或倒卵形的分生孢子，但后两者的环痕明显，能与白帚霉明显区分。

世界分布：中国、法国、意大利、荷兰、德国、美国、加拿大。

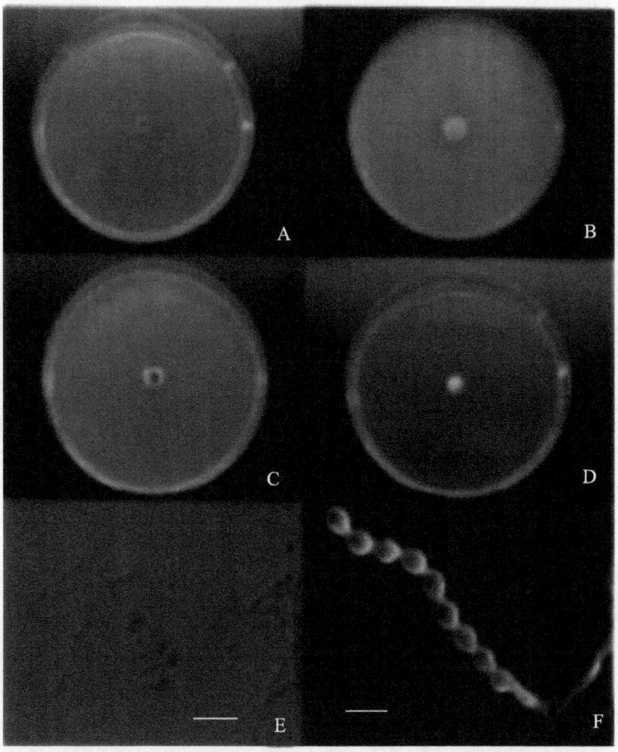

图 124 白帚霉 *Scopulariopsis candida* Vuill.（引自余进等，2008）

A～D. MEA、OA、PDA 和 SDA 培养基上的菌落；E、F. 产孢结构。标尺：E = 20 μm；F = 10 μm

黑帚霉 图 125

Scopulariopsis carbonaria F.J. Morton & G. Sm., Mycol. Pap. 86: 59, 1963. Liu, Taxonomic studies on soil dematiaceous Hyphomycetes from warm temperate zone of Eastern China p. 138, 2004.

讨论：刘会梅（2004）记载，该菌在 PDA 培养基上，菌丝体表生或埋生。分生孢子梗短小，淡褐色或褐色，4.5～6.5 × 2～3.5 μm。环痕梗常单生于气生菌丝、菌索或分生孢子梗上，也有时 2～4 个簇生，瓶形，基部稍膨大，宽 2.5～4 μm。分生孢子链生，卵形，基部平截，光滑或粗糙，成团时黑色或暗褐色，4～5.5 × 3.5～4.5 μm。

Morton 和 Smith（1963）根据环痕梗的特点将该属内的真菌分为四组，并将其与多拉霉属的环痕梗进行了比较，认为多拉霉属内的环痕梗很少圆柱形，一般基部膨大，然后骤窄形成环痕带，常呈锥形。该种由 Morton 和 Smith（1963）建立，分离自巴拿马的土壤、茶树的根际土壤及尘土中。

世界分布：中国、印度、日本、巴拿马。

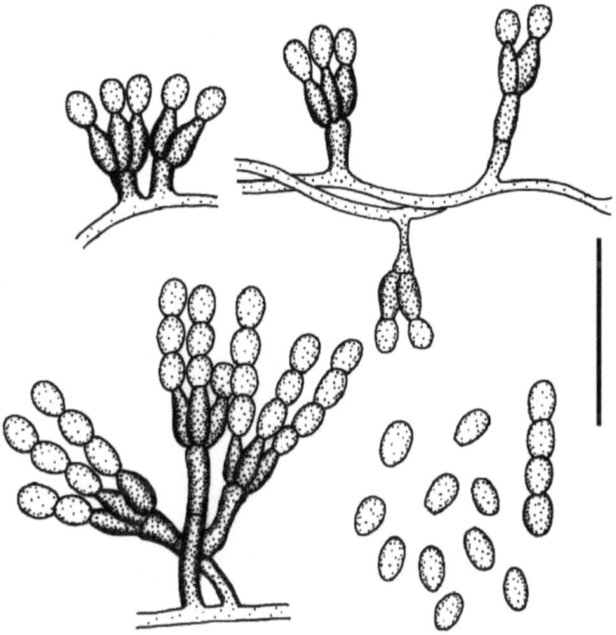

图 125 黑帚霉 *Scopulariopsis carbonaria* F.J. Morton & G. Sm.（引自刘会梅，2004）
分生孢子梗和分生孢子。标尺：25 μm

纸帚霉　图 126

Scopulariopsis chartarum (G. Sm.) F.J. Morton & G. Sm., Mycol. Pap. 86: 64, 1963. Yu *et al.*, Journal of Clinical Dermatology 9: 565, 2008.

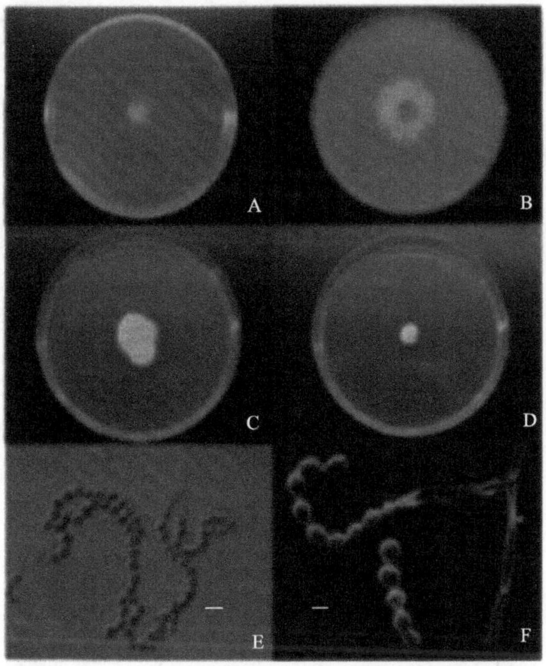

图 126　纸帚霉 *Scopulariopsis chartarum* (G. Sm.) F.J. Morton & G. Sm.（引自余进等，2008）
A～D. MEA、OA、PDA 和 SDA 培养基上的菌落；E、F. 产孢结构。标尺：E = 20 μm；F = 10 μm

讨论：余进等（2008）记载，该菌在 PDA 培养基上菌落局限，生长慢，灰色至灰黑色，短绒毛。产孢细胞多单个存在，也可 2～3 个成簇，烧瓶状，5～10 × 2.5～3.5 μm，环痕区较窄。分生孢子为深棕色，卵圆形，基底部有切痕，3～4 × 2.5～3.5 μm。

该种最初分离自腐烂的墙纸，但是紧接着又在松树林、雪松林、柚木林、小麦田等土壤里分离得到。

世界分布：中国、英国。

粗糙帚霉　图 127

Scopulariopsis crassa Z.F. Zhang, F. Liu & L. Cai, Persoonia 39: 25, 2017b.

讨论：Zhang 等（2017b）记载，该菌在 PDA 培养基上，毡状，平展，浅褐色。分生孢子梗着生于匍匐菌丝或聚集在分生孢子体上，分枝，无色或淡棕色，2～3.5 μm。产孢细胞着生于气生菌丝或在分生孢子梗上，1～3 个轮生，具环痕，无色，圆柱状或有时不规则，略弯曲，偶具隔膜，15～43 × 3～6 μm。分生孢子链生，光滑或有细小疣状物，球形或近球形，5～10.5 × 5～8.5 μm，基部平截。

基于多基因系统发育分析，粗糙帚霉与车叶草帚霉、白帚霉具有较近的关系。但对比形态特征，粗糙帚霉可通过产生较长的产孢细胞（粗糙帚霉 15～43 μm，车叶草帚霉 8～25 μm，白帚霉 11～22 μm）与上述两个种相区别。

世界分布：中国。

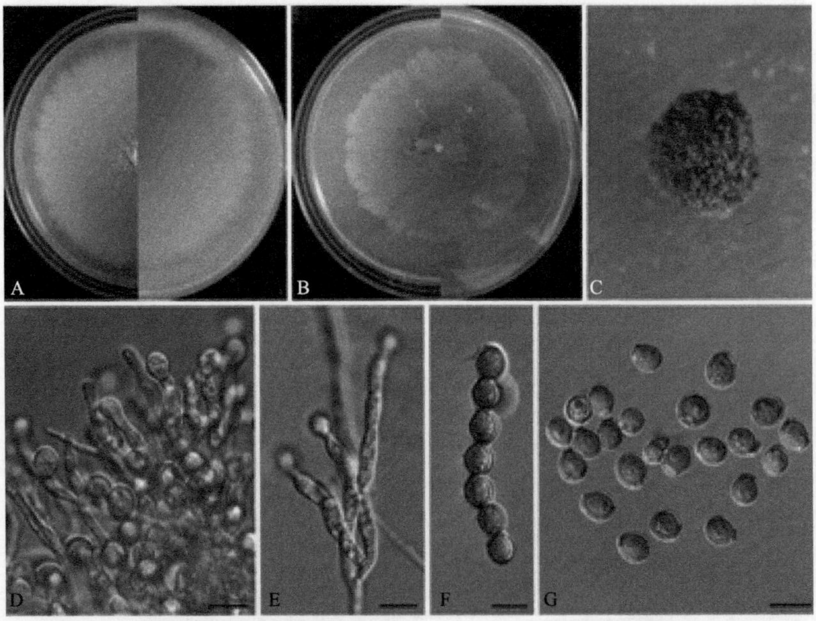

图 127　粗糙帚霉 *Scopulariopsis crassa* Z.F. Zhang, F. Liu & L. Cai（引自 Zhang *et al.*，2017b）

A、B. PDA 和 SNA 培养基上的菌落；C. 载孢体；D、E. 分生孢子梗；F、G. 分生孢子。标尺：D～G = 10 μm

番红花帚霉　图 128

Scopulariopsis croci J.F.H. Beyma, Antonie van Leeuwenhoek 10: 52, 1945. Liu, Taxonomic studies on soil dematiaceous Hyphomycetes from warm temperate zone of

Eastern China p. 139, 2004.

讨论：刘会梅（2004）记载，该菌在 PDA 培养基上，菌丝体表生或埋生。分生孢子梗缺或分化很不明显。环痕梗常侧生于菌丝，单生或很少情况下 2～3 个簇生在短的分生孢子梗上。环痕梗有两种类型，一种为基部窄，至中部膨大，继之骤然变窄，至环痕带宽 1～1.5 μm；第二种为基部稍膨大，至顶部逐渐变窄；一般长 4～10 μm。分生孢子球形或卵圆形，淡棕色，聚集在一起常呈橄榄色，4～5 × 3～4 μm。

该种与纸帚霉在分生孢子的形态及大小都非常近似，据 Morton 和 Smith（1963）描述，其主要区别在于同样培养条件下前者生长速度是后者的 2 倍。

世界分布：中国、荷兰。

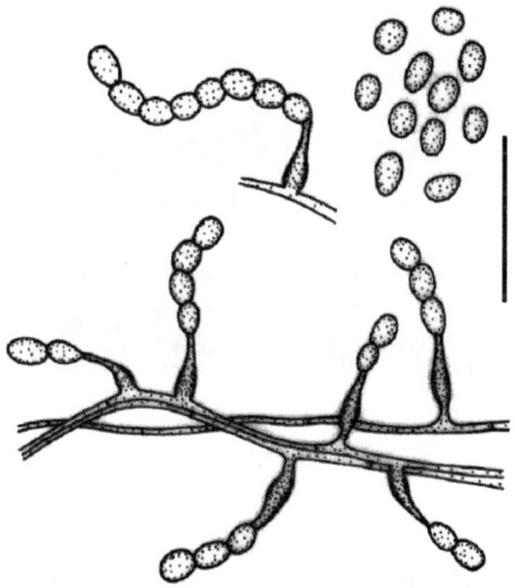

图 128　番红花帚霉 *Scopulariopsis croci* J.F.H. Beyma（引自刘会梅，2004）
分生孢子梗和分生孢子。标尺：25 μm

黄帚霉　图 129

Scopulariopsis flava (Sopp) F.J. Morton & G. Sm., Mycol. Pap. 86: 43, 1963. Yu *et al.*, Journal of Clinical Dermatology 9: 565, 2008.

讨论：余进等（2008）记载，该菌在 PDA 培养基上，菌落扩展，生长快，有皱褶，白色至浅褐色。产孢细胞单个或 2～3 个成簇，圆柱状，基底部稍膨大，5～25 × 2.5～3.5 μm，环痕不明显。分生孢子圆形或卵圆形，基底部有切迹，5～8 × 5～7 μm，表面粗糙。

黄帚霉与帚霉属其他种最主要的区别为分生孢子表面粗糙，菌落呈白色至浅褐色，生长速度快，分生孢子无色，圆形或卵圆形。

世界分布：中国、英国。

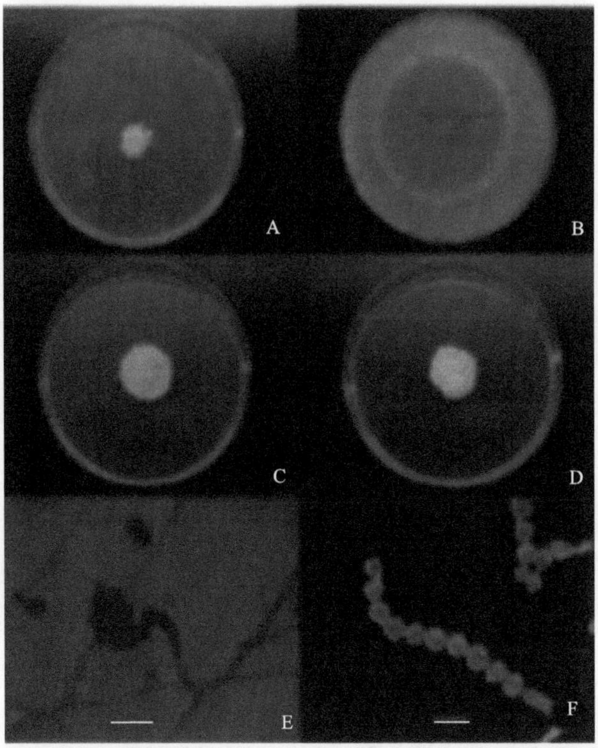

图 129　黄帚霉 *Scopulariopsis flava* (Sopp) F.J. Morton & G. Sm.（引自余进等，2008）

A～D. MEA、OA、PDA 和 SDA 培养基上的菌落；E、F. 产孢结构和分生孢子。标尺：E = 20 μm；F = 10 μm

暗色帚霉　图 130

Scopulariopsis fusca Zach, Öst. Bot. Z. 83: 174, 1934. Geng, A survey on soil dematiaceous hyphomycetes at species level from Tibetan Plateau p. 89, 2008.

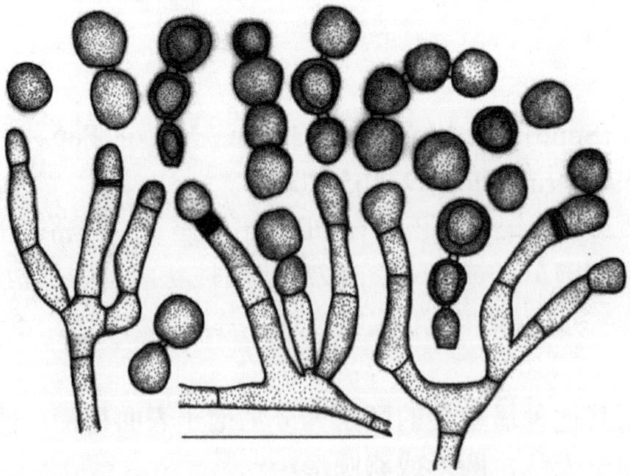

图 130　暗色帚霉 *Scopulariopsis fusca* Zach（引自耿月华，2008）

分生孢子梗和分生孢子。标尺：25 μm

讨论：耿月华（2008）记载，该菌在 PDA 培养基上，菌落生长较快，平展，近粉状，黄褐色或黑褐色，菌丝表生，部分埋生。产孢梗分化不明显，偶见产孢梗长度达

7.3～13 μm，通常侧生于菌丝或菌索上，单生或者 2～10 个簇生，环痕梗基部膨大，至顶部渐细，8～27 × 2.5～4.0 μm，基部膨大部分宽达 4～6 μm。分生孢子球形或者近球形，橄榄色或棕色，光滑，常链生，基部平截，5～8 × 5～7 μm。

暗色帚霉多见于土壤、纸张和奶酪等基质上，也有分离自空气的记录。本种真菌的环痕梗比较清晰，基部膨大，分枝较多。

世界分布：中国、法国。

长梗帚霉 图 131

Scopulariopsis longipes H.Q. Pan & T.Y. Zhang, Mycosystema 33(1): 2, 2014.

讨论：Pan 等（2014）记载，该菌在 PDA 培养基上，菌丝生长较快，平展，灰白色，表面放射状。菌丝体部分表生，部分埋生。产孢梗着生于气生或埋生菌丝上，直立或弯曲，无色或极浅褐色，长 15～44 μm。环痕梗单生或 2～3 个簇生于产孢梗顶端，光滑，近无色，3.5～8 × 1.5～2.5 μm，环痕区域不明显。分生孢子链生，褐色，球形或近球形，3.5～5 μm。

长梗帚霉具有相对较小的分生孢子（3.5～5 μm）可以与近似种暗色帚霉较大的分生孢子（5～8 μm）相区别。

世界分布：中国。

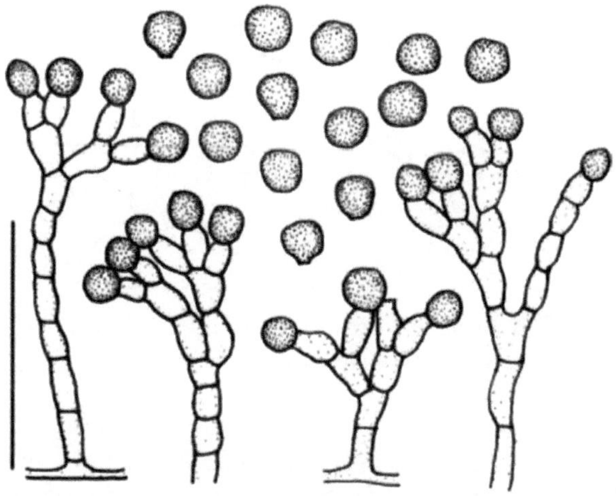

图 131 长梗帚霉 *Scopulariopsis longipes* H.Q. Pan & T.Y. Zhang（引自 Pan *et al.*，2014）

分生孢子梗和分生孢子。标尺：25 μm

马杜拉帚霉

Scopulariopsis maduramycosis Q.T. Chen, Chinese Medical Journal 99(5): 378, 1986.

讨论：该菌在 PDA 培养基上，菌落生长迅速。菌落呈浅灰棕色，平坦，短发状或偶呈粉状，具少许辐射状沟纹。菌丝上侧生单生的分生孢子梗（环孢子梗），通常平行着生或着生于菌丝短侧分枝顶端。环孢子梗长 15～20 μm，基部窄，顶部膨大。分生孢子着生于环痕梗末端，环痕梗末端具环痕，无色或浅棕色，末端有一个或一些环痕。分生孢子卵圆形，6.5～8 × 4.6～6 μm，无色至浅棕色，成串或链。

世界分布：中国。

雪白帚霉　图 132

Scopulariopsis nivea Demelius, Verh. Kaiserl.-Königl. Zool.-Bot. Ges. Wien 66: 490, 1916. Pan *et al*., Mycosystema 33(1): 2, 2014.

讨论：Pan 等（2014）记载，该菌在 PDA 培养基上，菌落生长慢，平展，白色至黑褐色。菌丝体表生或埋生。环痕梗通常单生于菌丝的两侧或菌索上，但更常见的是小段菌丝或菌索上 2～3 个环痕梗簇生，长 3～6.5 μm，膨大部分宽 2～3.5 μm，光滑，无色至浅棕色。分生孢子浅黄褐色至黄褐色，球形，单生或短链，光滑或粗糙，表面具细密刺状突起，3～5 μm。

雪白帚霉以分生孢子具疣状物或小刺，菌丝和分生孢子梗表面光滑，无色或浅棕色等特征与帚霉属其他种相区别。

世界分布：中国、西班牙、澳大利亚。

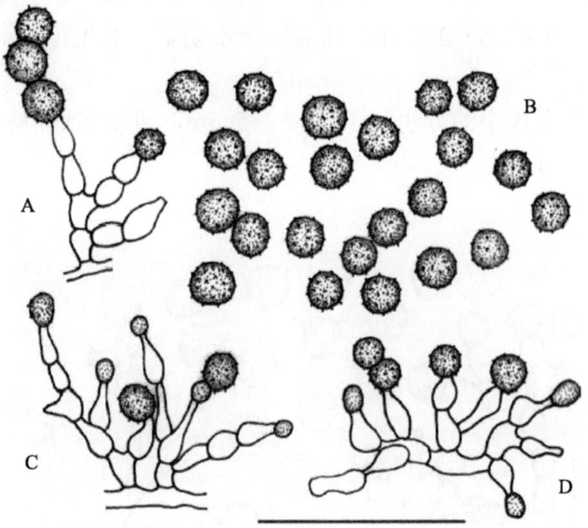

图 132　雪白帚霉 *Scopulariopsis nivea* Demelius（引自 Pan *et al.*，2014）

A、C、D. 分生孢子梗和分生孢子；B. 分生孢子。标尺：25 μm

细小帚霉　图 133

Scopulariopsis parvula F.J. Morton & G. Sm., Mycol. Pap. 86: 65, 1963. Geng, A survey on soil dematiaceous hyphomycetes at species level from Tibetan Plateau p. 91, 2008.

讨论：耿月华（2008）记载，该菌在 PDA 培养基上，菌落生长较慢，平展，近粉状，黄褐色或淡褐色。菌丝体表生，部分埋生。产孢梗分化不明显，偶见产孢梗长度达 7.3～13 μm。环痕梗通常侧生于菌丝或者菌索上，多单生，有时可见 2～3 个簇生，环痕梗通常有两种类型：一种是环痕梗基部膨大，至顶部渐细，长 4～13 μm，环痕区较长，为 3～7 μm；另一种为基部窄，中部膨大，至顶部渐细或者顶部膨大，淡褐色，无隔膜，长 6～13 μm，环痕区部分长 2～8 μm。分生孢子球形或者近球形，粗糙，偶见光滑，常链生，褐色，2.5～3.7 μm。

按照环痕梗的类型，细小帚霉应该属于布鲁姆帚霉组内。在该组真菌中，本种在分生孢子形态、大小方面与番红花帚霉有些相似，但后者孢子表面光滑，且明显偏大。

世界分布：中国、德国、加拿大。

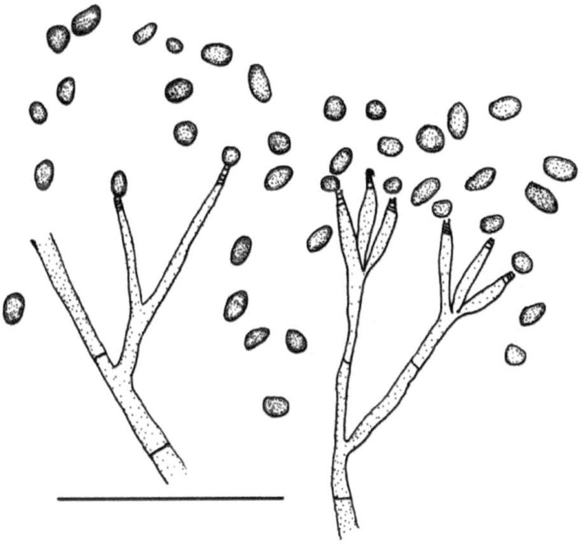

图 133　细小帚霉 *Scopulariopsis parvula* F.J. Morton & G. Sm.（引自耿月华，2008）

分生孢子梗和分生孢子。标尺：25 μm

青霉状帚霉　图 134

Scopulariopsis penicillioides H.Q. Pan, Y.L. Jiang, H.F. Wang & T.Y. Zhang, Mycosystema 33(1): 3, 2014.

讨论：Pan 等（2014）记载，该菌在 PDA 培养基上，菌落生长缓慢。菌落毛絮状，灰白色，表面褶皱。菌丝体部分表生，部分埋生。产孢梗顶生或侧生于菌丝上，顶端帚状分枝，浅褐色，光滑，环痕梗长 5～9.5 μm，基部稍膨大，宽 2.5～3.5 μm，环痕区域 1～2.5 × 1.5～3.5 μm。分生孢子浅褐色，顶端钝圆，基部平截，椭圆形，5～7 × 3～4 μm。

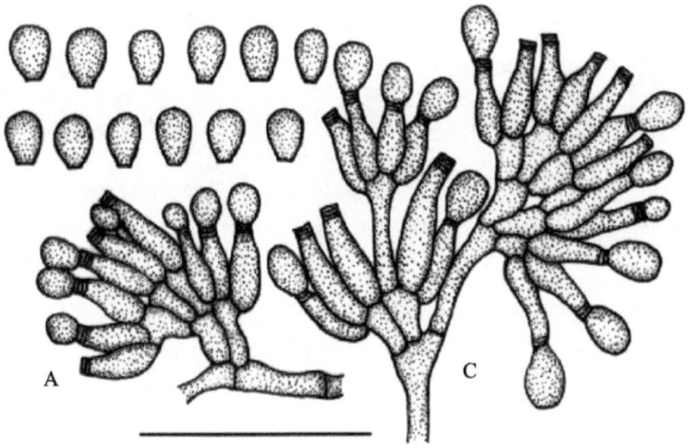

图 134　青霉状帚霉 *Scopulariopsis penicillioides* H.Q. Pan, Y.L. Jiang, H.F. Wang & T.Y. Zhang（引自 Pan *et al.*，2014）

A～C. 分生孢子梗和分生孢子。标尺：25 μm

在帚霉属中，具有帚状的分生孢子梗的物种有黑帚霉、纸帚霉和土生帚霉。青霉状帚霉与它们相区别的特征为其环痕梗具有不明显膨胀的基部和较大的分生孢子（5～7 × 3～4 μm）。黑帚霉菌落黑色，具有粗糙且厚壁的孢子；纸帚霉分生孢子卵圆形，具有烧瓶状环痕梗；而土生帚霉孢子椭圆形，具有短柱状环痕梗，且它们的分生孢子均比青霉状帚霉的小。

世界分布：中国。

具瘤帚霉 图 135

Scopulariopsis verrucifera H.F. Wang & T.Y. Zhang, Mycosystema 33(1): 4, 2014.

讨论：Pan 等（2014）记载，该菌在 PDA 培养基上，菌落平展，粉状，灰褐色至黑褐色。菌丝体大部分表生，少部分埋生。菌丝淡褐色至中度褐色，表面布满瘤突，形成菌丝束或菌索。分生孢子梗缺或分化不明显。环痕梗常侧生于菌丝或菌丝束上，偶见直接生于菌丝顶端，单生或 2～3 个簇生在短的分生孢子梗上。环痕梗基部窄至中部膨大，继之骤然变窄，膨大处宽 1.9～2.5 μm，产孢梗和环痕梗表面布满瘤突，黄褐色至中褐色，无隔膜。分生孢子球形或近球形，黄色至中褐色，粗糙，链生，3～5 μm。

具瘤帚霉与番红花帚霉在形态上具有相似性，均具有膨胀的环痕梗和球形至近球形的分生孢子。二者的区别为：在 25℃条件下培养两周后，具瘤帚霉的生长速度明显快于番红花帚霉；具瘤帚霉的菌丝、环痕梗和分生孢子表面具异物，而番红花帚霉的菌丝、环痕梗和分生孢子表面光滑。

世界分布：中国。

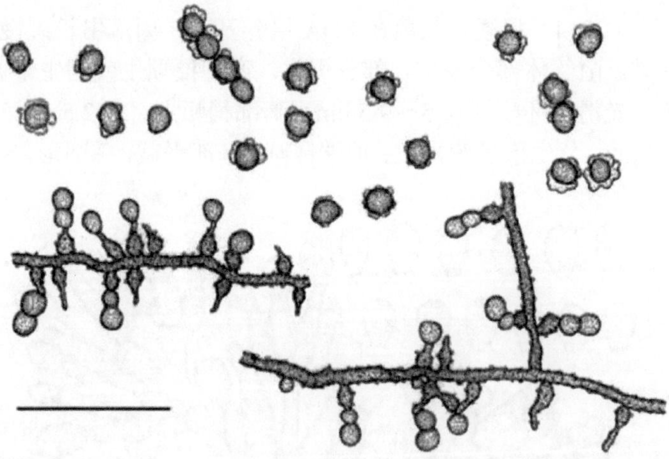

图 135　具瘤帚霉 *Scopulariopsis verrucifera* H.F. Wang & T.Y. Zhang（引自 Pan *et al.*，2014）
分生孢子梗和分生孢子。标尺：25 μm

云南帚霉 图 136

Scopulariopsis yunnanensis Q.T. Chen & C.L. Jiang, Acta Mycol. Sin. 4(3): 167, 1985.

讨论：姜成林等（1985）记载，该菌在 PDA 培养基上，菌落局限，绒状，中央稍呈绳束状，有稀疏的放射褶沟，浅白肉色至浅灰褐色。菌丝表生或部分埋生于基质中，直径 1～2.5 μm；产孢结构为不分枝单生的环痕瓶梗或为少数几个顶生聚成帚状的环痕瓶梗，2～8 × 1.5～2.5 μm。环痕瓶梗瓶形或柱形，可层生，基部膨大，直径 1.5～2.5 μm。老龄分生孢子大多数球形，直径 2～3 μm；少数为卵形，基部平截，2.8～3.3 × 2～2.5 μm，稍粗糙，近无色或微黄褐色，顶生，向基式生长成串，具 1～3 个侧生芽孔，由芽孔产生短侧生分生孢子链，以其末端和亲本分生孢子链相连接。

在分生孢子体形小且主要为球形的特征上，它与细小帚霉有些近似，但后者的环痕瓶梗不层生，无柱形环痕瓶梗，分生孢子体形较大（直径 2.5～3.7 μm）并且粗糙，不产生侧生芽孔，分生孢子链无接合的特征，故与本种有明显的区别。

世界分布：中国。

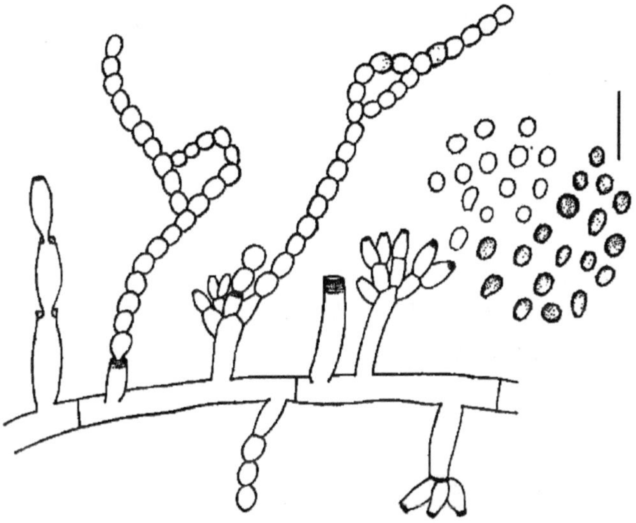

图 136 云南帚霉 *Scopulariopsis yunnanensis* Q.T. Chen & C.L. Jiang（引自姜成林等，1985）
分生孢子梗和分生孢子。标尺：5 μm

嗜热疣霉属 Thermothelomyces Y. Marín, Stchigel, Guarro & Cano Mycologia 107(3): 630, 2015

该属形态特征与毁丝霉属类似，主要根据多基因系统发育研究所建立，除建立该属后发表的新种外，其余成员均来自毁丝霉属，同时其研究应用现状可参考毁丝霉属。

模式种：嗜热疣霉 *Thermothelomyces thermophilus* (Apinis) Y. Marín, Stchigel, Guarro & Cano。

中国嗜热疣霉属分种检索表

1. 有性型不存在 ……………………………………………………………… 2
1. 有性型存在 ………………………………………………………………… 4
  2. 分生孢子具有明显瘤突、小刺 ……………………… **黄褐嗜热疣霉 *T. hinnuleus***
  2. 分生孢子光滑 …………………………………………………………… 3
3. 菌落淡棕色或红棕色，分生孢子卵圆形到椭圆形，3.2～5.4 × 2.2～3.2 μm ……………
   …………………………………………………………… **嗜热疣霉 *T. thermophilus***
3. 菌落黄褐色至灰褐色，分生孢子水滴状，4.8～7.2 × 3.0～5.0 μm ……………

……………………………………………………………………………………………水滴状嗜热疣霉 ***T. guttulatus***

4\. 分生孢子卵圆形、梨形，5.5～10.5 × 3.5～7.5 μm………………………弗格斯嗜热疣霉 ***T. fergusii***

4\. 分生孢子卵圆形到椭圆形，3.2～8.6 × 2.2～4.3 μm………………异宗嗜热疣霉 ***T. heterothallicus***

弗格斯嗜热疣霉 图 137

Thermothelomyces fergusii X. Wei Wang & Houbraken, Stud. Mycol. 101: 205, 2022.

= *Thielavia thermophila* Fergus & Sinden, Canad. J. Bot. 47: 1635, 1969.

= *Chrysosporium fergusii* Klopotek, Arch. Mikrobiol. 98: 366, 1974.

= *Myceliophthora fergusii* (Klopotek) Oorschot, Persoonia 9(3): 406, 1977.

= *Crassicarpon thermophilum* (Fergus & Sinden) Y. Marín, Stchigel, Guarro & Cano, Mycologia 107: 630, 2015.

在 PDA 培养基上，45～50℃培养，菌落初稀疏柔毛状伸长，白色，渐铺展生长，桃红奶油色，培养基中出现不同程度亮紫色（分泌紫红色素），后变黄至黄棕色，不均匀颗粒粉末状，以明显分散的小簇状分布，簇堆常黏至培养皿边缘和皿盖。菌丝具分枝和隔膜，直径 1.5～7.5 μm，无分化的分生孢子梗。产孢细胞多不分化，菌丝直接产胞，少数分化成小柄，偶见会形成坛形隆起，顶端着生多个分生孢子。分生孢子无色，壁薄，光滑，成熟时黄褐色，卵圆形、梨形，5.5～10.5 × 3.5～7.5 μm，顶生或侧生，单生，偶见 2 个或 3 个链生，易从菌丝上脱离，脱离后继续膨大成球形或亚球形，基部窄，明显平截。

研究菌株：陕西：渭南市，农田土，2013 年 4 月 10 日，韩燕峰，E201008H；山东：莱芜，2003 年 7 月，王冬梅。

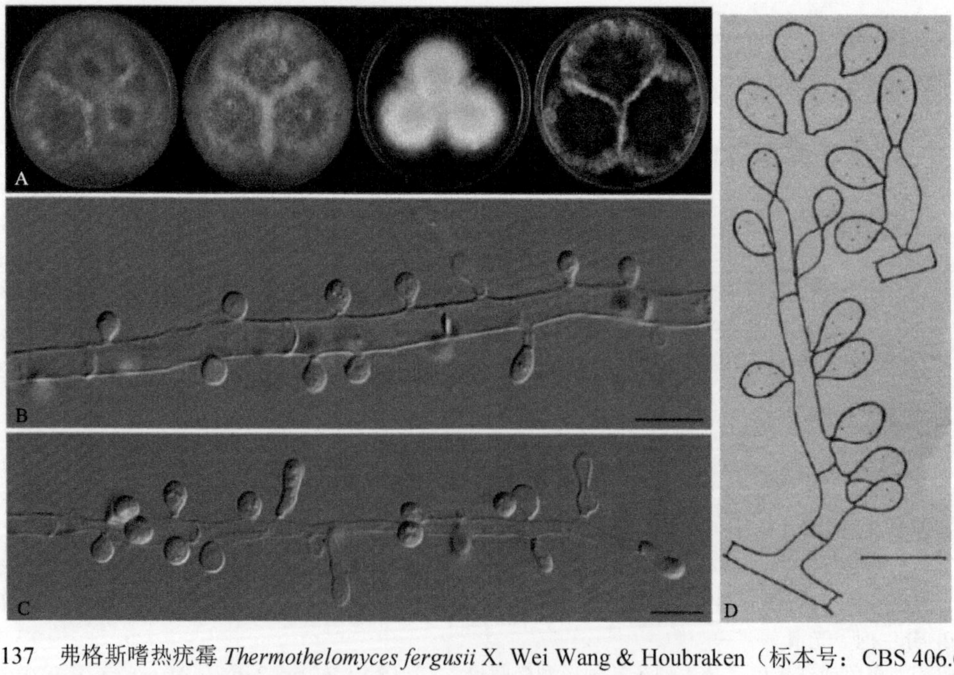

图 137 弗格斯嗜热疣霉 *Thermothelomyces fergusii* X. Wei Wang & Houbraken（标本号：CBS 406.69）

A. OA、CMA、MEA 和 PDA 培养基上的菌落；B～D. 产孢结构。标尺：B～D = 10 μm（A～C 引自 Wang *et al.*，2022；D 引自 van Oorschot，1980）

世界分布：中国、德国。

讨论：Fergus 和 Sinden（1969）将弗格斯嗜热疣霉的有性型描述为嗜热梭孢壳 *Thielavia thermophila* Fergus & Sinden。von Klopotek（1974）将其移至棒囊壳属，并根据分生孢子形态将其更名为 *Chrysosporium fergusii* Klopotek。van Oorschot（1977）将其无性型种移至毁丝霉属，命名为弗格斯毁丝霉。Marín-Felix 等（2015）基于 3 个位点的系统发育分析将该种组合至无效提议属粗糙壳属。Wang 等（2022）对标本进行了重新描述和鉴定后发表成一新种。该种可通过产孢细胞多不分化，菌丝直接产孢，少数分化成小柄，顶端着生多个分生孢子等特征明显区别于该属的其他物种（王冬梅等，2004）。

水滴状嗜热疣霉　图 138

Thermothelomyces guttulatus (Yu Zhang & L. Cai) Y. Marín, Stchigel, Guarro & Cano, Mycologia 107 (3): 630, 2015.

= *Myceliophthora guttulata* Yu Zhang & L. Cai, Mycol. Progr. 13(1): 168, 2014.

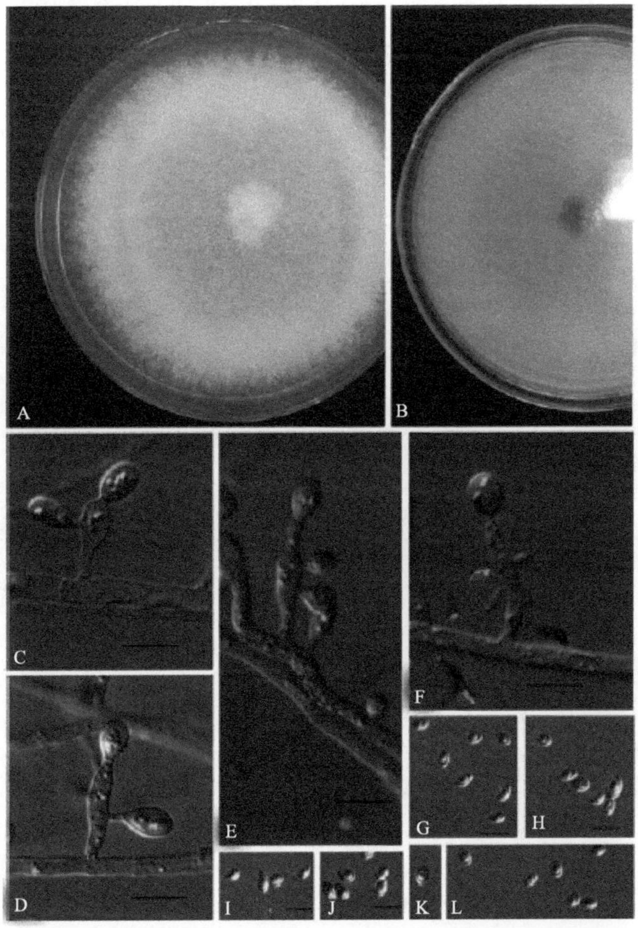

图 138　水滴状嗜热疣霉 *Thermothelomyces guttulatus* (Yu Zhang & L. Cai) Y. Marín, Stchigel, Guarro & Cano（标本号：HMAS 244238）（引自 Zhang *et al.*，2014）

A、B. PDA 培养基上的菌落；C～F. 分生孢子梗和产孢细胞；G～L. 分生孢子。标尺：C～L = 10 μm

在 PDA 培养基上，45℃培养 5 天，菌落直径达 77 mm，平展，最初白色，棉花状，然后变成黄褐色至灰褐色；圆形，边缘规则；背面淡黄褐色。菌丝分隔，无色或半透明，光滑，宽 1.6～3.9 μm；无球拍状菌丝。分生孢子顶生或侧生，1～4 个直接着生于菌丝或短柄或长柄或安瓿瓶形的膨大突起上，单胞，光滑，无色，近球形、倒卵圆形至梨形，大多数水滴状，4.8～7.2 × 3.0～5.0 μm。

研究菌株：贵州：贵阳市，森林公园，林下土，2014 年 3 月，韩燕峰，E38；湖南：莽山国家森林公园，林下土，HMAS 244238。

世界分布：中国。

讨论：水滴状嗜热疣霉形态上有别于该属其他物种在于：其不存在有性型，分生孢子光滑，近球形、倒卵圆形至梨形，以及膨大的产孢细胞（Zhang *et al.*，2014）。

异宗嗜热疣霉　图 139

Thermothelomyces heterothallicus (Klopotek) Y. Marín, Stchigel, Guarro & Cano, Mycologia 107 (3): 630, 2015.

≡ *Thielavia heterothallica* Klopotek, Arch. Mikrobiol. 107(2): 223,1976.

= *Myceliophthora heterothallica* (Klopotek) van den Brink & Samson, Fungal Diversity 52(1): 206, 2012. Han *et al.*, Microbiology China 43(9): 1961, 2016.

= *Corynascus heterothallicus* (Klopotek) Arx, Dreyfuss & Müller, Persoonia 12(2): 174, 1984.

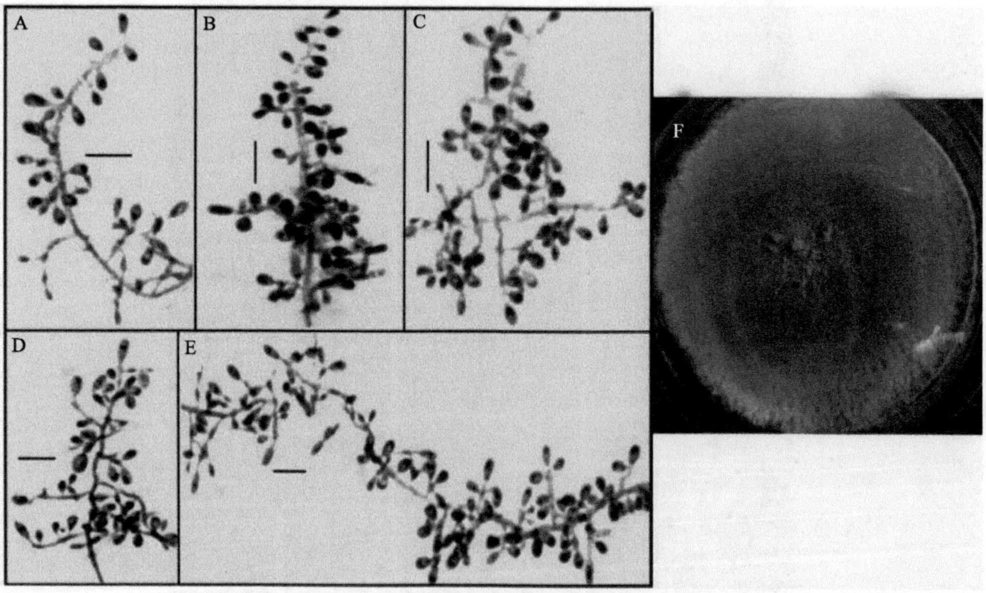

图 139　异宗嗜热疣霉 *Thermothelomyces heterothallicus* (Klopotek) Y. Marín, Stchigel, Guarro & Cano（标本号：GZAC G10）

A～E. 产孢结构和孢子；F. PDA 培养基上的菌落。标尺：A～E = 10 μm

在 PDA 培养基上，40℃培养 5 天，菌落直径 90 mm，平展，最初绒状，后变成毡状至粉状，中部棕红色，边缘白色，圆形，边缘规则；背面黄棕色。菌丝分隔，无色或半透明，光滑，宽 1.1～3.2 μm；无球拍状菌丝。分生孢子顶生、侧生，直接着生于菌丝或短柄或短的膨大突起或分枝上，大多数单生，少数 2～3 个成短链，单胞，壁稍厚，光滑，无色，成熟脱落后常呈淡黄棕色，卵圆形到椭圆形，3.2～8.6 × 2.2～4.3 μm；基痕宽 0.5～2 μm。无间生孢子，厚垣孢子未见。

研究菌株：重庆：歌乐山森林公园，林下土，2013 年 6 月 1 日，韩燕峰，GZAC G10。

世界分布：中国、澳大利亚。

讨论：形态上与异宗嗜热疣霉相似的种有嗜热疣霉、弗格斯嗜热疣霉、黄褐嗜热疣霉和水滴状嗜热疣霉。异宗嗜热疣霉的分生孢子比黄褐嗜热疣霉和弗格斯嗜热疣霉的小（韩燕峰等，2016），且黄褐嗜热疣霉分生孢子表面粗糙；水滴状嗜热疣霉的分生孢子单生和分生孢子表面具水滴状有异于异宗嗜热疣霉；嗜热疣霉的菌落质地和分生孢子着生方式与异宗嗜热疣霉有差异。

黄褐嗜热疣霉　图 140

Thermothelomyces hinnuleus (Awao & Udagawa) Y. Marín, Stchigel, Guarro & Cano, Mycologia 107(3): 630, 2015.

= *Myceliophthora hinnulea* Awao & Udagawa, Mycotaxon 16(2): 436, 1983. Zhao & Li, Journal of Fungal Research 6(3): 134, 2008.

在不同温度的 PDA 培养基上分别培养 5 天后，19℃时无生长现象；40℃时生长迅速，3 天满皿（直径= 90 mm），孢子大量产生；50℃时生长缓慢，直径约为 47 mm，第 10 天仍未见有孢子产生。菌落初白色，绒毛状，稀疏，中间厚，渐现黄白色粉末状，后整个菌落呈灰褐色。菌丝无色，直径 1～9 μm，具隔膜、分枝。产孢细胞顶生或侧生于菌丝，多分化成小柄或形成隆起，顶端生 1～4 个分生孢子。分生孢子单胞，顶生、侧生、单生或 2～3 个链生，多数长卵圆形，壁厚，易留痕脱落，脱落后继续膨大成纺锤形或椭圆形，初淡黄色，成熟后褐色，具明显瘤突、小刺，8.0～10 × 5～6.5 μm。未见有性态。

研究菌株：福建：福州市，动物园，绿地土，2015 年 6 月，王玉荣，E212005；海南：2008 年，赵春青，HSAUP-H6，HSAUP-SY10。

世界分布：中国、日本。

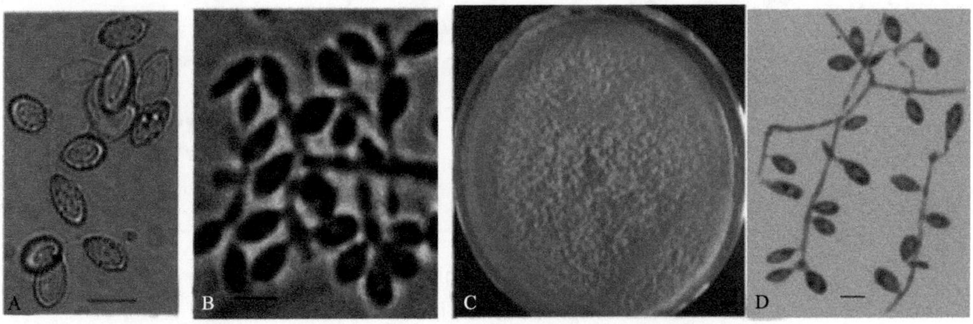

图 140　黄褐嗜热疣霉 *Thermothelomyces hinnuleus* (Awao & Udagawa) Y. Marín, Stchigel, Guarro & Cano（引自赵春青和李多川，2008）

A、B、D. 分生孢子和分生孢子梗；C. PDA 培养基上的菌落。标尺：A、B、D = 10 μm

讨论：黄褐嗜热疣霉是该属中唯一一种产生疣状、深色分生孢子的物种（Wang *et al.*，2022；赵春青和李多川，2008）。

嗜热疣霉 图 141

Thermothelomyces thermophilus (Apinis) Y. Marín, Stchigel, Guarro & Cano, Mycologia 107 (3): 630, 2015.

= *Myceliophthora thermophila* (Apinis) Oorschot, Persoonia 9(3): 403, 1977. Han *et al.*, Microbiology China 43(9): 1963, 2016.

= *Sporotrichum thermophilum* Apinis, Nova Hedwigia 5: 74, 1963.

= *Chrysosporium thermophilum* (Apinis) Klopotek, Arch. Mikrobiol. 98(4): 366, 1974.

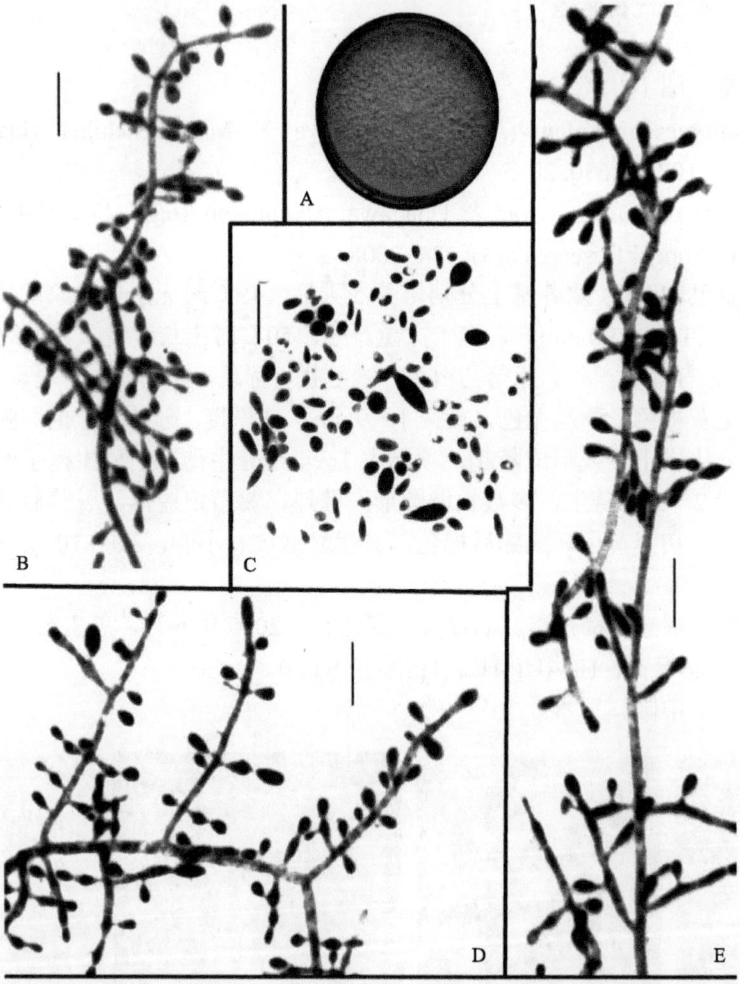

图 141 嗜热疣霉 *Thermothelomyces thermophilus* (Apinis) Y. Marín, Stchigel, Guarro & Cano（标本号：E20701H）

A. PDA 培养基上的菌落；B～E. 产孢结构和分生孢子。标尺：B～E = 10 μm

在 PDA 培养基上，40℃培养 4 天，菌落直径 65～85 mm，最初白色，后变为淡棕色或红棕色，初长绒状，后变成粉状或毯状，有时菌丝尖端结成团而形成网孔结构，有的菌株中央有液体分泌，边缘规则；背面奶酪色到浅褐色。菌丝分隔，无色，光滑，宽 1～6 μm。分生孢子顶生、侧生；单生或 2～3 个成短链；直接着生于菌丝或短柄或短的膨大突起或分枝上，大多数卵圆形到椭圆形，3.2～5.4 × 2.2～3.2 μm；少数长椭圆形到长倒卵圆形、棒状，6.5～9.7 × 2.2～5.4 μm；基痕宽 0.5～2.2 μm。

研究菌株：贵州：开阳县，大叶黄杨根系土样，2012 年 2 月 4 日，王玉荣，E20701H；海南：三亚市，三角梅根系土，2012 年 12 月，王玉荣，E60501H；湖南：张家界，树木根系土，2012 年 7 月，陈万浩，E40501H；贵州：修文县，贵州森林野生动物园，鸸鹋毛，2012 年 4 月，王玉荣，E31001H；贵州：修文县，贵州森林野生动物园，麻雀毛，2012 年 4 月，王玉荣，E30701H；海南：三亚市，河泥，2012 年 2 月，梁宗琦，E21401H；贵州：开阳县，白菜地土，2012 年 2 月，王玉荣，E21201H；贵州：开阳县，猪粪，2012 年 2 月，王玉荣，E21103H；贵州：开阳县，蚕豆土，2012 年 2 月，王玉荣，E20901H；贵州：开阳县，带鸡毛和猪毛土，2012 年 2 月，王玉荣，E20801H；陕西：太白，2003 年 7 月，王冬梅，HSAUPIII8002。

世界分布：中国、英国、美国、加拿大等。

讨论：嗜热疣霉在形态上与异宗嗜热疣霉较为相似，但可以通过其产孢结构加以区分（王冬梅等，2004；韩燕峰等，2016）。前者的有性型不存在；而后者可产生有性型，且为异宗配合（Wang *et al.*，2022）。

发癣菌属 **Trichophyton** Malmsten

Arch. Anat. Physiol. Wiss. Med., 1848: 14, 1848

菌落大多絮状，白色至浅黄色；背面奶油色、棕色、红色或紫罗兰色。菌丝壁薄，无色。大分生孢子和小分生孢子如果出现，末端着生或沿着未分枝的菌丝着生。大分生孢子，具有 2 个或多个细胞，壁薄或壁厚，表面光滑，无色，呈柱状、棍棒状或雪茄形。小分生孢子壁薄且表面光滑，无色，单胞，卵形、梨形或棍棒状。菌丝在配接后形成有性型，形如关节状。

模式种：断发发癣菌 *Trichophyton tonsurans* Malmsten。

中国发癣菌属分种检索表

1. 未见大、小分生孢子 …… 2
1. 大或小分生孢子可见 …… 4
   2. 大小分生孢子未见或偶现 …… 3
   2. 大小分生孢子未见 …… **许兰氏发癣菌 *T. schoenleinii***
3. 大分生孢子出现时呈菜豆状，小分生孢子出现时呈卵圆形至梨形 …… **疣状发癣菌 *T. verrucosum***
3. 未见分生孢子或出现泪滴状小分生孢子 …… **同心发癣菌 *T. concentricum***
   4. 大小分生孢子区分不明显 …… **土生发癣菌 *T. terrestre***
   4. 大小分生孢子区分明显 …… 5
5. 大分生孢子未见，可见小分生孢子 …… 6
5. 大小分生孢子均可见 …… 7

6. 小分生孢子丰富，梨形至棍棒状…………………………………………………苏丹发癣菌 ***T. soudanense***
6. 小分生孢子偶出现在培养时间长的区域………………………………………紫色发癣菌 ***T. violaceum***
7. 大分生孢子呈圆柱形至雪茄状…………………………………………………………………………8
7. 大分生孢子呈棍棒状或雪茄状…………………………………………………………………………9
8. 小分生孢子呈锥形至梨形……………………………………………………………红色发癣菌 ***T. rubrum***
8. 小分生孢子呈棍棒状或至近圆柱形…………………………………………断发发癣菌 ***T. tonsurans***
9. 小分生孢子呈卵圆形至梨形……………………………………………………阿耶罗发癣菌 ***T. ajelloi***
9. 小分生孢子呈球形………………………………………………………………须癣发癣菌 ***T. mentagrophytes***

阿耶罗发癣菌　图 142

Trichophyton ajelloi (Vanbreus.) Ajello, Sabouraudia 6(2): 148, 1968. Wang, Medical Mycology-Guide to Laboratory Examination p.182, 2005.

在 SDA 培养基上，菌落平展，粉状至绒状，米黄色至赭黄色，菌落背面淡黄色，琼脂中渗出深紫色的色素。菌丝分隔，光滑，宽 1.5～2 μm。具两种类型的孢子，大分生孢子无色，表面光滑，壁薄，雪茄状，含 8～12 个细胞，40～70 × 9～12 μm；小分生孢子具隔膜或未见，卵圆形至梨形，3～9 × 2～5 μm。

研究菌株：贵州：贵阳市，修文县，贵州森林野生动物园，绿地土，2010 年 8 月 10 日，韩燕峰，E30。

世界分布：中国、英国、法国、加拿大等。

讨论：阿耶罗发癣菌的鉴别特征主要为菌落形态，以及大、小分生孢子的特点，即小分生孢子卵圆形至梨形；大分生孢子多具隔膜、雪茄状，壁厚，光滑。

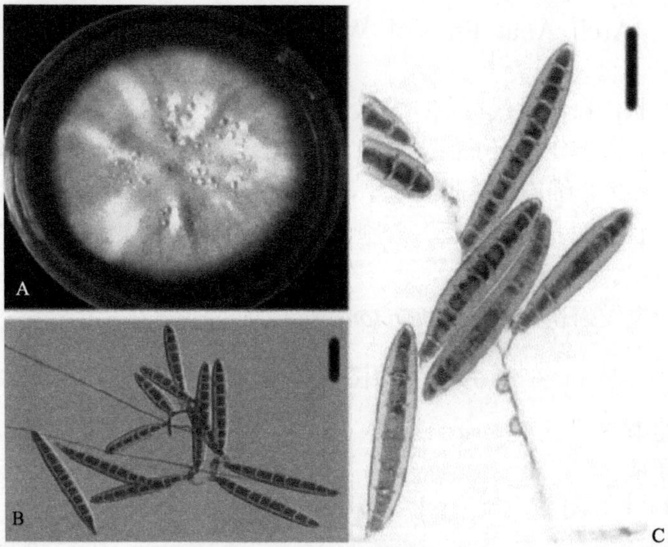

图 142　阿耶罗发癣菌 *Trichophyton ajelloi* (Vanbreus.) Ajello（引自王端礼，2005）
A. SDA 培养基上的菌落；B、C. 产孢结构和分生孢子。标尺：B、C = 25 μm

同心发癣菌　图 143

Trichophyton concentricum R. Blanch., Traité Path. Gen. (Paris) 2: 916, 1895. Wang, Medical Mycology-Guide to Laboratory Examination p. 172, 2005.

在 SDA 培养基上生长缓慢，菌落起初为白色，后变成米黄色、琥珀色、蜂蜜棕色至橘黄色，平滑至天鹅绒状，隆起、不规则，通常在琼脂上龟裂；菌落背面米黄色至棕色。分枝菌丝大量缠结，未见分生孢子或泪滴状小分生孢子出现。

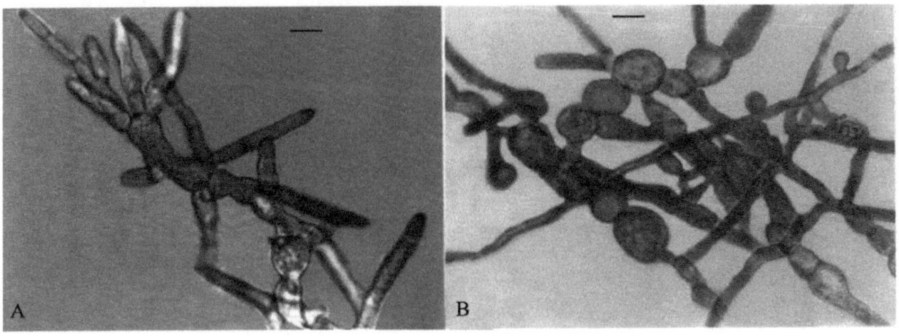

图 143　同心发癣菌 *Trichophyton concentricum* R. Blanch.（引自王端礼，2005）

A. 缠结的菌丝；B. 厚垣孢子。标尺：A、B = 10 μm

研究菌株：贵州：贵阳市，修文县，贵州森林野生动物园，绿地土，2010 年 8 月 10 日，韩燕峰。E32。

世界分布：中国、危地马拉、所罗门群岛。

讨论：同心发癣菌的典型特征为菌落形态，其生长需要维生素 B_1，可引起叠瓦状癣。分离菌株在培养基上生长缓慢，不规则，结合系统发育分析结果，鉴定该分离菌株为同心发癣菌。

须癣发癣菌　图 144

Trichophyton mentagrophytes (C.P. Robin) Sabour., Bouchard Traite de Pathologie Generale 2: 912, 1895. Wang, Medical Mycology-Guide to Laboratory Examination p. 164, 2005.

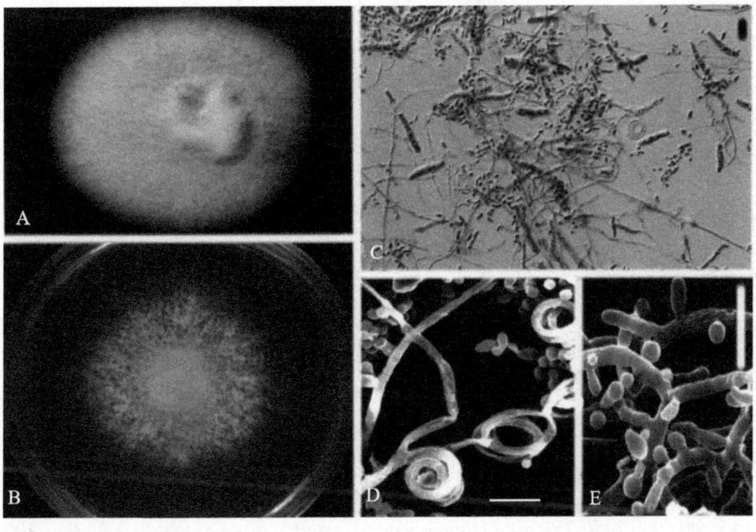

图 144　须癣发癣菌 *Trichophyton mentagrophytes* (C.P. Robin) Sabour.（引自王端礼，2005）

A、B. SDA 培养基上的菌落；C～E. 产孢结构和分生孢子。标尺：C = 25 μm；D、E = 10 μm

在 SDA 培养基上，菌落粉状至絮状，米黄色至淡黄色，粉状的菌落上偶现星形；菌落背面黄褐色至红棕色，偶现黄色至深褐色。大分生孢子 3～8 个细胞，表面光滑，壁薄，棍棒状至雪茄状，20～50 × 6～8 μm，常具隔膜；小分生孢子球形，直径 2 μm，无柄，密集排列，葡萄状簇生或与菌丝一起。螺旋菌丝经常出现，吊灯状结构和厚垣孢子偶尔出现。

研究菌株：福建：福州市，福州市动物园，绿地土，2014 年 8 月 10 日，韩燕峰，E20。

世界分布：中国、印度、德国、墨西哥。

讨论：须癣发癣菌的典型特征为菌落形态，集簇的小分生孢子和螺旋菌丝，典型的分生孢子。该菌适应性强，分布广泛（赵小东等，2004；刘艳等，2009；陈柏叡等，2010；胡小平等，2011）。

红色发癣菌 图 145

Trichophyton rubrum (Castell.) Sabour., British Journal of Dermatology 23: 389, 1911. Wang, Medical Mycology-Guide to Laboratory Examination p. 161, 2005.

在 SDA 培养基上，菌落绒毛状或絮状，白色，有时中间部分变成玫瑰色，菌落背面酒红色至橄榄色，有时呈黄色。大分生孢子大多数情况未见，出现时壁薄，大小差异小，圆柱状至雪茄状，40～55 × 6.0～7.5 μm，有不连接的趋势；小分生孢子锥形至梨形，3.0～5.5 × 2.0～3.5 μm，直接着生于菌丝，无小梗结构。偶仅现小分生孢子或大分生孢子，培养时皆不育。有些菌株常出现节孢子。

研究菌株：海南：三亚市，沙滩土，2010 年 7 月 10 日，韩燕峰，E15。

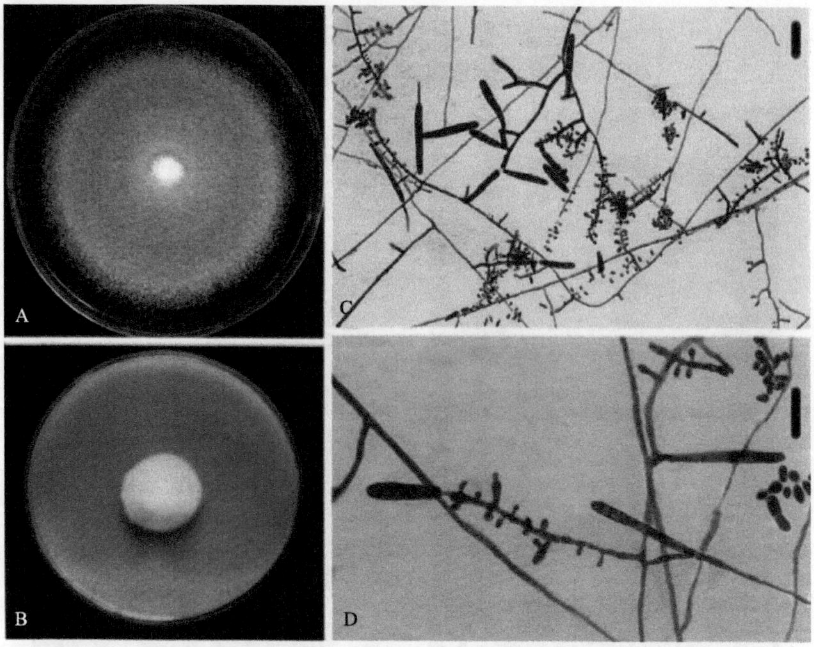

图 145 红色发癣菌 *Trichophyton rubrum* (Castell.) Sabour.（引自王端礼，2005）

A、B. SDA 培养基上的菌落；C、D. 产孢结构。标尺：C、D = 25 μm

世界分布：中国、英国。

讨论：红色发癣菌的典型特征为菌落形态及在培养基上产生红色色素，小分生孢子锥形至梨形，无柄，着生于未分化菌丝旁侧。

许兰氏发癣菌　图 146

Trichophyton schoenleinii (Lebert) Langeron & Miloch. ex Nann., Trattato di Micopatologia Umana (Firenze) 4: 198, 1934. Wang, Medical Mycology-Guide to Laboratory Examination p. 182, 2005.

在 SDA 培养基上，菌落生长相当缓慢，蜡状，之后变成绒状，折叠，随着培养时间的延长而堆积成脑形，经常在琼脂上开裂或龟裂，略带白色或米黄色，由于吊灯状结构的出现边缘呈羽状。菌落背面无色素或淡黄色。大分生孢子和小分生孢子通常未见，鹿角状菌丝具二歧状分枝，顶端膨大，出现在新鲜培养的菌落边缘，厚垣孢子丰富。

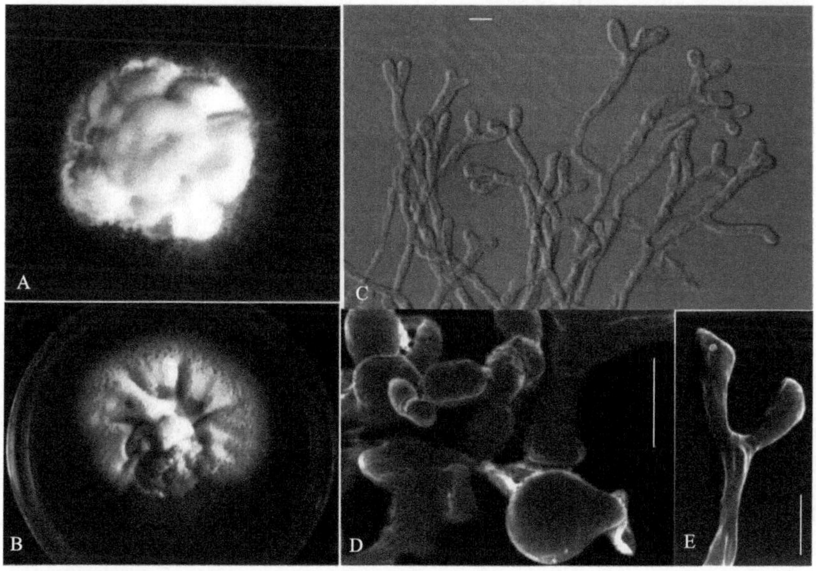

图 146　许兰氏发癣菌 *Trichophyton schoenleinii* (Lebert) Langeron & Miloch. ex Nann.（引自王端礼，2005）

A、B. SDA 培养基上的菌落；C～E. 产孢结构。标尺：C～E = 10 μm

研究菌株：海南：三亚市，沙滩土，2010 年 7 月 10 日，韩燕峰，E18。

世界分布：中国、日本、韩国、巴西。

讨论：许兰氏发癣菌的典型特征是菌落形态和鹿角状菌丝，分离菌株的形态特征与该种的典型形态特征基本一致，故鉴定该菌株为许兰氏发癣菌。

苏丹发癣菌　图 147

Trichophyton soudanense Joyeux, C. r. Seanc. Soc. Biol. 73: 15, 1912. Wang, Medical Mycology-Guide to Laboratory Examination p. 178, 2005.

在 SDA 培养基上生长缓慢，25℃培养 14 天，菌落起初表面光滑，后呈辐射状，菌丝堆积和折叠，直径达 30 mm。菌落中部黄金色、橘黄色或赭色，或杏黄色，至边缘逐渐消失；背面中心呈橘黄色至赭色，边缘呈浅橘黄色至深橘黄色。一些培养物变成玫瑰色至浅玫瑰色或葡萄酒色，表面呈青红色或深葡萄酒色；背面血色或红棕色，培养时间长后菌落表面轻微粉状；培养物通常因快速突变产生绒毛状浅葡萄酒色至白色菌丝体而布满整个菌落。菌丝无色，具隔膜，表面光滑，直径约 3 μm。起初成簇存在，大量分枝；分枝大多数呈锐角或钝角；锐角分枝着生于细胞顶端，钝角分枝出现在末端。直角分枝偶尔出现在细胞中部。生长时间长的菌丝膨大，直径约为 8.5 μm，有断裂形成独立细胞的趋势；末端或中间偶尔形成厚垣孢子；菌落中间由一个紧密交织的大量死亡和扭曲的菌丝及偶现的厚垣孢子组成。老的气生菌丝有断裂形成菌丝碎片和单独细胞的趋势。小分生孢子丰富，形成粉状至绒毛状菌落，梨形至棍棒状，2～10 × 1.5～3 μm，通常不具隔膜，偶现 1 个具隔膜，无柄或具短梗，着生于简单菌丝的侧面。曾经有报道，在培养物为大麦上时出现过大分生孢子。

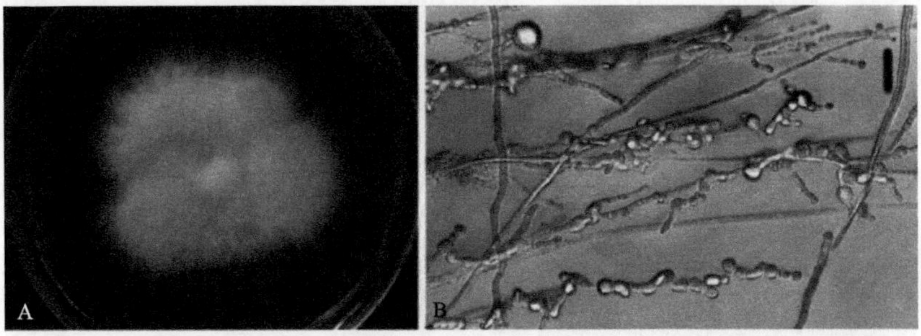

图 147　苏丹发癣菌 *Trichophyton soudanense* Joyeux（引自王端礼，2005）

A. SDA 培养基上的菌落；B. 产孢结构。标尺：B = 25 μm

研究菌株：贵州：修文县，贵州森林野生动物园，花坛土，2010 年 7 月 10 日，韩燕峰，E23。

世界分布：中国、英国、意大利、美国、新西兰。

讨论：苏丹发癣菌的典型特征为其菌落颜色和菌丝形态多样，分布比较广（Albanese *et al.*，1995）。

土生发癣菌　图 148

Trichophyton terrestre Durie & D. Frey, Mycologia 49: 401, 1957. Wang, Medical Mycology-Guide to Laboratory Examination p. 184, 2005.

在 SDA 培养基上，菌落平展，绒毛状至粉状，白色至浅米黄色或浅棕色；菌落背面浅灰色至淡黄色或赭色。大分生孢子与小分生孢子区分不明显，由 2～6 个细胞组成，表面光滑，壁薄，圆柱状至略棍棒状，9～50 × 4～5 μm。小分生孢子着生于具有直立分枝的浓密排列的分生孢子梗上，短圆柱状至短棍棒状，平截，4～6.5 × 1.5 μm。

研究菌株：贵州：修文县，贵州森林野生动物园，花坛土，2010 年 7 月 10 日，韩燕峰，E20。

世界分布：中国、印度、英国、美国。

讨论：土生发癣菌的典型特征主要是菌落特征和其产生的大、小分生孢子的特征。

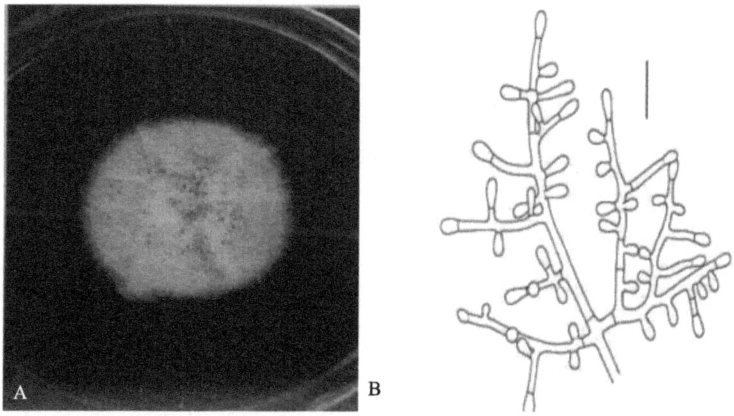

图 148 土生发癣菌 *Trichophyton terrestre* Durie & D. Frey

A. SDA 培养基上的菌落；B. 产孢结构。标尺：B = 10 μm（A 引自王端礼，2005；B 引自 MycoBank）

断发发癣菌 图 149

Trichophyton tonsurans Malmsten, Arch. Anat. Physiol. Wiss. Med., 1848: 14, 1848.

Wang, Medical Mycology-Guide to Laboratory Examination p. 167, 2005.

在 SDA 培养基上，菌落可变，大多数呈革状，径向或不规则沟纹，白色至浅灰色、淡黄色或浅棕色，有时中部浅粉色至浅橄榄色；菌落背面红褐色、黄色或棕色。小分生孢子大小可变，产生于松散丛生的分子孢子梗或变厚的菌丝末端，无柄，棍棒状或近圆柱状，有时呈气球状。大分生孢子出现时大小可变，壁薄，由 2～6 个细胞组成，圆柱状至雪茄状，10～65 × 4～12 μm，末端或中间形成大量的、膨大的厚垣孢子。

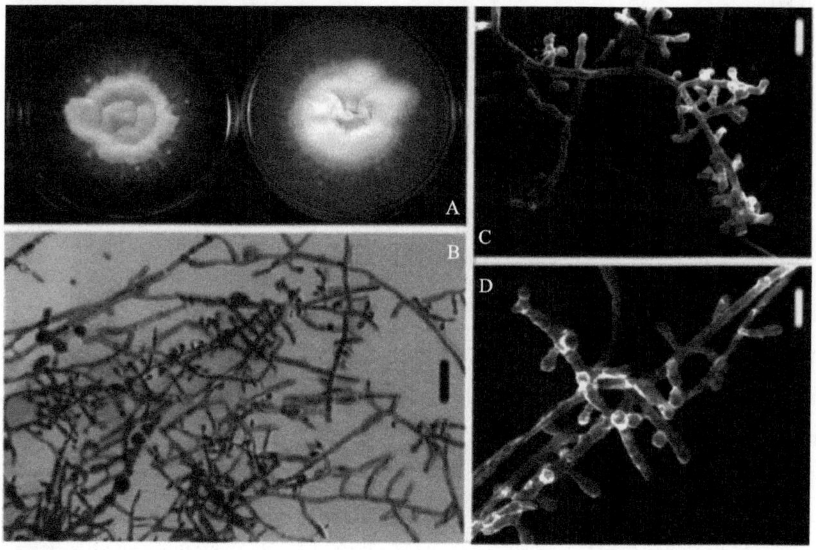

图 149 断发发癣菌 *Trichophyton tonsurans* Malmsten（引自王端礼，2005）

A. SDA 培养基上的菌落；B～D. 产孢结构及分生孢子。标尺：B～D = 25 μm

研究菌株：贵州：贵阳市，黔灵山公园，林下土，2016 年 7 月 5 日，韩燕峰，E24。

世界分布：中国、英国、美国、加拿大。

讨论：断发发癣菌的典型特征为其特征性小分生孢子和大分生孢子。分离菌株的形态特征与该种的典型特征基本一致，故分离菌株鉴定为断发发癣菌。

作者未观察的种

疣状发癣菌 图 150

Trichophyton verrucosum E. Bodin, Les Champignons Parasites de l'homme et des Animaux Domestiques p. 121, 1902. Wang, Medical Mycology-Guide to Laboratory Examination p. 176, 2005.

讨论：在 SDA 培养基上，菌落生长非常缓慢，堆积或呈纽扣状，或被埋生菌丝包围成羽毛状；菌落开始平滑，后呈绒毛状，米黄色至灰白色，有时带有鲑鱼色或淡黄色；背面淡米黄色至鲑鱼色。当大分生孢子出现时，由 4～7 个细胞组成，表面光滑，壁薄，菜豆状；当小分生孢子出现时，卵圆形至梨形。新鲜分离的菌株出现厚垣孢子，通常链生，有时出现在鹿角状菌丝末端，顶端不膨大。新鲜分离的菌株菌丝顶端膨大。

该菌的典型特征为独特的菌落形态，生长缓慢，且生长需要维生素 B_1 和肌醇，有鹿角状菌丝。

世界分布：中国、巴基斯坦、意大利、法国、澳大利亚。

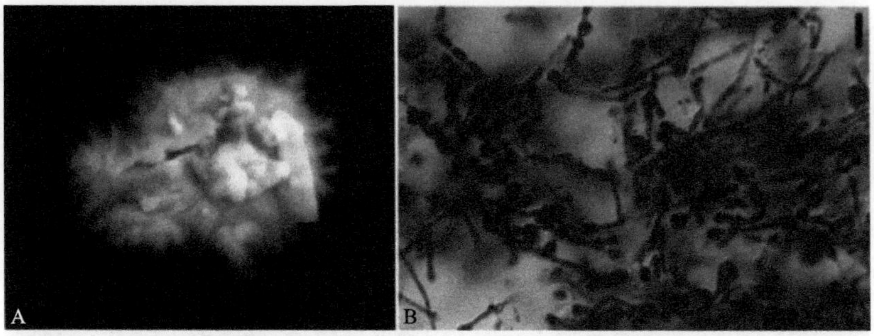

图 150　疣状发癣菌 *Trichophyton verrucosum* E. Bodin（引自王端礼，2005）

A. SDA 培养基上的菌落；B. 产孢结构。标尺：B = 25 μm

紫色发癣菌 图 151

Trichophyton violaceum Sabour. ex E. Bodin, Les Champignons Parasites de l'homme et des Animaux Domestiques: 113, 1902. Wang, Medical Mycology-Guide to Laboratory Examination p. 182, 2005.

讨论：在 SGA 培养基上，菌落生长缓慢，平滑，深葡萄酒色至血红色，具有辐射状沟纹，大多数埋生菌丝具有宽的辐射边缘。菌丝大量分枝，分枝大多数呈锐角，偶见直立和钝角；在菌落老的部位，末端或间生的球形厚垣孢子单生或成链，直径达 20 μm。菌落的中心部分，由死亡、膨大和扭曲的菌丝和厚垣孢子，无色或鲜红色。小分生孢子偶出现在老培养物的绒毛状区域，偶由 2～5 个细胞组成。在一些维生素 B_1 富集的培养基上生长的菌株，绒状气生菌丝出现类似棍棒状的大分生孢子。

紫色发癣菌的典型特征为紫色、具辐射状沟纹的菌落，无大、小分生孢子同时出现，

其生长需要维生素 B_1。

世界分布：中国、意大利、美国、澳大利亚、新西兰。

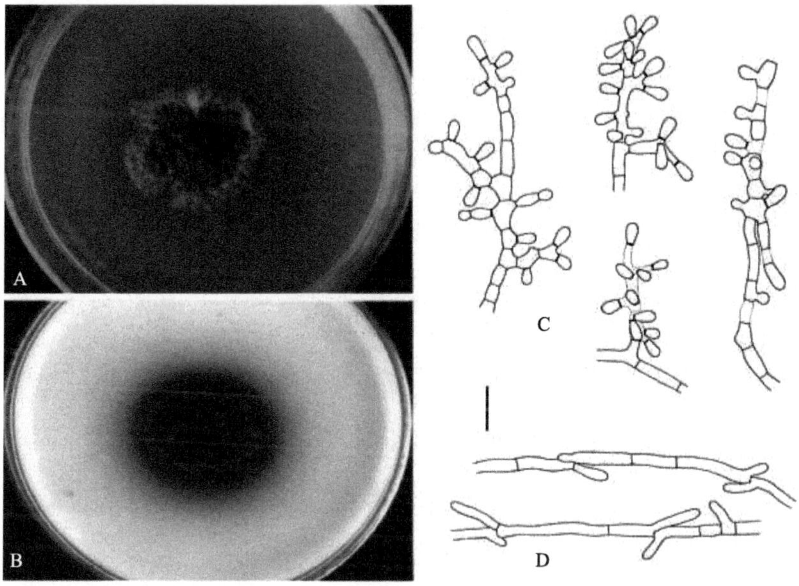

图 151　紫色发癣菌 *Trichophyton violaceum* Sabour. ex E. Bodin

A、B. SDA 培养基上的菌落；C、D. 产孢结构。标尺：C、D = 10 μm（A、B 引自王端礼，2005；C、D 引自 MycoBank）

参考文献

曹军, 郝林, 宋志文, 唐晓萌, 张成刚. 2001. 栖土曲霉生产角蛋白酶研究. 微生物学杂志 21(2): 11-12, 32.

曹艳云, 徐顺明. 2017. 犬小孢子菌病 20 例临床分析及文献回顾. 中国真菌学杂志 12(1): 29-33.

陈柏叡, 孙毅, 胡小平, 万喆, 李若瑜. 2010. 须癣毛癣菌的形态学及分子生物学鉴定. 中国真菌学杂志 5(6): 321-326.

陈剑, 易金玲, 刘丽, 尹颂超, 陈荣章, 李美容, 叶聪秀, 张玉清, 杨素莲, 陆春, 赖维. 2009. 红色毛癣菌金属蛋白酶 Metalloprotease 致病性研究. 中国热带医学 9(4): 605-607, 653.

陈立良, 钟金城, 陶金娇, 林定忠. 2017. 惠州地区浅部真菌病及病原菌种类分析. 甘肃医药 36(2): 122-123.

陈曦. 2010. 东北地区药用植物根际土壤真菌多样性的研究. 大连: 辽宁师范大学硕士学位论文.

陈先进, 沈永年, 吕桂霞, 刘维达. 2009. 红色毛癣菌对动物毛发降解的实验研究. 中国麻风皮肤病杂志 25(3): 166-168.

程永庆, 胡继兰, 戚长菁. 1992. 南极土壤真菌 C3368 菌株产生的抗生素: I. C3368 菌株的分类学、发酵及抗生素的生物活性. 中国抗生素杂志 17(6): 401-403.

戴玉成, 庄剑云. 2010. 中国菌物已知种数. 菌物学报 29(5): 625-628.

董龙龙. 2017. 南极真菌 *Geomyces* sp. WNF-15A 产红色素的发酵条件优化及诱变研究. 青岛: 青岛科技大学硕士学位论文.

冯涛, 王旭增, 王一非, 孙敏, 姚凌云, 徐志民. 2018. 葡萄园土壤中酿酒酵母的分离鉴定及其发酵葡萄酒香气成分分析. 食品科学 39(14): 213-220.

高云超, 朱文珊, 陈文新. 2001. 秸秆覆盖免耕土壤真菌群落结构与生态特征研究. 生态学报 21(10): 1704-1710.

郭志丽. 2007. 粉小孢子菌致头部脓癣 1 例. 中国真菌学杂志 2(5): 291-292.

韩龙. 2013. 耐冷玫红假裸囊菌 HD1031 的鉴定、所产纤维素酶的纯化及其性质的研究. 曲阜: 曲阜师范大学硕士学位论文.

韩龙, 石家骥, 石磊, 王有智, 钞亚鹏, 杨敬, 张国青, 钱世钧. 2013. 一株耐冷玫红假裸囊菌 HD1031 的鉴定及其所产纤维素酶的研究. 微生物学通报 40(6): 928-938.

韩燕峰, 梁宗琦, 梁建东. 2006. 中国金孢霉属两个新记录种. 菌物研究 4(1): 53-56.

韩燕峰, 邵秋雨, 杨娟, 张延威, 陈万浩, 梁建东, 梁宗琦. 2019. 来自土壤的金孢属四个新记录种. 微生物学通报 46(9): 2207-2213.

韩燕峰, 沈鑫, 梁建东, 梁宗琦. 2017a. 金孢属的分类研究进展及其分类特征. 山地农业生物学报 36(5): 1-5.

韩燕峰, 王成, 王玉荣, 陈万浩, 梁建东, 梁宗琦. 2016. 中国毁丝霉属两个新记录种. 微生物学通报 43(9): 1960-1965.

韩燕峰, 王玉荣, 梁建东, 梁宗琦. 2013. 来自青海农田土壤的金孢属一新种. 菌物学报 32(4): 606-611.

韩燕峰, 郑欢, 张芝元, 陈万浩, 罗韵, 王玉荣, 梁宗琦. 2017b. 广西和海南的赛多孢属2新种. 菌物学报 36(2): 145-153.

何苏琴, 金秀琳, 罗进仓, 王春明. 2011. *Pseudogymnoascus roseus* 的生物学特性及液体发酵培养基的筛选. 微生物学通报 38(9): 1371-1376.

胡小平, 万喆, 王晓红, 李若瑜. 2011. 我国代表地区须癣毛癣菌复合体的分子鉴定与分型研究. 中国真菌学杂志 6(2): 70-76.
胡以仁. 1986. 云南水稻土中的若干小型真菌. 云南农业大学学报 1(1): 23-27.
黄悦华. 2009. 西藏部分地区土壤淡色丝孢菌物种多样性研究. 泰安: 山东农业大学硕士学位论文.
姜成林, 徐丽华, 陈庆涛. 1985. 帚霉属的一个新种. 真菌学报 4(3): 167-170.
金滨滨. 2014. 南极红色素的成分分析和性质研究. 成都: 成都中医药大学硕士学位论文.
居哈尔·米吉提, 于世荣, 普雄明. 2009. NB-UVB 治疗 30 例银屑病临床疗效观察. 新疆医学 39(1): 43-45.
冷文川, 王玲玲, 卫灿东, 杨剑, 金奇. 2005. 红色毛癣菌分泌性蛋白酶的分析. 微生物学报 45(4): 601-605.
李奥兰. 2015. 一株拮抗马铃薯晚疫病菌菌株的分离鉴定及抑菌性初步分析. 泰安: 山东农业大学硕士学位论文.
李彩霞, 刘维达. 2011. 我国大陆近年儿童头癣流行情况的回顾分析. 中国真菌学杂志 6(2): 77-82.
李欣, 韩淑梅, 张芝元, 韩燕峰, 梁宗琦. 2022. 嗜角蛋白真菌的界定、研究方法及其应用价值. 微生物学通报 49(1): 292-305.
李云海, 王晓丽, 卢小玲, 于豪冰, 刘小宇. 2017. 极地真菌 *Geomyces* sp. 3-1 次级代谢产物的研究. 天然产物研究与开发 29(4): 590-594.
李增智. 2015. 我国利用真菌防治害虫的历史、进展及现状. 中国生物防治学报 31(5): 699-711.
李治建, 古力娜•达吾提, 斯拉甫•艾白, 安惠霞, 肖威. 2009. 高效液相色谱法分析红色毛癣菌麦角甾醇的含量. 石河子大学学报: 自然科学版 27(1): 73-76.
梁晨, 吕国忠. 2002. 辽宁省农田作物根围的真菌 II. 沈阳农业大学学报 33(4): 266-269.
梁建东, 韩燕峰, 梁宗琦. 2007a. 一类嗜角蛋白真菌: 金孢属的应用价值评介. 菌物研究 5(2): 113-118.
梁建东, 韩燕峰, 梁宗琦. 2008. 中国金孢属一新记录种. 菌物学报 27(3): 447-451.
梁建东, 韩燕峰, 梁宗琦. 2013. 1 株低温地丝霉菌的鉴定及其分解角蛋白与纤维素特性. 贵州农业科学 41(8): 135-138. [Liang JD, Han YF, Liang ZQ. 2013. Identification of a halophilic *Geomyces* strain and its decomposing feature for keratin and cellulose. *Guizhou Agricultural Sciences* 41(8): 135-138.]
梁建东, 韩燕峰, 梁宗琦, 张纪伟. 2007b. 产角蛋白酶耐热金孢菌的初筛研究. 菌物研究 5(4): 210-213.
梁宗琦. 1991. 一个分离自冬虫夏草的金孢霉新种. 真菌学报 10(1): 50-56.
林娟, 叶秀云, 曹泽丽, 谢范英, 徐美爱. 2015. “红茶菌” 中微生物的分离及纯菌混合发酵生产. 中国食品学报 15(2): 39-48.
林燕荣. 2010. 高校足球教学潜在安全因素及其对策研究. 无锡职业技术学院学报 9(5): 90-93.
刘丽, 赖维. 2016. 不同底物体外诱导不同基因型红色毛癣菌产生角蛋白酶的活性比较. 临床皮肤科杂志 45(2): 90-92.
刘鑫, 侯若琳, 金珊珊, 戚梦, 胡开辉, 郑明锋, 傅俊生. 2018. 药用真菌粗毛纤孔菌的分子甄别及其发酵液抗乳腺癌活性研究. 菌物学报 37(2): 215-225.
刘艳, 冉玉平, 张瑞峰, 代亚玲. 2009. 须癣毛癣菌所致脓癣病发的扫描和透射电镜观察及菌体外酶活性分析: 2 例报告. 中国真菌学杂志 4(3): 173-176.
刘云龙, 何永宏, 谢超, 何永明, 曾垄钢, 鲁海菊. 2005. 云南大围山自然保护区的丝孢菌(III). 云南农业大学学报 20(1): 23-26.
罗挺, 刘素琴, 陈希亮. 1996. 贵阳地区羊毛状小孢子菌性真菌病 183 例分析. 中华皮肤科杂志 29(5): 369.
罗文, 周世宁, 钟英长. 1993. 羽毛粉降解菌的分离及特性. 微生物学通报 20(4): 209-212.
罗韵, 陈万浩, 王垚, 韩燕峰, 梁宗琦. 2016. 一株产黑色素的地丝霉新种. 菌物学报 35(2): 123-130.

吕珊珊, 侯运华, 闫孟节, 钟耀华. 2016. 工业真菌高效产酶突变技术与高产机制. 中国生物工程杂志 36(3): 111-119.
马越娥, 顾俊瑛, 高飞, 刘至昱, 许辉, 李秀丽. 2013. 30 例犬小孢子菌致体癣病例的临床分析. 中国真菌学杂志 8(2): 90-92.
毛叶红, 何蒙文, 郑岳臣, 连昕, 曾敬思. 2017. 1960-2015 年湖北及周边地区头癣病原菌变化情况分析. 中国真菌学杂志 12(2): 98-105.
慕东艳. 2012. 黑龙江省药用植物根际土壤真菌多样性的研究. 大连: 辽宁师范大学硕士学位论文.
潘好芹. 2009. 长江、黄河源头地区土壤中暗色丝孢菌物种多样性研究. 泰安: 山东农业大学博士学位论文.
齐志国, 张铁鹰, 董杰丽, 杜家菊. 2012. 角蛋白降解菌分离、鉴定及其降解机制研究. 饲料工业 33(17): 19-24.
冉玉平, 杨琴, 庄凯文, 唐教清, 徐小茜, 游紫梦, 肖慧, 冉昕, 代亚玲. 2017. 常见皮肤真菌病临床及真菌形态学检查. 临床检验杂志 35(10): 750-757.
邵力平, 沈瑞祥, 张素轩, 项存悌, 谭松山. 1983. 真菌分类学. 北京: 中国林业出版社.
邵秋雨, 张芝元, 董醇波, 邹晓, 韩燕峰, 梁宗琦. 2019. 分离自土壤的裸囊菌科真菌中国及大陆新记录种. 山地农业生物学报 38(3): 16-21.[Shao QY, Zhang ZY, Dong CB, Zou X, Han YF, Liang ZQ. 2019. Newly recorded species of the Fungal genus Arthrodermataceae isolated from the soil. *Journal of Mountain Agriculture and Biology* 38(3): 16-21.]
沈鑫, 张芝元, 郑欢, 邹晓, 韩燕峰, 梁宗琦. 2018. 富角蛋白有机物对医院绿地真菌群落组成的影响. 菌物学报 37(8): 999-1005.
宋伟. 2008. 中国秦岭太白山及周围地区土壤中的暗色丝孢菌多样性研究. 泰安: 山东农业大学硕士学位论文.
孙晓红, 洪霞, 帅丽华, 宋秋荷, 朱钰力, 黄龙. 2013. 石膏样小孢子菌致儿童脓癣 20 例鉴定与临床分析. 实验与检验医学 31(3): 223-225.
谭盈盈, 冯继勤, 郑平. 2005. 拟青霉菌株 Z-1 降解羽毛角蛋白特性的研究. 浙江大学学报 31(1): 82-87.
佟盼琢, 杨晓亮, 冯琴, 王冰心, 杨云, 谢红霞. 2011. 盐酸小檗碱联合抗真菌药物对犬小孢子菌体外抗菌活性的研究. 临床皮肤科杂志 40(7): 391-393.
王冬梅, 李多川, 赵春青. 2004. 中国毁丝霉属嗜热型真菌两新记录种. 菌物研究 2(2): 47-49.
王冬梅, 王峰祥, 高红. 2007. 嗜热真菌的分类研究: 半知菌. 菌物研究 5(1): 10-13.
王端礼. 2005. 医学真菌学: 实验室检验指南. 北京: 人民卫生出版社: 1-525.
王娜. 2012. 烟草根际土壤真菌多样性的研究. 大连: 辽宁师范大学硕士学位论文.
王澎, 徐英春, 窦洪涛, 谢秀丽, 王辉. 2007. 尖端赛多孢和多育赛多孢所致的深部真菌感染 2 例并文献复习. 中国真菌学杂志 2(4): 210-213. [Wang P, Xu YC, Dou HT, Xie XL, Wang H. 2007. Invasive *Scedosporium apiospermum* and *Scedosporium prolificans* infections: case report and review. *Chinese Journal of Mycology* 2(4): 210-213.]
王垚, 王玉荣, 罗韵, 陈万浩, 韩燕峰, 梁宗琦. 2014. 一株在低温条件下产纤维素酶的草生帚霉. 山地农业生物学报 33(3): 50-54. [Wang Y, Wang YR, Luo Y, Chen WH, Han YF, Liang ZQ. 2014. A cold-adapted cellulase-producing strain of *Scopulariopsis hibernica*. *Journal of Mountain Agriculture and Biology* 33(3): 50-54.]
王玉荣, 陈万浩, 王垚, 韩燕峰, 梁宗琦. 2013. 一株能强降解鸡毛的帚霉菌株的分离鉴定. 山地农业生物学报 32(4): 295-298. [Wang YR, Chen WH, Wang Y, Han YF, Liang ZQ. 2013. Isolation and identification of *Scopulariopsis brevicaulis* strain decomposing feather duster. *Journal of Mountain Agriculture and Biology* 32(4): 295-298.]
王玉荣, 王垚, 罗韵, 梁建东, 陈万浩, 韩燕峰, 梁宗琦. 2014. 橘黄赛多孢, 一种潜在的致病有丝分裂产孢真菌. 微生物学通报 41(10): 2090-2093.
吴润标, 李前勇, 李奥, 刘威, 曹平, 张德志. 2017. 山羊疣状毛癣菌鉴定与敏感天然药物的体外筛选试验. 吉林农业大学学报 40(3): 358-363.
肖媛媛, 王爱平, 曹存巍, 万哲, 李若瑜. 2008. 皮肤癣菌体外蛋白水解酶活性测定. 中国真菌学杂志 3(1): 5-7.
谢美华, 陈华红, 陈玉红, 周红杰, 张春花. 2013. 几种普洱茶发酵微生物对茶多酚代谢的初步研究. 食品科技 38(5): 82-85.
徐宇, 杨国玲. 2016. 犬小孢子菌的致病机制和治疗研究进展. 中国真菌学杂志 11(1): 62-64.
杨虹, 高志琴, 陈健, 李民, 戴鹤骏, 杨连娟. 2013. 犬小孢子菌致皮肤癣菌病 162 例回顾分析. 中国真菌学杂志 8(4): 217-219.
杨连, 曹永洪, 杨培龙, 黄火清. 2015. 角蛋白酶的研究与应用前景. 生物技术进展 5(1): 29-34.
于晓朋, 赵新强. 2015. 角蛋白酶研究进展. 中国果菜 35(2): 77-81.
余进, 李若瑜, 王晓红, 万喆, 王端礼. 2008. 分离自甲真菌病的帚霉形态学研究. 临床皮肤科杂志 9: 564-567. [Yu J, Li RY, Wang XH, Wan Z, Wang DL. 2008. Morphological study on *Scropulariopsis* spp. isolated from nails of patients with onychomycosis. *Journal of Clinical Dermatology* 9: 564-567.]
虞伟衡, 章强强. 2008. 石膏样小孢子菌致须癣 1 例. 中国真菌学杂志 3(1): 29-30.
曾毅, 李忠琴, 许小平. 2004. 角蛋白酶的研究进展. 海峡药学 16(6): 10-13.
张芳芳, 刘金鹏, 鄢秋龙, 戴安安, 杨国玲. 2016. 须癣毛癣菌烯醇化酶基因的克隆、原核表达与纯化. 中国皮肤性病学杂志 30(6): 574-578.
张文州, 许嵘. 2014. 食药用真菌多糖的研究进展. 食品工业科技 35(15): 395-399.
张小鸣, 于英瑶, 葛新红, 王琪. 2012. 宁夏 203 例儿童头癣病原菌分析. 宁夏医科大学学报 34(1): 66-68.
张延威, 陈万浩, 邹晓, 韩燕峰, 梁宗琦. 2016. 来自山西鸟窝的金孢属一新种. 菌物学报 35(11): 1337-1343.
张英, 金敏. 2016. 425 例皮肤浅部真菌病的病原菌分析. 中国微生态学杂志 28(6): 719-721.
张颖, 罗庆录. 2007. 皮肤癣菌外分泌角蛋白酶的研究进展. 中国真菌学杂志 2(3): 190-192.
张影, 郑宝勇. 2015. 石膏样小孢子菌致伪膜性阴囊癣一例. 实用皮肤病学杂志 8(4): 310-311.
张芝元. 2018. 特定生境中角蛋白降解真菌物种多样性. 贵阳: 贵州大学硕士学位论文.
张芝元, 郑欢, 韩燕峰, 陈万浩, 梁宗琦. 2017a. 川、黔、渝几所医院土生角蛋白降解真菌多样性分析. 菌物学报 36(11): 1504-1513.
张芝元, 郑欢, 韩燕峰, 陈万浩, 梁宗琦. 2017b. 不同宠物狗体表一些部位角蛋白降解真菌物种多样性. 菌物学报 36(5): 563-572.
张芝元, 邹晓, 韩燕峰, 梁宗琦. 2017c. 贵州小孢子菌: 小孢子菌属一新种. 菌物学报 36(5): 535-541.
赵春青, 李多川. 2008. 中国大陆嗜热真菌 2 新记录种. 菌物研究 6(3): 133-135.
赵明月, 李治建, 古力娜·达吾提, 斯拉甫·艾白. 2012. 红色毛癣菌中羊毛甾醇含量的高效液相色谱法分析. 时珍国医国药 23(12): 2980-2981.
赵小东, 付萌, 王冬青, 孙林潮, 张海龙, 刘红娟, 刘玉峰. 2004. 角蛋白与毛发诱导须癣毛癣菌产生角蛋白酶的比较. 中国美容医学 13(5): 519-521.
赵小东, 卢涛, 刘玉峰. 2005. 角蛋白酶研究进展. 中国美容医学 14(5): 647-649.
郑飞, 孟歌, 安琪, 田雪梅, 司静. 2017. 药用真菌桑黄液体培养过程中的抗氧化活性研究. 菌物学报 36(1): 98-111.
周帼萍, 沙涛, 程立忠, 丁骅孙. 2002. 云南省部分地区产菊粉酶霉菌的初步调查. 生物技术 12(2): 24-25.
朱敬先. 2017. 常见儿童真菌性皮肤病的诊断与治疗. 皮肤科学通报 34(5): 522-530.
朱小红, 张海平, 尹兴平, 华海康. 2014. 3 例不同临床表现的石膏样小孢子菌感染. 中国真菌学杂志

9(3): 172-174.

Abarca ML, Castellá G, Martorell J, Cabañes FJ. 2010. *Chrysosporium guarroi* sp. nov. a new emerging pathogen of pet green iguanas (*Iguana iguana*). *Medical Mycology* 48: 365-372.

Abbas Soleymani MD, Mohammad Hoseini MD, Hamze Sharifi MS. 2015. Species diversity of keratinophilic fungi in various soil type of Babol Medical University's Hospitals' yard. *International Journal of Applied Science and Technology* 5(3): 55-59.

Abbott SP, Sigler L. 2001. Heterothallism in the Microascaceae demonstrated by three species in the *Scopulariopsis brevicaulis* series. *Mycologia* 93: 1211-1220.

Abbott SP, Sigler L, Currah RS. 1998. *Microascus brevicaulis* sp. nov., the teleomorph of *Scopulariopsis brevicaulis*, supports placement of *Scopulariopsis* with the Microascaceae. *Mycologia* 90: 297-302.

Abneuf MA, Krishnan A, Aravena MG, Pang KL, Convey P, Mohamad-Fauzi N, Alias SA. 2016. Antimicrobial activity of microfungi from maritime Antarctic soil. *Czech Polar Reports* 6(2): 141-154.

Abrantes RA, Refojo N, Hevia AI, Fernández J, Isla G, Córdoba S, Dávalos MF, Lubovich S, Maldonado I, Davel GO, Stchigel AM. 2021. *Scedosporium* spp. from clinical setting in Argentina, with the proposal of the new pathogenic species *Scedosporium americanum*. *Journal of Fungi* 7(3): 160.

Achterman RR, White TC. 2012. A foot in the door for dermatophyte research. *PLoS Pathogens* 8(3): e1002564.

Adams PR. 1997. Growth and amylase production in *Sporotrichum thermophile* Apinis. *Biotechnology and Applied Biochemistry* 26(3): 169-170.

Adetunji CO, Makanjuola OR, Arowora K, Afolayan SS, Adetunji JB. 2012. Production and application of keratin-based organic fertilizer from microbially hydrolyzed feathers to cowpea (*Vigna unguiculata*). *International Journal of Scientific & Engineering Research* 3(12): 164-172.

Afzal AJ, Ali S, Latif F, Rajoka MI, Siddiqui KS. 2005. Innovative kinetic and thermodynamic analysis of a purified superactive xylanase from *Scopulariopsis* sp. *Applied Biochemistry and Biotechnology* 120(1): 51-70.

Ainsworth GC, Georg LK. 1954. Nomenclature of the faviform trichophytons. *Mycologia* 46(1): 9-11.

Ajello L. 1953. The dermatophyte, *Microsporum gypseum*, as a saprophyte and parasite. *Journal of Investigative Dermatology* 21(3): 157-171.

Ajello L. 1956. Soil as natural reservoir for human pathogenic fungi. *Science* 123(3203): 876-879.

Ajello L. 1968. A taxonomic review of the dermatophytes and related species. *Sabouraudia* 6(2): 147-159.

Ajello L. 1974. Natural history of the dermatophytes and related fungi. *Mycopathologia et Mycologia applicata* 53(1-4): 93-110.

Ajello L. 1977. Milestones in the history of medical mycology: the dermatophytes. In: Iwata K. Recent Advances in Medical and Veterinary Mycology. Tokyo: University of Tokyo.

Ajello L, Kuttin ES, Beemer AM, Kaplan W, Padhye A. 1977. Occurrence of *Histoplasma capsulatum* Darling, 1906 in Israel, with a review of the current status of histoplasmosis in the Middle East. *The American Journal of Tropical Medicine and Hygiene* 26(1): 140-147.

Ajello L, Runyon L. 1953. Abortive "perithecial" production by *Phialophora verrucosa*. *Mycologia* 45(6): 947-950.

Ajello L, Zeidberg LD. 1951. Isolation of *Histoplasma capsulatum* and *Allescheria boydii* from soil. *Science* 113(2945): 662-663.

Albanese G, Cintio RD, Crippa D, Galbiati G. 1995. *Trichophyton soudanense* in Italy: *Trichophyton soudanense* in Italien. *Mycoses* 38(5-6): 229-230.

Aldridge DC, Davies AB, Jackson MR, Turner WB. 1974. Pentaketide metabolites of the fungus *Phialophora lagerbergii*. *Journal of the Chemical Society, Perkin Transactions* 1: 1540-1541.

Ali TH, Ali NH, Mohamed LA. 2011. Production, purification and some properties of extracellular keratinase from feathers-degradation by *Aspergillus oryzae* nrrl-447. *Journal of Applied Sciences in Environmental Sanitation* 6(2): 123-136.

Al-Musallam A, Tan CS. 1989. *Chrysosporium zonatum*, a new keratinophilic fungus. *Persoonia* 14(1): 69-71.

Anane S, Al-Yasiri MHY, Normand AC, Ranque S. 2015. Distribution of keratinophilic fungi in soil across Tunisia: a descriptive study and review of the literature. *Mycopathologia* 180(1): 1-8.

Anbu P, Gopinath SCB, Hilda A, Annadurai G. 2005. Purification of keratinase from poultry farm isolate-*Scopulariopsis brevicaulis* and statistical optimization of enzyme activity. *Enzyme and Microbial Technology* 36(5-6): 639-647.

Anbu P, Gopinath SCB, Hilda A, Lakshmipriya T, Annadurai G. 2007. Optimization of extracellular keratinase production by poultry farm isolate *Scopulariopsis brevicaulis*. *Bioresource Technology* 98(6): 1298-1303.

Anbu P, Gopinath SCB, Hilda A, Mathivanan N, Annadurai G. 2006. Secretion of keratinolytic enzymes and keratinolysis by *Scopulariopsis brevicaulis* and *Trichophyton mentagrophytes*: regression analysis. *Canadian Journal of Microbiology* 52(11): 1060-1069.

Apinis AE. 1964. Revision of British Gymnoascaceae. *Mycological Papers* 96: 1-56.

Apodaca G, McKerrow JH. 1989. Regulation of *Trichophyton rubrum* proteolytic activity. *Infection and Immunity* 57(10): 3081-3090.

Apodaca G, McKerrow JH. 1990. Expression of proteolytic activity by cultures of *Trichophyton rubrum*. *Journal of Medical and Veterinary Mycology* 28(2): 159-171.

April TM, Abbott SP, Foght JM, Currah RS. 1998. Degradation of hydrocarbons in crude oil by the ascomycete *Pseudallescheria boydii* (Microascaceae). *Canadian Journal of Microbiology* 44(3): 270-278.

April TM, Foght JM, Currah RS. 1999. Hydrocarbon-degrading filamentous fungi isolated from flare pit soils in northern and western Canada. *Canadian Journal of Microbiology* 46(1): 38-49.

Aquino RSS, Silveira SS, Pessoa WFB, Rodrigues A, Andrioli JL, Delabie JHC, Fontana R. 2012. Filamentous fungi vectored by ants (Hymenoptera: Formicidae) in a public hospital in north-eastern Brazil. *Journal of Hospital Infection* 83(3): 200-204.

Awao T, Udagawa SI. 1983. A new thermophilic species of *Myceliophthora*. *Mycotaxon* 16(2): 436-440.

Ayer WA, Trifonov LS. 1994. Anthraquinones and a 10-Hydroxyanthrone from *Phialophora alba*. *Journal of Natural Products* 57(2): 317-319.

Babot ED, Rico A, Rencoret J, Kalum L, Lund H, Romero J, Del Río JC, Martínez AT, Gutiérrez A. 2011. Towards industrially-feasible delignification and pitch removal by treating paper pulp with *Myceliophthora thermophile* laccase and a phenolic mediator. *Bioresource Technology* 102(12): 6717-6722.

Badhan AK, Chadha BS, Kaur J, Saini HS, Bhat MK. 2007. Production of multiple xylanolytic and cellulolytic enzymes by thermophilic fungus *Myceliophthora* sp. IMI 387099. *Bioresource Technology* 98: 504-510.

Bahuguna S, Kushwaha RK. 1992. Antifungal activity of some *Chrysosporium* spp. *Biomedical Letters* 47: 99-103.

Bainier G. 1907. Mycothèque de l'École de Pharmacie, XIV. *Scopulariopsis* (*Penicillium proparte*) genre nouveau de mucédinées. *Bulletin Trimestriel de la Société Mycologique de France* 23(2): 98-105.

Bakos L, Bonamigo RR, Pisani AC, Mariante JC, Mallmann R. 1996. Scutular favus-like tinea cruris et pedis in a patient with AIDS. *Journal of the American Academy of Dermatology* 34: 1086-1087.

Barnett HL, Barry B. 1998. Illustrated genera of imperfect fungi. Minneapolis: Burgess Publishing Company.

Barratt SR, Ennos AR, Greenhalgh M, Robson GD, Handley PS. 2003. Fungi are the predominant micro-organisms responsible for degradation of soil-buried polyester polyurethane over a range of soil water holding capacities. *Journal of Applied Microbiology* 95(1): 78-85.

Bary A. 1854. Ueber die Entwickelung und den Zusammenhang von *Aspergillus glaucus* und Eurotium. *Botanische Zeitung* 12: 425-434.

Basu M. 1985. *Myceliophthora indica* in a new thermophilic species from India. *Nova Hedwigia* 40: 85-90.

Baudet EARF. 1930. Sur une teigne trichophytique du dromadaire produite par une espèce nouvelle de *Grubyella, G. langeroni* n. sp. *Annales De Parasitologie Humaine et Comparee* 8(3-4): 411-418.

Beeson WT, Iavarone AT, Hausmann CD, Cate JH, Marletta MA. 2011. Extracellular aldonolactonase from *Myceliophthora thermophila*. *Applied and Environmental Microbiology* 77(2): 650-656.

Beguin H, Larcher G, Nolard N, Chabasse D. 2005. *Chrysosporium chiropterorum* sp. nov., isolated in France, resembling *Chrysosporium* state of *Ajellomyces capsulatus* (*Histoplasma capsulatum*). *Medical Mycology* 43(2): 161-169.

Bell RG. 1978. Comparative virulence and immunodiffusion analysis of *Petriellidium boydii* (Shear) Malloch strains isolated from feedlot manure and a human mycetoma. *Canadian Journal of Microbiology* 24(7): 856-863.

Bergero R, Girlanda M, Varese GC, Intili D, Luppi AM. 1999. Psychrooligotrophic fungi from Arctic soils of Franz Joseph Land. *Polar Biology* 21: 361-368.

Berka RM, Schneider P, Golightly EJ, Brown SH, Madden M, Brown KM, Xu F. 1997. Characterization of the gene encoding an extracellular laccase of *Myceliophthora thermophila* and analysis of the recombinant enzyme expressed in *Aspergillus oryzae*. *Applied and Environmental Microbiology* 63(8): 3151-3157.

Bertrand S, Schumpp O, Bohni N, Monod M, Gindro K, Wolfender JL. 2013. *De novo* production of metabolites by fungal co-culture of *Trichophyton rubrum* and *Bionectria ochroleuca*. *Journal of Natural Products* 76(6): 1157-1165.

Bhat KM, Maheshwari R. 1987. *Sporotrichum thermophile* Growth, Cellulose Degradation, and Cellulase Activity. *Applied and Environmental Microbiology* 53: 2175-2182.

Blank F, Ng AS, Just G. 1966. Metabolites of pathogenic Fungi: V. Isolation and tentative structures of vioxanthin and vippurpurin, two colored metabolites of *Trichophyton violaceum*. *Canadian Journal of Chemistry* 44(23): 2873-2879.

Blehert DS, Hicks AC, Behr M, Meteyer CU, Berlowski-Zier BM, Buckles EL, Okoniewski JC. 2009. Bat white-nose syndrome: an emerging fungal pathogen? *Science* 323(5911): 227.

Błyskal B. 2009. Fungi utilizing keratinous substrates. *International Biodeterioration & Biodegradation* 63(6): 631-653.

Bodin E. 1907. A new fungus of favus (*Achorion gypseum*) [Sur un nouveau champignon du favus (*Achorion gypseum*)]. *Annales de Dermatologie et de Syphiligraphie* 8: 585-602.

Bosso L, Scelza R, Testa A, Cristinzio G, Rao MA. 2015. Depletion of pentachlorophenol contamination in an agricultural soil treated with *Byssochlamys nivea, Scopulariopsis brumptii* and urban waste compost: a laboratory microcosm study. *Water, Air, & Soil Pollution* 226(6): 183.

Bosso L, Scelza R, Varlese R, Meca G, Testa A, Rao MA, Cristinzio G. 2016. Assessing the effectiveness of *Byssochlamys nivea* and *Scopulariopsis brumptii* in pentachlorophenol removal and biological control of two *Phytophthora* species. *Fungal Biology* 120(4): 645-653.

Bourbeau PAUL, McGough DA, Fraser H, Shah N, Rinaldi MG. 1992. Fatal disseminated infection caused by *Myceliophthora thermophila*, a new agent of mycosis: case history and laboratory characteristics. *Journal of Clinical Microbiology* 30(11): 3019-3023.

Bowman BH, Taylor JW. 1993. Molecular phylogeny of pathogenic and non-pathogenic Onygenales. In: Reynolds DR, Taylor JW. The Fungal Holomorph: Mitotic, Meiotic and Pleomorphic Speciation in Fungal Sytematics. Wallingford: CAB International.

Bowman BH, White TJ, Taylor JW. 1996. Human pathogeneic fungi and their close nonpathogenic relatives. *Molecular Phylogenetics and Evolution* 6: 89-96.

Brandelli A. 2008. Bacterial keratinases: useful enzymes for bioprocessing agroindustrial wastes and beyond. *Food and Bioprocess Technology* 1(2): 105-116.

Brandsberg JW, Weeks RJ, Hill WB, Piggott WR. 1969. A study of fungi found in association with *Histoplasma capsulatum*: three bird roosts in SE Missouri, USA. *Mycopathologia* 38(1): 71-81.

Brandt ME, Gaunt D, Iqbal N, McClinton S, Hambleton S, Sigler L. 2005. False-positive *Histoplasma capsulatum* Gen-Probe chemiluminescent test result caused by a *Chrysosporium* species. *Journal of Clinical Microbiology* 43(3): 1456-1458.

Brasch J, Müller S, Gräser Y. 2015. Unusual strains of *Microsporum audouinii* causing tinea in Europe. *Mycoses* 58(10): 573-577.

Brouta F, Descamps F, Fett T, Losson B, Gerday C, Mignon B. 2001. Purification and characterization of a 43.5 kDa keratinolytic metalloprotease from *Microsporum canis*. *Medical Mycology* 39(3): 269-275.

Brown AH, Smith G. 1957. The genus *Paecilomyces* Bainier and its perfect stage *Byssochlamys* Westling. *Transactions of the British Mycological Society* 40(1): 17-89.

Buckley KF, Dobson AD. 1998. Extracellular ligninolytic enzyme production and polymeric dye decolourization in immobilized cultures of *Chrysosporium lignorum* CL1. *Biotechnology Letters* 20: 301-306.

Bukhtojarov FE, Ustinov BB, Salanovich TN, Antonov AI, Gusakov AV, Okunev ON, Sinitsyn AP. 2004. Cellulase complex of the fungus *Chrysosporium lucknowense*: isolation and characterization of endoglucanases and cellobiohydrolases. *Biochemistry* (*Moscow*) 69(5): 542-551.

Bulter T, Alcalde M, Sieber V, Meinhold P, Schlachtbauer C, Arnold FH. 2003. Functional expression of a fungal laccase in *Saccharomyces cerevisiae* by directed evolution. *Applied Environmental Microbiology* 69: 987-995.

Burge MN, Isaac I. 1974. *Phialophora asteris*, causal agent of Aster wilt. *Transactions of the British Mycological Society* 62(2): 367-376.

Burnside CE. 1928. Saprophytic fungi associated with the honeybee. *Michigan Academy of Science* 8: 59-86.

Burtseva YV, Verigina NS, Sova VV, Pivkin MV, Zvyagintseva TN. 2003. Filamentous marine fungi as producers of *O*-glycosylhydrolases: β-1,3-glucanase from *Chaetomium indicum*. *Marine Biotechnology* 5(4): 349-359.

Cabello MN. 1989. Deuteromycotina from Antarctica: new species of hyphomycetes from Danco coast, Antarctic Peninsula. *Mycotaxon* 36(1): 91-94.

Cai J, Yang J, Du Y, Fan L, Qiu Y, Li J, Kennedy JF. 2006. Purification and characterization of chitin deacetylase from *Scopulariopsis brevicaulis*. *Carbohydrate Polymers* 65(2): 211-217.

Cain RF. 1952. Studies of fungi imperfecti. I. *Phialophora*. *Canadian Journal of Botany* 30(3): 338-343.

Călin M, Dracea O, Raut I, Vasilescu G, Doni MB, Arsene ML, Alexandrescu E, Aruxandei DC, Jecu L, Lazar V. 2016. *In vitro* biodegradation of keratinized substrates by keratinophilic fungi. *Scientific Bulletin* Series F, 20: 248-253.

Călin M, Raut I, Doni M, Alexandrescu E, Macovescu G, Arsene ML, Gurban AM, Vasilescu G, Jecu L. 2019. The potential of keratinolytic fungi for biotechnological applications in leather manufacture.

Revista de Chimie 70(9): 3152-3157.

Canevascini G, Borer P, Dreyer JL. 1991. Cellobiose dehydrogenases of *Sporotrichum* (*Chrysosporium*) *thermophile*. *European Journal of Biochemistry* 198(1): 43-52.

Cano J, Guarro J. 1994. Studies on keratinophilic fungi III. *Chrysosporium siglerae* sp. nov. *Mycotaxon* 51: 75-79.

Čapek A, Fassatiova O, Hanč O. 1975. Progesterone transformations as a biochemical feature of species of the genus *Scopulariopsis*. *Folia Microbiologica* 20(6): 517-518.

Caretta G, Piontelli E. 1977. *Microsporum magellanicum* and *Cunninghamella antarctica*, new species isolated from australic and antarctic soil of Chile. *Sabouraudia* 15(1): 1-10.

Carmichael JW. 1962. *Chrysosporium* and some other aleuriosporic hyphomycetes. *Canadian Journal of Botany* 40: 1137-1173.

Carmo ES, Belem LF, Catão RM. 2007. Microbiota fúngica presente em diversos setores de um hospital público em Campina Grande-PB. *Revista Bŕasileira de Análises Clínicas* 39: 213-216.

Carrión AL. 1942. Chromoblastomycosis. *Mycologia* 34(4): 424-441.

Castellani A. 1939. Brief note on three new species of *Trichophyton*. *T. batonrougei*, *T. guzzonii*, *T. tenuishypha*, and a little known species *T. louisianicum*. *Journal of Tropical Medicine and Hygiene* 42: 373-378.

Castellani A, Chalmers AJ. 1919. Manual of tropical medicine. New York: Williams, Wood, and Co.: 2110-2150.

Cejp K, Milko AA. 1966. Genus *Pseudogymnoascus* Radio (Gymnoascaceae). *Èeská Mykologie* 20: 160-163.

Chabasse D. 1988. Taxonomic study of keratinophilic fungi isolated from soil and some mammals in France. *Mycopathologia* 101(3): 133-140.

Chabert J, Nicot J. 1966. Aerobiology notes: density of *U. spores* in the air of Rabat (Morocco). Predominance of cynodontis. *Bulletin Trimestriel de la Societe Mycologique de France* 82(4): 569-581.

Chen WH, Zeng GP, Luo Y, Liang ZQ, Han YF. 2017. Morphological traits and molecular analysis for *Geomyces fujianensis* sp. nov. from China. *Mycosphere* 8(1): 38-43.

Chen Z, Bi XL, Gu J, Xu XG, Wang Y, Wu JH. 2013. White paint dot-like lesions of the scrotum: *Microsporum gypseum* infection. *Australasian Journal of Dermatology* 54(4): 95-96.

Cheng S, Chong L. 2002. A prospective epidemiological study on tinea pedis and onychomycosis in Hong Kong. *Chinese Medical Journal* 115(6): 860-865.

Childress MK, Dragone NB, Young BD, Adams BJ, Fierer N, Quandt CA. 2025. Three new *Pseudogymnoascus* species (Pseudeurotiaceae, Thelebolales) described from antarctic soils. *IMA Fungus* 16: e142219.

Choi JS, Graser Y, Walther G, Peano A, Symoens F, de Hoog S. 2012. *Microsporum mirabile* and its teleomorph *Arthroderma mirabile*, a new dermatophyte species in the *M. cookei* clade. *Medical Mycology* 50(2): 161-169.

Choi KS, Lee SH, Hong SP, Lee HD, Bae DK, Choi C. 1998. Characteristics and action pattern of α-galactosidase from *Scopulariopsis brevicaulis* in Korean traditional Meju. *Applied Biological Chemistry* 41(7): 489-495.

Chong LM, Sheridan JE. 1982. Mycoflora of barley (*Hordeum vulgare* L.) seed in New Zealand. *New Zealand Journal of Botany* 20(2): 187-189.

Claus H, Faber G, König H. 2002. Redox-mediated decolorization of synthetic dyes by fungal laccases. *Applied Microbiology and Biotechnology* 59(6): 672-678.

Claußen M, Schmidt S. 1998. Biodegradation of phenol and *p*-cresol by the hyphomycete *Scedosporium apiospermum. Research in Microbiology* 149(6): 399-406.

Clauβen M, Schmidt S. 1999. Biodegradation of phenylbenzoate and some of its derivatives by *Scedosporium apiospermum. Research in Microbiology* 150(6): 413-420.

Cole GT, Kendrick B. 1973. Taxonomic studies of *Phialophora. Mycologia* 65(3): 661-688.

Conant NF. 1937. The occurrence of a human pathogenic fungus as a saprophyte in nature. *Mycologia* 29(5): 597-598.

Conant NF, Smith DT, Baker RD, Callaway JL, Martin DS. 1954. Manual of Clinical Mycology. Philadelphia: W. B. Saunders Co. 2: 240-253.

Cooke WB, Kabler P. 1955. Isolation of potentially pathogenic fungi from polluted water and sewage. *Public Health Reports* 70(7): 689.

Cooney DG, Emerson R. 1964. Thermophilic fungi. an account of their biology, activities, and classification. *Quarterly Review of Biology* 6(2): 292.

Corda ACJ. 1833. Deutschlands Flora, Abt. III. *Die Pilze Deutschlands* 3: 65-96.

Cortez KJ, Roilides E, Quiroz-Telles F, Meletiadis J, Antachopoulos C, Knudsen T, Buchanan W, Milanovich J, Sutton DA, Fothergill A, Rinaldi MG, Shea YR, Zaoutis T, Kottilill S, Walsh TJ. 2008. Infections caused by *Scedosporium* spp. *Clinical Microbiology Reviews* 21(1): 157-197.

Costantin J. 1892. Sur quelques maladies du blanc de champignon. *Cr Hebd Séanc Acad Sci Paris* 114: 849-851.

Crous PW, Cowan DA, Maggs-Kölling G, Yilmaz N, Larsson E, Angelini C, Groenewald JZ. 2020. Fungal Planet description sheets: 1112-1181. *Persoonia* 45: 251.

Crous PW, Wingfield MJ, Guarro J, Cheewangkoon R, van der Bank M, Swart WJ, Stchigel AM, Cano-Lira JF, Roux J, Madrid H, Damm U, Wood AR, Shuttleworth LA, Hodges CS, Munster M, de Jesús Yáñez-Moralcs M, Zúñiga-Estrada L, Cruywagen EM, de Hoog GS, Silvera C, Najafzadeh J, Davison EM, Davison PJN, Barrett MD, Barrett RL, Manamgoda DS, Minnis AM, Kleczewski NM, Flory SL, Castlebury LA, Clay K, Hyde KD, Maússe-Sitoe SND, Chen S, Lechat C, Hairaud M, Lesage-Meessen L, Pawłowska J, Wilk M, Śliwińska-Wyrzychowska A, Mętrak M, Wrzosek M, Pavlic-Zupanc D, Maleme HM, Slippers B, Mac Cormack WP, Archuby DI, Grünwald NJ, Tellería MT, Dueñas M, Martín MP, Marincowitz S, de Beer ZW, Perez CA, Gené J, Marin-Felix Y, Groenewald JZ. 2013. Fungal Planet description sheets 154-213. *Persoonia-Molecular Phylogeny and Evolution of Fungi* 31: 188-296.

Crous PW, Wingfifield MJ, Richardson DM, Le Roux JJ, Strasberg D, Edwards J, Roets F, Hubka V, Taylor PWJ, Heykoop M, Martín MP, Moreno G, Sutton DA, Wiederhold NP, Barnes CW, Carlavilla JR, Gené J, Giraldo A, Guarnaccia V, Guarro J, Hernández-Restrepo M, Kolařík M, Manjón JL, Pascoe IG, Popov ES, Sandoval-Denis M, Woudenberg JHC, Acharya K, Alexandrova AV, Alvarado P, Barbosa RN, Baseia IG, Blanchette RA, Boekhout T, Burgess TI, Cano-Lira JF, Čmoková A, Dimitrov RA, Dyakov MY, Dueñas M, Dutta AK, Raventós FE, Fedosova AG, Fournier J, Gamboa P, Gouliamova DE, Grebenc T, Groenewald M, Hanse B, Hardy GESTJ, Held BW, Jurjević Ž, Kaewgrajang T, Latha KPD, Lombard L, Luangsa-ard JJ, Lysková P, Mallátová N, Manimohan P, Miller AN, Mirabolfathy M, Morozova OV, Obodai M, Oliveira NT, Ordóñez ME, Otto EC, Paloi S, Peterson SW, Phosri C, Roux J, Salazar WA, Sánchez A, Sarria GA, Shin HD, Silva BDB, Silva GA, Smith MTH, Souza-Motta CM, Stchigel AM, Stoilova-Disheva MM, Sulzbacher MA, Telleria MT, Toapanta C, Traba JM, Valenzuela-Lopez N, Watling R, Groenewald JZ. 2016. Fungal Planet description sheets: 400-468. *Persoonia* 36: 316-458.

Crous PW, Wingfield MJ, Lombard L, Roets F, Swart WJ, Alvarado P, Carnegie AJ, Moreno G, Luangsa-ard J, Thangavel R, Alexandrova AV, Baseia IG, Bellanger JM, Bessette AE, Bessette AR,

Delapeña-Lastra S, García D, Gené J, Pham THG, Heykoop M, Malysheva E, Malysheva V, Martín MP, Morozova OV, Noisripoom W, Overton BE, Rea AE, Sewall, BJ, Smith, ME, Smyth, CW, Tasanathai K, Visagie CM, Adamčík S, Alves A, Andrade JP, Aninat MJ, Araújo RVB, Bordallo JJ, Boufleur T, Baroncelli R, Barreto RW, Bolin J, Cabero J, Caboň M, Cafà G, Caffot MLH, Cai L, Carlavilla JR, Chávez R, Decastro RRL, Delgat L, Deschuyteneer D, Dios MM, Domínguez LS, Evans HC, Eyssartier G, Ferreira BW, Figueiredo CN, Liu F, Fournier J, Galli-Terasawa LV, Gil-Durán C, Glienke C, Gonçalves MFM, Gryta H, Guarro J, Himaman W, Hywel-Jones N, Iturrieta-González I, Ivanushkina NE, Jargeat P, Khalid AN, Khan J, Kiran M, Kiss L, Kochkina GA, Kolařík M, Kubátová A, Lodge DJ, Loizides M, Luque D, Manjón JL, Marbach PAS, Massolajr NS, Mata M, Miller AN, Mongkolsamrit S, Moreau PA, Morte A, Mujic A, Navarro-Ródenas A, Németh MZ, Nóbrega TF, Nováková A, Olariaga I, Ozerskaya SM, Palma MA, Petters-Vandresen DAL, Piontelli E, Popov ES, Rodríguez A, Requejo Ó, Rodrigues ACM, Rong IH, Roux J, Seifert KA, Silva BDB, Sklenář F, Smith JA, Sousa JO, Souza HG, Desouza JT, Švec K, Tanchaud P, Tanney JB, Terasawa F, Thanakitpipattana D, Torres-Garcia D, Vaca I, Vaghefi N, Vaniperen AL, Vasilenko OV, Verbeken A, Yilmaz N, Zamora JC, Zapata M, Jurjevi Ž, Groenewald JZ. 2019. Fungal planet description sheets: 951-1041. *Persoonia* 43: 223-425.

Currah RS. 1985. Taxonomy of the Onygenales: Arthrodermataceae, Gymnoascaceae, Myxotrichaceae, and Onygenaceae. *Mycotaxon* 24: 1-216.

Curzi M. 1930. Una nuova specie di Microascus. *Bolletino della Stazione di Patologia Vegetale di Roma* 10: 302-309.

Curzi M. 1931. Rapporti fra i generi Microascus Zukal e *Scopulariopsis* Bainier. *Bolletino della Stazione di Patologia Vegetale di Roma* 11: 55-60.

Da Silva M, Umbuzeiro GA, Pfenning LH, Canhos VP, Espoito E. 2003. Filamentous fungi isolated from estuarine sediments contaminated with industrial discharges. *Soil and Sediment Contamination: an International Journal* 12(3): 345-356.

Dabrowa N, Landau JW, Newcomer VD, Plunkett OA. 1964. A survey of tide-washed coastal areas of southern California for fungi potentially pathogenic to man. *Mycopathologia et Mycologia Applicata* 24(2): 137-150.

Dai YC, Cui BK, Si J, He SH, Hyde KD, Yuan HS, Liu XY, Zhou LW. 2015. Dynamics of the worldwide number of fungi with emphasis on fungal diversity in China. *Mycological Progress* 14: 62.

Daniels G. 1953. The digestion of human hair keratin by *Microsporum canis* Bodin. *Journal of General Microbiology* 8: 289-294.

Davies JS, Wellman AM, Zajic JE. 1973. Hypomycetes utilizing natural gas. *Canadian Journal of Microbiology* 19(1): 81-85.

Dayanandan A, Kanagaraj J, Sounderraj L, Govindaraju R, Rajkumar GS. 2003. Application of an alkaline protease in leather processing: an ecofriendly approach. *Journal of Cleaner Production* 11(5): 533-536.

de Hoog GS, Dukik K, Monod M, Packeu A, Stubbe D, Hendrickx M, Kupsch C, Stielow JB, Freeke J, Göker M, Rezaei-Matehkolaei A, Mirhendi H, Gräser Y. 2017. Toward a novel multilocus phylogenetic taxonomy for the dermatophytes. *Mycopathologia* 182(1-2): 5-31.

de Hoog GS, Guarro J, Gene J, Figueras MJ. 2000a. Atlas of Clinical Fungi. 2nd edition. Netherlands, Utrecht: Centraalbureau voor Schimmelcultures (CBS).

de Hoog GS, Marvin-Sikkema FD, Lahpoor GA, Gottschall JC, Prins RA, Guého E. 1994. Ecology and physiology of the emerging opportunistic fungi *Pseudallescheria boydii* and *Scedosporium prolificans*: Ökologie und Physiologie der opportunistischen Pilze *Pseudallescheria boydii* und *Scedosporium prolificans*. *Mycoses* 37(3-4): 71-78.

de Hoog GS, Mayser P, Haase G, Horre R, Horrevorts AM. 2000b. A new species, *Phialophora europaea*, causing superficial infections in humans. *Mycoses* 43(11/12): 409-416.

de Hoog GS, Weenink XO, van den Ende AHGG. 1999. Taxonomy of the *Phialophora verrucosa* complex with the description of two new species. *Studies in Mycology* 43: 107-121.

de Stefano M, Nicoletti R, Raimo F. 2002. Antagonism of *Scopulariopsis candida* against *Phytophthora nicotianae*: ultrastructural evidence. *Journal of Plant Pathology* 84: 180.

Deacon JW, Scott DB. 1983. *Phialophora zeicola* sp. nov., and its role in the root rot-stalk rot complex of maize. *Transactions of the British Mycological Society* 81(2): 247-262.

Deepak C, Shahnawaz MOHD, Butt MI, Sanjay S. 2017. Isolation and characterization of protease from *Pseudogymnoascus* sp. strain BPF6. *Trends in Biosciences* 10(42): 8852-8859.

Del-Cid A, Ubilla P, Ravanal MC, Medina E, Vaca I, Levicán G, Chávez R. 2014. Cold-active xylanase produced by fungi associated with Antarctic marine sponges. *Applied Biochemistry and Biotechnology* 172(1): 524-532.

Deshmukh SK. 2004. Isolation of dermatophytes and other keratinophilic fungi from the vicinity of salt pan soils of Mumbai, India. *Mycopathologia* 157: 265-267.

Deshmukh SK, Mandeel QA, Verekar SA. 2008. Keratinophilic fungi from selected soils of Bahrain. *Mycopathologia* 165(3): 143-147.

Deshmukh SK, Verekar SA. 2006. Keratinophilic fungi from the vicinity of meteorite crater soils of Lonar (India). *Mycopathologia* 162(4): 303-306.

Destino L, Sutton DA, Helon AL, Havens PL, Thometz JG, Willoughby RE, Chusid MJ. 2006. Severe osteomyelitis caused by *Myceliophthora thermophila* after a pitchfork injury. *Annals of Clinical Microbiology and Antimicrobials* 5(1): 21.

Dodge CW. 1935. Medical Mycology. St. Louis: The CV Mosby Company: 850.

Dominik T. 1967. *Chrysosporium* Corda. *Zeszyty Naukowe Wyzszej Szkoly Rolniczej* 24: 37-66.

Dominik T, Majchrowicz I. 1970. Further contribution to the knowledge of keratinolytic and keratinophilic soil fungi of the region of Szczecin—keratinolytic and keratinophilic fungi in the immediate surroundings of cattle. *Ekologia Polska* 18(29): 571-611.

Domsch KH, Gams W, Anderson TH. 1980. Compendium of Soil Fungi. Volume 1. London: Academic Press Ltd.

Domsch KH, Gams W, Anderson TH. 1993. Compendium of Soil Fungi (Reprint of 1980 Ed. Verlag). London: Academic Press Ltd.

Donk MA. 1974. Check List of European Polypores. Amsterdam: North-Holland Publishing.

Donoghue AJO, Knudsen GM, Beekman C, Perry JA, Johnson AD, DeRisi JL, Craik CS, Bennett RJ. 2015. Destructin-1 is a collagen-degrading endopeptidase secreted by *Pseudogymnoascus destructans*, the causative agent of white-nose syndrome. *Proceedings of the National Academy of Sciences of the United States of America* 112(24): 7478-7483.

Doyer CM. 1927. Meded. Phytopathological Laboratory 'Willie Commelin Scholten' 10: 32.

Duncan SM, Minasaki R, Farrell RL, Thwaites JM, Held BW, Arenz BE, Blanchette RA. 2008. Screening fungi isolated from historic Discovery Hut on Ross Island, Antarctica for cellulose degradation. *Antarctic Science* 20(5): 463-470.

Eidam E. 1880. Beitrag zur Kenntnis der Gymnoasceen. *Beiträge zur Biologie der Pflanzen* 3: 267-305.

Ellis MB. 1971. More dematiaceous hyphomycetes. *Mycologia* 83(69): 439.

Ellis MB. 1976. More dematiaceous hyphomycetes. England, Kew, Surrey: Commonwealth Mycological Institute.

Elnaggar M, Liu Z, Ebada S, Ashour M, Ebrahim W, Singab A, Proksch P. 2015. New xanthone and sesquiterpene derivatives from *Scopulariopsis* sp., a marine-derived fungus isolated from the Red Sea hard coral *Stylophora* sp. *Planta Medica* 81: 183.

Elnaggar MS, Ebada SS, Ashour ML, Ebrahim W, Müller WEG, Mándi A, Kurtán T, Singab A, Lin WH, Liu Z, Proksch P, Proksch P. 2016. Xanthones and sesquiterpene derivatives from a marine-derived fungus *Scopulariopsis* sp. *Tetrahedron* 72(19): 2411-2419.

Elnaggar MS, Ebada SS, Ashour ML, Ebrahim W, Singab A, Lin WH, Liu Z,Proksc P, Proksch P. 2017. Two new triterpenoids and a new naphthoquinone derivative isolated from a hard coral-derived fungus *Scopulariopsis* sp. *Fitoterapia* 116: 126-130.

Emmons CW. 1934. Dermatophytes: natural grouping based on the form of the spores and accessory organs. *Archives of Dermatology and Syphilology* 30(3): 337-362.

Emmons CW. 1944. *Allescheria boydii* and *Monosporium apiospermum*. *Mycologia* 36(2): 188-193.

Emmons CW, Carrión AL. 1937. Sporulation of the *Phialophora* Type in Hormodendrum. *Mycologia* 29(3): 327-333.

English MP, Stockdale PM. 1968. *Trichophyton proliferans* sp. nov., a human pathogen. *Sabouraudia* 6(3): 267-270.

Erne JB, Walker MC, Strik N, Alleman AR. 2007. Systemic infection with *Geomyces organisms* in a dog with lytic bone lesions. *Journal of the American Veterinary Medical Association* 230(4): 537-540.

Fakhfakh-Zouari N, Haddar A, Hmidet N, Frikha F, Nasri M. 2010. Application of statistical experimental design for optimization of keratinases production by *Bacillus pumilus* A1 grown on chicken feather and some biochemical properties. *Process Biochemistry* 45(5): 617-626.

Farina C, Gamba A, Tambini R, Beguin H, Trouillet JL. 1998. Fatal aortic *Myceliophthora thermophila* infection in a patient affected by cystic medial necrosis. *Medical Mycology* 36(2): 113-118.

Fenice M, Selbmann L, Zucconi L, Onofri S. 1997. Production of extracellular enzymes by Antarctic fungal strains. *Polar Biology* 17: 275-280.

Feng CH. 2012. The primary study on the diversity of fungi from sediment from Yellow Sea and Yangtze Estuary. Qingdao: Master's Thesis of Ocean University of China. [冯春辉. 2012. 黄海、长江口沉积物中真菌属种多样性初步研究. 青岛: 中国海洋大学硕士学位论文.]

Fergus CL, Sinden JW. 1969. A new thermophilic fungus from mushroom compost: *Thielavia thermophila* spec. nov. *Canadian Journal of Botany* 47: 1635-1637.

Figueroa L, Jiménez C, Rodríguez J, Areche C, Chávez R, Henríquez M, Cruz MDL, Díaz C, Segade YR, Vaca I. 2014. 3-Nitroasterric acid derivatives from an antarctic sponge-derived *Pseudogymnoascus* sp. Fungus. *Journal of Natural Products* 78(4): 919-923.

Fike JM, Kollipara R, Alkul S, Stetson CL. 2018. Case report of onychomycosis and tinea corporis due to *Microsporum gypseum*. *Journal of Cutaneous Medicine and Surgery* 22(1): 94-96.

Fisher JF, Shadomy S, Teabeaut JR, Woodward J, Michaels GE, Newman MA, White E, Cook P, Seagraves A, Yaghmai F, Rissing JP. 1982. Near-drowning complicated by brain abscess due to *Petriellidium boydii*. *Archives of Neurology* 39(8): 511-513.

Floersheim GL, Looser R, Grundmann HP, Meyer JC. 1982. Inhibition of growth of human cancers by extracts from *Trichophyton verrucosum*. *The Lancet* 319(8274): 708-710.

Frank J, Crous PW, Groenewald JZ, Oertel B, Hyde KD, Phengsintham P, Schroers HJ. 2010. *Microcyclospora* and *Microcyclosporella*: novel genera accommodating epiphytic fungi causing sooty blotch on apple. *Persoonia* 24(1): 93-105.

Fredenhagen A, Petersen F, Tintelnot-Blomley M, Rösel J, Mett H, Hug P. 1997. Semicochliodinol A and B: inhibitors of HIV-1 protease and EGF-R protein tyrosine kinase related to asterriquinones produced by the fungus *Chrysosporium merdarium*. *The Journal of Antibiotics* 50(5): 395-401.

Fuller L. 2009. Changing face of tinea capitis in Europe. *Current Opinion in Infectious Diseases* 22(2):

115-118.
Gabriel KT, Neville JJ, Pierce GE, Cornelison CT. 2019. Lipolytic Activity and the Utilization of Fatty Acids Associated with Bat Sebum by *Pseudogymnoascus destructans*. *Mycopathologia* 184: 625-636.
Gams W. 2016. Recent Changes in Fungal Nomenclature and Their Impact on Naming of Microfungi. In: Li DW. Biology of Microfungi. Switzerland: Springer International Publishing.
Gams W, Holubová-Jechová V. 1976. Chloridium and some other dematiaceous Hyphomycetes growing on decaying wood. *Studies in Mycology* 13: 1-99.
Gams W, McGinnis MR. 1983. *Phialemonium*, a new anamorph genus intermediate between *Phialophora* and *Acremonium*. *Mycologia* 75(6): 977-987.
Gan GG, Kamarulzaman A, Goh KY, Ng KP, Na SL, Soo-Hoo TS. 2002. Non-sporulating *Chrysosporium*: an opportunistic fungal infection in a neutropenic patient. *The Medical Journal of Malaysia* 57(1): 118-122.
Ganaie MA, Sood S, Rizvi G, Khan TA. 2010. Isolation and Identification of Keratinophilic Fungi from Different Soil Samples in Jhansi City (India). *Plant Pathology Journal* 9(4): 194-197.
Gao B, Mao Y, Zhang L, He L, Wei D. 2016. A novel saccharifying α-amylase of Antarctic psychrotolerant fungi *Geomyces pannorum*: Gene cloning, functional expression, and characterization. *Starch-Stärke* 68(1-2): 20-28.
Gargas A, Trest MT, Christensen M, Volk TJ, Blehert DS. 2009. *Geomyces destructans* sp. nov. associated with bat white-nose syndrome. *Mycotaxon* 108(1): 147-154.
Gawas-Sakhalkar P, Singh SM. 2011. Fungal community associated with Arctic moss, *Tetraplodon mimoides* and its rhizosphere: bioprospecting for production of industrially useful enzymes. *Current Science* 100(11): 1701-1705.
Gené J, Guarro J, Ulfig K, Vidal P, Cano J. 1994. Studies on keratinophilic fungi. II. *Chrysosporium pilosum* sp. nov. *Mycotaxon* 50: 107-113.
Geng YH. 2008. A survey on soil dematiaceous hyphomycetes at species level from Tibetan Plateau. Tai'an: Master's Thesis of Shandong Agricultural University. [耿月华. 2008. 西藏高原地区土壤中的暗色丝孢菌物种多样性考察. 泰安: 山东农业大学硕士学位论文.]
Geng YH, Zhang TY. 2009. Soil dematiaceous hyphomycetes from Changtang Steppe in northern Tibet. *Mycosystema* 28(5): 660-663.
Georg LK. 1950. The relation of nutrition to the growth and morphology of *Trichophyton faviforme*. *Mycologia* 42(6): 693-716.
Georg LK. 1956. Studies on *Trichophyton tonsurans*. I. The taxonomy of *T. tonsurans*. *Mycologia* 48(1): 65-82.
Georg LK, Maechling EH. 1949. *Trichophyton mentagrophytes*, variety nodular, *a* mutant with brilliant orange-red pigment isolated in nine cases of ringworm of the skin and nails. *Journal of Investigative Dermatology* 13(6): 339-350.
Gern RM, Furlan SA, Ninow JL, Jonas R. 2001. Screening for microorganisms that produce only endo-inulinase. *Applied Microbiology and Biotechnology* 55(5): 632-635.
Gianni C, Caretta G, Romano C. 2003. Skin infection due to *Geomyces pannorum* var. *pannorum*. *Mycoses* 46(9-10): 430-432.
Gilgado F, Cano J, Gené J, Guarro J. 2005. Molecular phylogeny of the *Pseudallescheria boydii* species complex: proposal of two new species. *Journal of Clinical Microbiology* 43(10): 4930-4942.
Gilgado F, Cano J, Gené J, Sutton DA, Guarro J. 2008. Molecular and phenotypic data supporting distinct species statuses for *Scedosporium apiospermum* and *Pseudallescheria boydii* and the proposed new species *Scedosporium dehoogii*. *Journal of Clinical Microbiology* 46(2): 766-771.

Gock MA, Hocking AD, Pitt JI, Poulos PG. 2003. Influence of temperature, water activity and pH on growth of some xerophilic fungi. *International Journal of Food Microbiology* 2481: 11-19.

Goddard HN. 1913. Can fungi living in agricultural soil assimilate free nitrogen? *Botanical Gazette* 56: 249-305.

Gokulshankar S, Ranjithsingh AJA, Ranjith MS, Ranganathan S, Palaniappan R. 2005. Role of *Chrysosporium keratinophillum* in the parasitic evolution of dermatophytes. *Mycoses* 48(6): 442-446.

Gräser Y, Kuijpers AFA, Presber W, Hoog GD. 1999. Molecular taxonomy of *Trichophyton mentagrophytes* and *T. tonsurans*. *Medical Mycology* 37(5): 315-330.

Gray LE, Gardner HW, Weisleder D, Leib M. 1999. Production and toxicity of 2,3-dihydro-5-hydroxy-2-methyl-4H-1-benzopyran-4-one by *Phialophora gregata*. *Phytochemistry* 50(8): 1337-1340.

Gruby D. 1843. Recherches sur la nature, le siege et le developpement du porrigo decalvans ou phytoalopecie. *Comptes Rendus de I'Academie Sciences* 17: 301-303.

Guarro J, Kantarcioglu AS, Horr R, Luis Rodriguez-Tudela J, Estrella MC, Berenguer J, de Hoog, GS. 2006. *Scedosporium apiospermum*: changing clinical spectrum of a therapy-refractory opportunist. *Medical Mycology* 44(4): 295-327.

Guarro J, Punsola L, Cano J. 1985. *Myceliophthora vellerea* (*Chrysosporium asperatum*) anamorph of *Ctenomyces serratus*. *Mycotaxon* 23: 419-427.

Gueho E, de Hoog GS. 1991. Taxonomy of the medical species of *Pseudallescheria* and *Scedosporium*. *Journal de Mycologie Médicale* 1(1): 3-9.

Gugnani HC, Paliwal-Joshi A, Rahman H, Padhye AA, Singh TSK, Das TK, Khanal B, Bajaj R, Rao S, Chukhani R. 2007. Occurrence of pathogenic fungi in soil of burrows of rats and of other sites in bamboo plantations in India and Nepal. *Mycoses* 50(6): 507-511.

Gusakov AV, Sinitsyn AP, Markov AV, Sinitsyna OA, Ankudimova NV, Berlin AG. 2001. Study of protein adsorption on indigo particles confirms the existence of enzyme-indigo interaction sites in cellulase molecules. *Journal of Biotechnology* 87(1): 83-90.

Gusakov AV, Sinitsyn AP, Salanovich TN, Bukhtojarov FE, Markov AV, Ustinov BB, van Zeijl C, Punt P, Burlingame R. 2005. Purification, cloning and characterisation of two forms of thermostable and highly active cellobiohydrolase I (Cel7A) produced by the industrial strain of *Chrysosporium lucknowense*. *Enzyme and Microbial Technology* 36(1): 57-69.

Han YF, Ge W, Zhang ZY, Liang JD, Chen WH, Huang JZ, Liang ZQ. 2022. Morphological and phylogenetic characterisations reveal nine new species of *Chrysosporium* (Onygenaceae, Onygenales) in China. *Phytotaxa* 539(1): 1-16.

Han YF, Wang YR, Liang JD, Liang ZQ. 2013. A new species of the genus *Chrysosporium* from the farmland soil of Qinghai province. *Mycosystema* 32: 606-611.

Han Z, Kautto L, Nevalainen H. 2017. Secretion of proteases by an opportunistic fungal pathogen *Scedosporium aurantiacum*. *PLoS One* 12(1): e0169403.

Hantschke D. 1969. Morphology and biology of *Trichophyton mentagrophytes* (Robin) Blanchard var. *goetzii* var. *nova*. *Mykosen* 12(2): 97-104.

Harmsen D, Schwinn A, Weig M, Bröcker EB, Heesemann J. 1995. Phylogeny and dating of some pathogenic keratinophilic fungi using small subunit ribosomal RNA. *Journal of Medical and Veterinary Mycology* 33: 299-303.

Harrington TC, McNew DL. 2003. Phylogenetc analysis places the *Phialophora*-like anamorph genus *Cadophora* in the Helotiales. *Mycotaxon* 87: 141-152.

Hatakeyama Y, Takeda H, Ooi T, Kinoshita S. 1996. Kinetic Parameters of β-Fructofuranosidase from

Scopulariopsis brevicaulis. *Journal of Fermentation and Bioengineering* 81(6): 518-523.
Havlickova B, Czaika VA, Friedrich M. 2008. Epidemiological trends in skin mycoses worldwide. *Mycoses* 51(S4): 2-15.
Hawksworth DL. 1983. A key to the lichen-forming, parasitic, parasymbiotic and saprophytic fungi occurring on lichens in the British Isles. *The Lichenologist* 15(1): 1-44.
Hawksworth DL. 2012. Global species numbers of fungi: are tropical studies and molecular approaches contributing to a more robust estimate? *Biodiversity and Conservation* 21(9): 2425-2433.
Hawksworth DL, Kirk PM, Sutton BC, Pegler DN. 1995. Ainsworth and Bisby's: Dictionary of the Fungi. 8th Edition. Wallingford: CAB International.
Hayakawa Y, Adachi H, Kim JW, Shin-ya K, Seto H. 1998. Adenopeptin, a new apoptosis inducer in transformed cells from *Chrysosporium* sp. *Tetrahedron* 54(52): 15871-15878.
Hayashi S, Naitoh K, Matsubara S, Nakahara Y, Nagasawa Z, Tanabe I, Kusaba K, Tadano J, Nishimura K, Sigler L. 2002. Pulmonary colonization by *Chrysosporium zonatum* associated with allergic inflammation in an immunocompetent subject. *Journal of Clinical Microbiology* 40(3): 1113-1115.
Haybrig P, Dania G, Josepa G, Josepa C. 2013. *Phialemoniopsis*, a new genus of Sordariomycetes, and new species of *Phialemonium* and *Lecythophora*. *Mycologia* 105(2): 398-421.
Hayes MA. 2012. The *Geomyces* fungi: ecology and distribution. *BioScience* 62(9): 819-823.
He JW, Chen GD, Gao H, Yang F, Li XX, Peng T, Guo LD, Yao XS. 2012. Heptaketides with antiviral activity from three endolichenic fungal strains *Nigrospora* sp., *Alternaria* sp. and *Phialophora* sp. *Fitoterapia* 83(6): 1087-1091.
Hedayati MT, Mohseni-Bandpi A, Moradi S. 2004. A survey on the pathogenic fungi in soil samples of potted plants from Sari hospitals, Iran. *Journal of Hospital Infection* 58(1): 59-62.
Heidemann S, Monod M, Gräser Y. 2010. Signature polymorphisms in the internal transcribed spacer region relevant for the differentiation of zoophilic and anthropophilic strains of *Trichophyton interdigitale* and other species of *T. mentagrophytes sensu lato*. *British Journal of Dermatology* 162(2): 282-295.
Hennebert GL. 1974. *Lomentospora prolificans*, a new hyphomycete from greenhouse soil. *Mycotaxon* 1: 50.
Henríquez M, Vergara K, Norambuena J, Beiza A, Maza F, Ubilla P, Darias MJ. 2014. Diversity of cultivable fungi associated with Antarctic marine sponges and screening for their antimicrobial, antitumoral and antioxidant potential. *World Journal of Microbiology and Biotechnology* 30(1): 65-76.
Hoarau G, Miquel J, Picot S. 2016. Kerion Celsi caused by *Microsporum gypseum*. *Journal of Pediatrics* 178: 296-297.
Hocking AD, Pitt JI. 1988. Two new species of xerophilic fungi and a further record of *Eurotium halophilicum*. *Mycologia* 80(1): 82-88.
Hocquette A, Grondin M, Bertout S, Mallié M. 2005. Les champignons des genres *Acremonium*, *Beauveria*, *Chrysosporium*, *Fusarium*, *Onychocola*, *Paecilomyces*, *Penicillium*, *Scedosporium* et *Scopulariopsis* *responsables* de hyalohyphomycoses. *Journal de Mycologie Médicale* 15(3): 136-149.
Hoshino Y, Ivanova VB, Yazawa K, Ando A, Mikami Y, Zaki SM, Karam AZ, Youssef YA, Gräfe U. 2002. Queenslandon, a new antifungal compound produced by *Chrysosporium queenslandicum*: production, isolation and structure elucidation. *The Journal of Antibiotics* 55(5): 516-519.
Hu KC, Xu MY, Li HJ, Yuan J, Tang G, Xu J, Yang DP, Lan WJ. 2016. Discovery of aromadendrane anologues from the marine-derived fungus *Scedosporium dehoogii* F41-4 by NMR-guided isolation. *RSC Advances* 6(97): 94763-94770.
Hua GR, Dong D, Jun WS, Qiang F, Bin B, Hui WW. 2017. Structural characteristics and fibrinolytic property of rediscovering rare compounds from polar-derived strain of marine fungi *Geomyces pannorum*. *Der Chemica Sinica* 8(5): 471-477.

Huang LH, Xu MY, Li HJ, Li JQ, Chen YX, Ma WZ, Li YP, Xu J, Yang DP, Lan WJ. 2017. Amino acid-directed strategy for inducing the marine-derived fungus *Scedosporium apiospermum* F41-1 to Maximize Alkaloid Diversity. *Organic Letters* 19(18): 4888-4891.

Hubálek Z, Hornich M. 1977. Experimental infection of white mouse with *Chrysosporium* and *Paecilomyces*. *Mycopathologia* 62(3): 173-178.

Hubka V, Dobiášová S, Dobiáš R, Kolařík M. 2014. *Microsporum aenigmaticum* sp. nov. from *M. gypseum* complex, isolated as a cause of tinea corporis. *Medical Mycology* 52(4): 387-396.

Hughes SJ. 1951. Studies on micro-fungi. XI. Some hyphomycetes which produce phialides. Mycological Papers No. 45. England, Kew: Comm. Mycol. Inst.

Hughes SJ. 1953. Conidiophores, conidia, and classification. *Canadian Journal of Botany* 31(5): 577-659.

Hughes SJ. 1958. Revisiones hyphomycetum aliquot cum appendice de nominibus rejiciendis. *Canadian Journal of Botany* 36: 727-836.

Hütter R. 1958. Untersuchungen über die Gattung Pyrenopeziza Fuck. *Journal of Phytopathology* 33: 1-54.

Ilkit M. 2010. Favus of the scalp: an overview and update. *Mycopathologia* 170(3): 143-154.

Ismail AMS, Housseiny MM, Abo-Elmagd HI, El-Sayed NH, Habib M. 2012. Novel keratinase from *Trichoderma harzianum* MH-20 exhibiting remarkable dehairing capabilities. *International Biodeterioration & Biodegradation* 70: 14-19.

Issakainen J, Jalava J, Hyvönen J, Sahlberg N, Pirnes T, Campbell CK. 2003. Relationships of *Scopulariopsis* based on LSU rDNA sequences. *Medical Mycology* 41: 31-42.

Ivanova LG. 1983. *Trichophyton sarkisovii* Ivanova et Polyakov sp. nov.: a new species of pathogenic fungus isolated from dermatomycosis of camel. *Mikologiya i Fitopatologiya* 17: 363-367.

Ivanova VB, Hoshino Y, Yazawa K, Ando A, Mikami Y, Zaki SM, Graefe U. 2002. Isolation and structure elucidation of two new antibacterial compounds produced by *Chrysosporium queenslandicum*. *The Journal of Antibiotics* 55(10): 914-918.

Iwatsu T, Miyaji M, Okamoto S. 1981. Isolation of *Phialophora verrucosa* and *Fonsecaea pedrosoi* from nature in Japan. *Mycopathologia* 75(3): 149-158.

Iwatsu T, Nishimura K, Miyaji M. 1985. Spontaneous disappearance of cutaneous sporotrichosis: report of two cases. *International Journal of Dermatology* 24(1): 524-525.

Iwatsu T, Udagawa S. 1985. A new species of *Phialophora* from Japan. *Mycotaxon* 24: 387-393.

Jain N, Sharma M. 2012. Biodiversity of keratinophilic fungal flora in university campus, Jaipur, India. *Iranian Journal of Public Health* 41(11): 27-33.

Jain PC, Agrawal SC. 1980. A note on the keratin decomposing capability of some fungi. *Transactions of the Mycoligical Societu of Japan* 21: 513-517.

Jain PC, Deshmukh SK, Agrawal SC. 1993. *Chrysosporium gourii* Jain, Deshmukh & Agrawal sp. nov. *Mycoses* 36: 77-79.

Janda-Ulfig K, Ulfig K, Cano J, Guarro J. 2008. A study of the growth of *Pseudallescheria boydii* isolates from sewage sludge and clinical sources on tributyrin, rapeseed oil, biodiesel oil and diesel oil. *Annals of Agricultural and Environmental Medicine* 15(1): 45-49.

Jiang YL, Wang Y. 2010. Two new species of *Phialophora* from Guizhou Province. *Mycosystema* 29(6): 783-785.

Johri BN, Alurralde JD, Klein J. 1990. Lipase production by free and immobilized protoplasts of *Sporotrichum* (*Chrysosporium*) *thermophile* Apinis. *Applied Microbiology and Biotechnology* 33(4): 367-371.

Joyeux C. 1912. Sur le *Trichophyton soudanense* sp. nov. *Comptes Redus des Seances de la Societe de Biologie et de Ses Filiales* 72: 15-16.

Jung H, Xu F, Li K. 2002. Purification and characterization of laccase from wood-degrading fungus *Trichophyton rubrum* LKY-7. *Enzyme and Microbial Technology* 30(2): 161-168.

Kacinová J, Tancinová D, Medo J. 2014. Production of extracellular keratinase by *Chrysosporium tropicum* and *Trichophyton ajelloi*. *The Journal of Microbiology, Biotechnology and Food Sciences* 3: 103-106.

Kadhim SK, Al-Janabi JK, Al-Hamadani AH. 2015. *In vitro*, determination of optimal conditions of growth and proteolytic activity of clinical isolates of *Trichophyton rubrum*. *Journal of Contemporary Medical Sciences* 1(3): 9-19.

Kandemir H, Dukik K, de Melo Teixeira M, Stielow JB, Delma FZ, AlHatmi AMS, Ahmed SA, Ilkit M, de Hoog GS. 2022. Phylogenetic and ecological reevaluation of the order Onygenales. *Fungal Diversity* 115: 1-72.

Kane J, Scott JA, Summerbell RC, Diena B. 1992. *Trichophyton krajdenii* sp. nov.: an anthropophilic dermatophyte. *Mycotaxon* 45: 307-316.

Katapodis K, Vršanská M, Kekos D, Nerinckx W, Biely P, Claeyssens M, Macris BJ, Christakopoulos P. 2003. Biochemical and catalytic properties of an endoxylanase purified from the filtrate of *Sporotrichum thermophile*. *Carbohydrate Research* 338: 1881-1890.

Kaur G, Kumar S, Satyanarayana T. 2004. Production, characterization and application of a thermostable polygalacturonase of a thermophilic mould *Sporotrichum thermophile* Apinis. *Bioresource Technology* 94(3): 239-243.

Khardenavis AA, Kapley A, Purohit HJ. 2009. Processing of poultry feathers by alkaline keratin hydrolyzing enzyme from *Serratia* sp. HPC 1383. *Waste Management* 29(4): 1409-1415.

Khizhnyak SV, Tausheva IV, Berezikova AA, Nesterenko EV, Rogozin D. 2003. Psychrophilic and psychrotolerant heterotrophic microorganisms of middle Siberian karst cavities. *Russian Journal of Ecology* 34 (4): 231-235.

Kiffer E, Morelet M. 1999. The Deuteromycetes-Mitosporic Fungi: classification and generic keys. Boca Raton: CRC Press.

Kinderlerer JL, Hatton PV, Chapman AJ, Rose ME. 1988. Essential oil produced by *Chrysosporium xerophilum* in coconut. *Phytochemistry* 27(9): 2761-2763.

Kingma FHVBT. 1943. Beschreibung der im centraalbureau voor schimmelcultures vorhandenen arten der gattungen *Phialophora* Thaxter und *Margarinomyces* Laxa, nebst Schüssel zu ihrer Bestimmung. *Antonie van Leeuwenhoek* 9(1): 51-76.

Kirk PW. 1967. A comparison of saline tolerance and sporulation in marine and clinical isolates of *Allescheria boydii* Shear. *Mycopathologia et Mycologia Applicata* 33(1): 65-75.

Kirk PM, Cannon PF, David JC, Stalpers JA. 2001. Ainsworth & Bisby's Dictionary of the Fungi. 9th edition. Wallingford: CAB International.

Klimek-Ochab M, Mucha A, Żymańczyk-Duda E. 2014. 2-Aminoethylphosphonate utilization by the cold-adapted *Geomyces pannorum* P11 strain. *Current Microbiology* 68(3): 330-335.

Kohno J, Hirano N, Sugawara K, Nishio M, Hashiyama T, Nakanishi N, Komatsubara S. 2001. Structure of TMC-69, a new antitumor antibiotic from *Chrysosporium* sp. TC 1068. *Tetrahedron* 57: 1731-1735.

Krasny L, Strohalm M, Bouchara JP, Sulc M, Lemr K, Barreto-Bergter E, Havlicek V. 2011. *Scedosporium* and *Pseudallescheria* low molecular weight metabolites revealed by database search. *Mycoses* 54(S3): 37-42.

Krempl-Lamprecht L. 1965. ber das Vorkommen von Pilzen aus der Gattung *Chrysosporium* auf der Haut und Diskussion ihrer systematischen Stellung. Berlin/Heidelberg/New York: Springer Verlag.

Krishnan A, Alias SA, Wong CMVL, Pang KL, Convey P. 2011. Extracellular hydrolase enzyme production by soil fungi from King George Island, Antarctica. *Polar Biology* 34(10): 1535-1542.

Krunic AL, Cetner A, Tesic V, Janda WM, Worobec S. 2007. A typical favic invasion of the scalp by *Microsporum canis*: report of a case and review of reported cases caused by *Microsporum* species. *Mycoses* 50(2): 156-159.

Kunert J. 1989. Biochemical mechanism of keratin degradation by the actinomycete *Streptomyces fradiae* and the fungus *Microsporum gypseum*: a comparison. *Journal of Basic Microbiology* 29(9): 597-604.

Kuroda K, Yoshida M, Uosaki Y, Ando K, Kawamoto I, Oishi E, Onuma H, Yamada KJ, Matsuda Y. 1993. AS-183, a novel inhibitor of acyl-CoA: cholesterol acyltransferase produced by *Scedosporium* sp. SPC-15549. *The Journal of Antibiotics* 46(8): 1196-1202.

Kushwaha RKS, Shrivastava JN. 1989. A new species of *Chysosporium*. *Current Science* 58(17): 970-971.

Kushwaha RKS, Guarro J. 2000. Biology of Dermatophytes and other Keratinophilic Fungi. Bilbao: Revista Iberoamericana de Micología.

Kushwaha RKS. 2000. The genus *Chrysosporium*, its physiology and biotechnological potential Kushwaha RKS, Guarro J1 Biology of Dermatophytes and other Keratinophilic Fungi. Bilbao: Revista Iberoamericana de Micología.

Laakkonen J, Sundell J, Soveri T. 1998. Lung parasites of least weasels in Finland. *Journal of Wildlife Diseases* 34(4): 816-819.

Labuda R, Bernreiter A, Hochenauer D, Kubátová A, Kandemir H, Schüller C. 2021. Molecular systematics of *Keratinophyton*: the inclusion of species formerly referred to *Chrysosporium* and description of four new species. *IMA Fungus* 12(1): 1-21.

Lackner M, de Hoog GS, Yang L, Moreno LF, Ahmed SA, Andreas F, Kaltseis J, Nagl M, Lass-Flörl C, Risslegger B, Rambach G, Speth C, Robert V, Buzina W, Chen SR, Bouchara JP, Cano-Lira JF, Guarro J, Gené J, Silva FF, Haido R, Haase G, Havlicek V, Garcia-Hermoso D, Meis JF, Hagen F, Kirchmair M, Rainer J, Schwabenbauer K, Zoderer M, Meyer W, Gilgado F, Schwabenbauer K, Vicente VA, Piecková E, Regenermel M, Rath PM, Steinmann J, Alencar XW, Symoens F, Tintelnot K, Ulfig K, Velegraki A, Tortorano AM, Giraud S, Mina S, Rigler-Hohenwarter K, Hernando FL, Ramirez-Garcia A, Pellon A, Kaur J, Bergter EB, de Meirelles JV, da Silva ID, Delhaes L, Alastruey-Izquerdo A, Li RY, Lu QY, Moussa T, Almaghrabi O, Al-Zahrani H, Okada G, Deng SW, Liao WQ, Zeng JS, Issakainen J, Lopes LCL. 2014. Proposed nomenclature for *Pseudallescheria*, *Scedosporium* and related genera. *Fungal Diversity* 67(1): 1-10.

Lackner M, Klaassen CH, Meis JF, van den Ende AHGG, de Hoog GS. 2012. Molecular identification tools for sibling species of *Scedosporium* and *Pseudallescheria*. *Medical Mycology* 50(5): 497-508.

Lagerberg T, Lundberg G, Melin E. 1927. Biological and practical researches into Blueing in Pine and Spruce. *Svenska Skogsvårdsföreningens Tidskrift* 25(2-4): 145-272.

Lange L, Huang Y, Busk PK. 2016. Microbial decomposition of keratin in nature: a new hypothesis of industrial relevance. *Applied Microbiology and Biotechnology* 100(5): 2083-2096.

Langeron M, Milochevitch S. 1930a. Metabolites of pathogenic fungi. VIII. 1-3 Floccosin and floccosic acid, two metabolites from *Epidermophyton floccosum*. *Annales de Parasitologie Humaine et Comparée* 8: 489.

Langeron M, Milochevitch S. 1930b. Morphologie des dermatophytes sur milieux naturels et milieux à base de polysaccharides. Essai de classification (Deuxième mémoire). *Annales de Parasitologie Humaine et Comparée* 8(5): 465-508.

Larcher G, Cimon B, Tronchin G, Chabasse D, Bouchara JP. 1996. A 33 kDa serine proteinase from *Scedosporium apiospermum*. *Biochemical Journal* 315(1): 119-126.

Le Gal M, Mangenot MF. 1956. Contribution à l'étude des Mollisioidées: 1. Note préliminaire: les formes conidiennes. *Revue de Mycologie* 21(1): 3-13.

Le Gal M, Mangenot MF. 1961. Contribution à l'étude des Mollisioldées: 4. *Revue de Mycologie* 26(5): 263-331.

Le Nours J, Ryttersgaard C, Lo Leggio L, Qstergaard PR, Borchert TV, Christensen LLH, Larsen S. 2003. Structure of two fungal β-1,4-galactanases: Searching for the basis for temperature and pH optimum. *Protein Science* 12(6): 1195-1204.

Lebasque J. 1934. Recherches morphologiques et biologiques sur les *Trichophyton mégasporés* du cheval et du boeuf. *Annales de Parasitologie Humaine et Comparée* 12(5): 418-444.

Lemaire JM, Ponchet J. 1963. *Phialophora radicicola* Cain, forme conidienne du *Linocarpon cariceti* B. et Br. *Compte rendu hebdomadaire des séances de l'Académie d'Agriculture de France* 49: 1067-1069.

Li CL. 2013. Diversity of marine-derived fungi from seaweeds in intertidal zone of Qingdao and sediments of Bohai Sea. Qingdao: Master's Thesis of Ocean University of China. [李长林. 2013. 青岛潮间带藻生与渤海沉积物真菌多样性研究. 青岛: 中国海洋大学硕士学位论文.]

Li X, Zhang ZY, Chen WH, Liang JD, Huang JZ, Han YF, Liang ZQ. 2022c. A new species of *Arthrographis* (Eremomycetaceae, Dothideomycetes), from the soil in Guizhou, China. *Phytotaxa* 538 (3): 175-181.

Li X, Zhang ZY, Chen WH, Liang JD, Liang ZQ, Han YF. 2022a. *Keratinophyton chongqingense* sp. nov. and *Keratinophyton sichuanense* sp. nov., from soil in China. *International Journal of Systematic and Evolutionary Microbiology* 72(8): 1-9.

Li X, Zhang ZY, Ren YL, Liang ZQ, Han YF. 2022b. Diversity and functional analysis of soil culturable microorganisms using a keratin baiting technique. *Microbiology* 91(5): 542-552.

Li Y, Sun B, Liu S, Jiang L, Liu X, Zhang H, Che Y. 2008. Bioactive asterric acid derivatives from the Antarctic ascomycete fungus *Geomyces* sp. *Journal of Natural Products* 71(9): 1643-1646.

Li Y, Xiao J, de Hoog GS, Wang X, Wan Z, Yu J, Liu W, Li R. 2017b. Biodiversity and human-pathogenicity of *Phialophora verrucosa* and relatives in Chaetothyriales. *Persoonia: Molecular Phylogeny and Evolution of Fungi* 38: 1-19.

Li Z, Zeng GP, Ren J, Han YF. 2017a. *Chrysosporium leigongshanense* sp. nov. from Guizhou Province, China. *Mycosphere* 8(8): 1030-1037.

Li Z, Zhang YW, Chen WH, Han YF. 2019. Morphological traits and molecular analysis for two new *Chrysosporium* species from Fujian Province, China. *Phytotaxa* 400(5): 257-264.

Li ZQ, Cui CQ. 1989. Study on the psychrophilic/psychrotrophic microorganisms III. *Geomyces laevis*, a new species of *Geomyces*. *Acta Mycoiogica Sinica* 8(1): 47-50.

Liang JD, Han YF, Du W, Liang ZQ, Li ZQ. 2009a. *Chrysosporium linfenense*: a new *Chrysosporium* species with keratinolytic activity. *Mycotaxon* 110: 65-71.

Liang JD, Han YF, Zhang JW, Du W, Liang ZQ, Li ZZ. 2011. Optimal culture conditions for keratinase production by a novel thermophilic *Myceliophthora thermophila* strain GZUIFR-H49-1. *Journal of Applied Microbiology* 110(3): 871-880.

Liang ZQ, Han YF, Chu HL, Fox RT. 2009b. Studies on the genus *Paecilomyces* in China V. *Taifanglania* gen. nov. for some monophialidic species. *Fungal Diversity* 34: 69-77.

Listemann H. 1973. *Trichophyton candelabreum* nov. spec. *Castellania* 1: 53-54.

Liu HM. 2004. Taxonomic studies on soil dematiaceous Hyphomycetes from warm temperate zone of Eastern China. Tai'an: Doctoral Dissertation of Shandong Agricultural University. [刘会梅. 2004. 中国东部暖温带地区土壤中暗色丝孢菌分类研究. 泰安: 山东农业大学博士学位论文.]

Liu HM, Zhang TY. 2004. A preliminary report of soil dematiaceous hyphomycetes from the Yellow River Delta I. *Mycosystema* 23(3): 77-78.

Liu YL, Xi PG, He XL, Jiang ZD. 2013. *Phialophora avicenniae* sp. nov., a new endophytic fungus in

Avicennia marina in China. *Mycotaxon* 124(1): 31-37.

Lloret L, Hollmann F, Eibes G, Feijoo G, Moreira MT, Lema JM. 2012. Immobilisation of laccase on Eupergit supports and its application for the removal of endocrine disrupting chemicals in a packed-bed reactor. *Biodegradation* 23(3): 373-386.

Locquin-Linard M. 1982. *Pseudogymnoascus dendroideus* Locquin-Linard, nouvelle espèce de Gymnascale (Ascomycètes) coprophile d'Afrique du Nord. *Cryptogamie Mycologie* 3: 409-414.

Lousberg RJ, Tirilly Y, Moreau M. 1976. Isolation of (－)-cryptosporiospin, a chlorinated cyclopentenone fungitoxic metabolite *from Phialophora asteris* f. sp. *helianthi*. *Experientia* 32(3): 331-332.

Lousberg RC, Tirilly Y. 1976. Structure of furasterin, a chlorinated metabolite from the fungus *Phialophora asteris* (Downson) Burge et Isaac. *Experientia* 32(11): 1394-1395.

Lu S, Xi LY, Zhang JM, Lu CM. 2009. Pseudomembranous-like tinea of the scrotum: report of six cases. *Mycoses* 52(3): 282-284.

Lukassen MB, Saei W, Sondergaard TE, Tamminen A, Kumar A, Kempken F, Wiebe MG, Sørensen JL. 2015. Identification of the scopularide biosynthetic gene cluster in *Scopulariopsis brevicaulis*. *Marine Drugs* 13(7): 4331-4343.

Lupa S, Seneczko F, Jeske J, Głowacka A, Ochęcka-Szymańska A. 1999. Epidemiology of dermatomycoses of humans in central Poland. Part IV. Onychomycosis due to dermatophytes. *Mycoses* 42(11-12): 657-659.

Lydolph MC, Jacobsen J, Arctander P, Gilbert MTP, Gilichinsky DA, Hansen AJ, Willerslev E, Lange L. 2005. Beringian paleoecology inferred from permafrost-preserved fungal DNA. *Applied and Environmental Microbiology* 71(2): 1012-1017.

MacCarthy L. 1925. Contribution a l'étude des épidermomycoses avec prés-entation de six parasites nouveaux. *Ann Dermatol Syphiligr* VI 6: 19-54.

Mahariya S, Sharma M. 2018. Fungal succession on keratinous hair and nail baits of human origin. *Mycopathologia* 183: 631-635.

Maheshwari R, Bharadwaj G, Bhat MK. 2000. Thermophilic fungi: their physiology and enzymes. *Microbiology and Molecular Biology Reviews* 64(3): 461-488.

Malloch D, Salkin I. 1984. A new species of *Scedosporium* associated with osteomyelitis in humans. *Mycotaxon* 21(10-11): 247-255.

Malviya HK, Rajak RC, Hasija SK. 1992. Purification and partial characterization of two extracellular keratinases of *Scopulariopsis brevicaulis*. *Mycopathologia* 119(3): 161-165.

Mangenot F. 1952. Recherches méthodiques sur les champignons de certains bois en décomposition. *Revue Générale de Botanique* 59: 381.

Marín-Felix Y, Stchigel AM, Miller AN, Guarro J, Cano-Lira JF. 2015. A re-evaluation of the genus *Myceliophthora* (Sordariales, Ascomycota): its segregation into four genera and description of *Corynascus fumimontanus* sp. nov. *Mycologia* 107: 619-632.

Matsushima K, Matsushima T. 1996. Matsushima Mycological Memoirs No. 9. Matsushima Fungus Collection, Kobe, Japan.

Matsushima T. 1983. Matsushima Mycological Memoirs No. 3. Matsushima Fungus Collection, Kobe, Japan.

Matsushima T. 1993. Matsushima Mycological Memoirs No. 7. Matsushima Fungus Collection, Kobe, Japan.

Mckittrick J, Chen PY, Bodde SG, Yang W, Novitskaya EE, Meyers MA. 2012. The structure, functions, and mechanical properties of keratin. *Journal of the Minerals Metals & Materials Society* 64(4): 449-468.

McNeill J. 2012. Guidelines for requests for binding decisions on application of the Code. *Taxon* 61: 477-478.

Medlar EM. 1915. A new fungus *Phialophora verrucosa*, pathogenic for man. *Mycologia* 7: 200-203.

Millar KR. 1990. A new species of *Phialophora* from lake sediment. *Mycologia* 82(5): 647-650.

Milochevitch S. 1931. Sur un cas de trichophytie produit par une espèce nouvelle de *Trichophyton*, *T. langeroni* n. sp. *Annales de Parasitologie Humaine et Comparée* 9(5): 456-461.

Milochevitch S. 1935. Une nouvelle espèce pathogène de *Ctenomyces*, *Ctenomyces bossæ* n. sp. *Annales de Parasitologie Humaine et Comparée* 13(6): 559-567.

Minnis AM, Lindner DL. 2013. Phylogenetic evaluation of *Geomyces* and allies reveals no close relatives of *Pseudogymnoascus destructans*, comb. nov., in bat hibernacula of eastern North America. *Fungal Biology* 117(9): 638-649.

Miranda MF, de Brito AC, Zaitz C, de Carvalho TN, Carneiro FR. 1998. *Microsporum gypseum* infection showing a white-paint-dot appearance. *International Journal of Dermatology* 37: 956-957.

Mitra SK. 1998. Dermatophytes isolated from selected ruminants in India. *Mycopathologia* 142(1): 13.

Mohanty SS, Prakash S. 2002. Efficacy of *Chrysosporium lobatum* against larvae of malaria vector, *Anopheles stephensi* in the laboratory. *Current Science* 1585-1588.

Mohanty SS, Prakash S. 2004. Extracellular metabolites of *Trichophyton ajelloi* against *Anopheles stephensi* and *Culex quinquefasciatus* larvae. *Current Science* 86(2): 323-325.

Moon SK, Shin YM, Shin DH, Choi JS, Kim KH. 2007. Two cases of dermatophytosis in patients with psoriasis. *Korean Journal Medical Mycology* 12(1): 18-22.

Moore M, de Almeida FP. 1936. New organisms of chromomycosis. *Annals of the Missouri Botanical Garden* 23(4): 543-552.

Moore RT. 2001. Hot fungi from Chernobyl. *Mycologist* 15(2): 63-64.

Moreau C. 1963. Morphologie comparee de quelques *Phialophora* et variations du *P. cinerescens* (Wr.) van Beyma. *Revue de Mycologie (Paris)* 28: 260-276.

Morio F, Fraissinet F, Gastinne T, Pape PL, Delaunay J, Sigler L, Gibas CFC, Miegeville M. 2011. Invasive *Myceliophthora thermophila* infection mimicking invasive aspergillosis in a neutropenic patient: a new cause of cross-reactivity with the *Aspergillus galactomannan* serum antigen assay. *Medical Mycology* 49(8): 883-886.

Mörner T, Avenäs A, Mattsson R. 1999. Adiaspiromycosis in a European beaver from Sweden. *Journal of Wildlife Diseases* 35(2): 367-370.

Morton FJ, Smith G. 1963. The genera *Scopulariopsis* Bainier, *Microascus* Zukal, and *Doratomyces* Corda. *Mycological Papers* 86: 1-96.

Mou XF, Liu X, Xu RF, Wei MY, Fang YW, Shao CL. 2018. Scopuquinolone B, a new monoterpenoid dihydroquinolin-2(1H)-one isolated from the coral-derived *Scopulariopsis* sp. fungus. *Natural Product Research* 32(7): 773-776.

Mouchacca J. 2000. Thermophilic fungi and applied research: a synopsis of name changes and synonymies. *World Journal of Microbiology & Biotechnology* 16: 881-888.

Mudau MM. 2006. The production, purification and characterization of endo-1,4-β-mannanase from newly isolated strains from *Scopulariopsis candida*. Bloemfontein: Doctoral dissertation of University of the Free State.

Mudau MM, Setati ME. 2008. Partial purification and characterization of endo-β-1,4-mannanases from *Scopulariopsis candida* strains isolated from solar salterns. *African Journal of Biotechnology* 7(13): 2279-2285.

Müller E, von Arx JA. 1982. *Pseudogymnoascus alpinus* sp. nov. *Sydowia* 35:135-137.

Murugesan AG, Prabu CS, Selvakumar C. 2009. Biolarvicidal activity of extracellular metabolites of the keratinophilic fungus *Trichophyton mentagrophytes* against larvae of *Aedes aegypti*: a major vector for Chikungunya and dengue. *Folia Microbiologica* 54(3): 213-216.

Nalli Y, Mirza DN, Wani ZA, Wadhwa B, Mallik FA, Raina C, Chaubey A, Riyaz-Ul-Hassan S, Ali A. 2015. Phialomustin A-D, new antimicrobial and cytotoxic metabolites from an endophytic fungus, *Phialophora mustea*. *RSC Advances* 5(115): 95307-95312.

Nannizzi A. 1927. Ricerche sull'origine saprofitica dei funghi delle tigne. 2. *Gymnoascus gypseum* sp. n. forma ascofora del *Sabouraudites* (*Achorion*) *gypseum*. (Bodin) Ota et langeron. *Atti dell'Accademia delle Scienze di Siena detta de'Fisiocritici* 10: 89-97.

Nannizzi A. 1934. Repertorio sistematico dei miceti dell'uomo e degli animali. Italy, Siena: SA Poligrafica Meini Siena.

Naranjo L, Urbina H, Sisto AD, Leon V. 2007. Isolation of autochthonous non-white rot fungi with potential for enzymatic upgrading of Venezuelan extra-heavy crude oil. *Biocatalysis and Biotransformation* 25(2-4): 341-349.

Nenoff P, Herrmann J, Gräser Y. 2007. *Trichophyton mentagrophytes* sive interdigitale? A dermatophyte in the course of time. *JDDG: Journal der Deutschen Dermatologischen Gesellschaft* 5(3): 198-202.

Neveu-Lemaire. 1921. Precis de Paresitologie humaine. Paris: Baillere et fils.

Ng AS, Just G. 1969. Metabolites of pathogenic fungi. VII. On the structure and stereochemistry of xanthomegnin, vioxanthin, and viopurpurin, pigments from *Trichophyton violaceum*. *Canadian Journal of Chemistry* 47: 1223.

Niu LH, Li Y, Xu L, Wang P, Zhang W, Wang C, Cai W, Wang L. 2017. Ignored fungal community in activated sludge wastewater treatment plants: diversity and altitudinal characteristics. *Environmental Science and Pollution Research International* 24(4): 4185-4193.

Niyonzima FN, More S. 2014. Purification and properties of detergent-compatible extracellular alkaline protease from *Scopulariopsis* spp. *Preparative Biochemistry and Biotechnology* 44(7): 738-759.

Nováková A, Kolařík M. 2010. *Chrysosporium speluncarum*, a new species resembling *Ajellomyces capsulatus*, obtained from bat guano in caves of temperate Europe. *Mycological Progress* 9: 253-260.

Orr GF. 1979. The genus *Pseudogymnoascus*. *Mycotaxon* 8: 165-173.

Orr GF, Kuehn HH, Plunkett OA. 1963. The genus *Gymnoascus* Baranetzky. *Mycopathologia et Mycologia Applicata* 21: 1-18.

Padhye AA, Carmichael JW. 1971. The genus *Arthroderma* Berkeley. *Canadian Journal of Botany* 49(9): 1525-1540.

Pan HQ. 2004. Species diversity of soil dematiaceous Hyphomycetes from Yangtse River and Yellow River source area of China. Tai'an: Doctoral Dissertation of Shandong Agricultural University. [潘好琴. 2004. 长江、黄河源头地区土壤中暗色丝孢菌物种多样性研究. 泰安: 山东农业大学博士学位论文.]

Pan HQ, Wang HF, Jiang YL, Zhang TY. 2014. *Scopulariopsis*: three new species and a key to species from soils in China. *Mycosystema* 33(1): 1-6.

Pannkuk EL, Risch TS, Savary BJ. 2015. Isolation and identification of an extracellular subtilisin-like serine protease secreted by the bat pathogen *Pseudogymnoascus destructans*. *PLoS One* 10(3): e0120508.

Paré JA, Coyle KA, Sigler L, Maas AK, Mitchell RL. 2006. Pathogenicity of the *Chrysosporium* anamorph of *Nannizziopsis vriesii* for veiled chameleons (*Chamaeleo calyptratus*). *Medical Mycology* 44(1): 25-31.

Parish CA, Cruz MDL, Smith SK, Zink D, Baxter J, Tucker-Samaras S, Collado J, Platas G, Bills G, Diez MT, Vicente F, Peláez F, Wilson K. 2009. Antisense-guided isolation and structure elucidation of pannomycin, a substituted cis-decalin from *Geomyces pannorum*. *Journal of Natural Products* 72(1):

59-62.

Patel S, Aglawe V, Jaget S, Arya P, Tiwari M. 2016. Epidemiology of pathogenic fungi from private hospital of Jabalpur, India. *International Journal Life Science Scientific Research* 2(6): 729-732.

Perdelli F, Cristina ML, Sartini M, Spagnolo AM, Dallera M, Ottria G. 2006. Fungal contamination in hospital environments. *Infection Control & Hospital Epidemiology* 27(1): 44-47.

Pereira L, Bastos C, Tzanov T, Cavaco-Paulo A, Gübitz GM. 2005. Environmentally friendly bleaching of cotton using laccases. *Environmental Chemistry Letters* 3: 66-69.

Pereira MM, Silva BA, Pinto MR, Barreto-Bergter E, dos Santos ALS. 2009. Proteins and peptidases from conidia and mycelia of *Scedosporium apiospermum* strain HLPB. *Mycopathologia* 167(1): 25-30.

Peterson S W, Sigler L. 1998. Molecular genetic variation in *Emmonsia crescens* and *Emmonsia parva*, etiologic agents of adiaspiromycosis, and their phylogenetic relationship to *Blastomyces dermatitidis* (*Ajellomyces dermatitidis*) and other systemic fungal pathogens. *Journal of Clinical Microbiology* 36(10): 2918-2925.

Pitt JI, Lantz H, Pettersson OV, Leong SLL. 2013. *Xerochrysium* gen. nov. and *Bettsia*, genera encompassing xerophilic species of *Chrysosporium*. *IMA Fungus* 4: 229-241.

Płaza G, Łukasik W, Ulfig K. 1998. Effect of cadmium on growth of potentially pathogenic soil fungi. *Mycopathologia* 141: 93-100.

Pollacci G, Nannizzi A. 1930. I miceti patogeni dell'uomo e degli animali, 10. Siena, Tip. San Bernardino, Bologna: Cappelli.

Priyanka S, Prakash S. 2003. Laboratory efficacy tests for fungal metabolites of *Chrysosporium tropicum* against *Culex quinquefasciatus*. *Journal of the American Mosquito Control Association* 19(4): 404-407.

Prochnau A, de Almeida HL, Souza PRM, Vetoratto G, Duquia RP, Defferari R. 2005. Scutular tinea of the scrotum: report of two cases. *Mycoses* 48: 162-164.

Qiu WY, Yao YF, Zhu YF, Zhang YM, Zhou P, Jin YQ, Zhang B. 2005. Fungal spectrum identified by a new slide culture and *in vitro* drug susceptibility using E test in fungal keratitis. *Current Eye Research* 30(12): 1113-1120.

Qoraan İ, Yasemin ÖZ, Bulur İ. 2017. The Identification of an Uncommon Dermatophyte as an Agent of Tinea Corporis, Microsporum Ferrugineum/Tinea Corporis' in Nadir bir Etkeni olarak Microsporum Ferrugineum'un İdentifikasyonu. *Osmangazi Tıp Dergisi* 39(2): 70-74.

Raillo A. 1929. Beiträge zur Kenntnis der Bodenpilze. Centralblatt für Bakteriologie, Parasitenkunde und Infektionskrankheiten. *Zweite Abtheilung* 78: 515-524.

Raina D, Gupta P, Khanduri A. 2016. A first case of *Microsporum ferrugineum* causing tinea corporis in Uttarakhand. *Annals of Tropical Medicine and Public Health* 9(5): 351.

Rainer J, Kaltseis J. 2010. Diversity in *Scedosporium dehoogii* (Microascaceae): *S. deficiens* sp. nov. *Sydowia* 62(1): 137-147.

Randhawa HS, Sandhu RS. 1964. *Keratinophyton terreum* gen. nov., sp. nov., a keratinophilic fungus from soil in India. *Sabouraudia* 3: 251-256.

Réblová M, Untereiner WA, Réblová K. 2013. Novel evolutionary lineages revealed in the Chaetothyriales (Fungi) based on multigene phylogenetic analyses and comparison of its secondary structure. *PLoS One* 8(5): e63547.

Ren N, Liu JJ, Yang DL, Peng YZ, Hong J, Liu X, Zhao NN, Zhou J, Luo YT. 2016. Indentification of vincamine indole alkaloids producing endophytic fungi isolated from *Nerium indicum*, Apocynaceae. *Microbiological Research* 192: 114-121.

Rice AV, Currah RS. 2006. Two new species of *Pseudogymnoascus* with *Geomyces* anamorphs and their phylogenetic relationship with *Gymnostellatospora*. *Mycologia* 98(2): 307-318.

Rippon JW, Carmichael JW. 1976. Petriellidiosis (Allescheriosis): four unusual cases and review of literature. *Mycopathologia* 58(2): 117-124.

Roilides E, Sigler L, Bibashi E, Katsifa H, Flaris N, Panteliadis C. 1999. Disseminated infection due to *Chrysosporium zonatum* in a patient with chronic granulomatous disease and review of non-*Aspergillus* fungal infections in patients with this disease. *Journal of Clinical Microbiology* 37(1): 18-25.

Rosa LH, Almeida Vieira MDL, Santiago IF, Rosa CA. 2010. Endophytic fungi community associated with the dicotyledonous plant *Colobanthus quitensis* (Kunth) Bartl. (Caryophyllaceae) in Antarctica. *FEMS Microbiology Ecology* 73(1): 178-189.

Roy SK, Dey SK, Raha SK, Chakrabarty SL. 1990. Purification and properties of an extracellular endoglucanase from *Myceliophthora thermophila* D-14 (ATCC 48104). *Microbiology* 136: 1967-1971.

Roy SK, Raha SK, Sadhukhan RK, Chakrabarty SL. 1991. Purification and characterization of extracellular β-glucosidase from *Myceliophthora thermophila*. *World Journal of Microbiology and Biotechnology* 7: 613-618.

Sabouraud R. 1910. Maladies du cuir chevelu. III. Les cryptogamiques. Les teignes. Paris: Masson.

Saccardo PA. 1888. Sylloge Fungorum. Berlin, Parisii: Sumptibus auctoris.

Saccardo PA. 1901. Sylloge Fungorum. Berlin, Parisii: Octave Doin edidit 8 place de l'Odéon 8.

Saccardo PA. 1911. Sylloge fungorum ommium hucusque cognitorum. *Borntragen* (Leipzig) 4: 282.

Sadhukhan R, Roy SK, Raha SK, Manna S, Chakrabarty SL. 1992. Induction and regulation of alpha-amylase synthesis in a cellulolytic thermophilic fungus *Myceliophthora thermophila* D14 (ATCC 48104). *Indian Journal of Experimental Biology* 30(6): 482-486.

Sadhukhan RK, Manna S, Roy SK, Chakrabarty SL. 1990. Thermostable amylolytic enzymes from a cellulolytic fungus *Myceliophthora thermophila* D14 (ATCC 48 104). *Applied Microbiology and Biotechnology* 33(6): 692-696.

Samson RA. 1972. Notes on *Pseudogymnoascus*, *Gymnoascus* and related genera. *Acta Botanica Neerlandica* 21: 517-527.

Sandoval-Denis M, Gené J, Sutton DA, Cano-Lira JF, de Hoog GS, Decock CA, Wiederhold NP, Guarro J. 2016. Redefining *Microascus*, *Scopulariopsis* and allied genera. *Persoonia: Molecular Phylogeny and Evolution of Fungi* 36: 1-36.

Santos VL, Heilbuth NM, Braga DT, Monteiro AS, Linardi VR. 2003. Phenol degradation by a *Graphium* sp. FIB4 isolated from industrial effluents. *Journal of Basic Microbiology* 43(3): 238-248.

Sarao LK, Arora M, Sehgal VK. 2010. Use of *Scopulariopsis acremonium* for the production of cellulase and xylanase through submerged fermentation. *African Journal of Microbiology Research* 4(14): 1506-1509.

Schmitt EC, Camozzi S, Vigo G, Tadini G. 1996. Tinea corporis resembling dermatophyte colonies on Sabouraud's agar in a patient with the human immunodeficiency virus. *Archives of Dermatology* 132: 233-234.

Schoch CL, Seifert KA, Huhndorf S, Robert V, Spouge JL, Levesque CA, Chen W, Fungal Barcoding Consortium. 2012. Nuclear ribosomal internal transcribed spacer (ITS) region as a universal DNA barcode marker for fungi. *Proceedings of the National Academy of Sciences of the United States of America* 109(16): 6241-6246.

Schol-Schwarz MB. 1970. Revision of the genus *Phialophora* (Moniliales). *Persoonia* 6(1): 59-94.

Schwartz IS, Emmons CW. 1968. Subcutaneous cystic granuloma caused by a fungus of wood pulp (*Phialophora richardsiae*). *American Journal of Clinical Pathology* 49(4): 500-505.

Shao CL, Xu RF, Wei MY, She ZG, Wang CY. 2013. Structure and absolute configuration of fumiquinazoline L, an alkaloid from a gorgonian-derived *Scopulariopsis* sp. fungus. *Journal of Natural*

Products 76(4): 779-782.

Sharaf EF, Khalil NM. 2011. Keratinolytic activity of purified alkaline keratinase produced by *Scopulariopsis brevicaulis* (Sacc.) and its amino acids profile. *Saudi Journal of Biological Sciences* 18(2): 117-121.

Sharma KD, Bhaitacharjee S, Bhadauria S. 1994. *Chrysosporium racemosus* sp. nov. *Indian Phytopathology* 46(4): 404.

Sharma R, Rajak RC, Pandey AK, Gräser Y. 2006. Internal Transcribed Spacer (ITS) of rDNA of appendaged and non-appendaged strains of *Microsporum gypseum*, reveals *Microsporum appendiculatum*, as its synonym. *Antonie van Leeuwenhoek* 89(1): 197-202.

Sharma R, Shouche YS. 2020. Diversity of onygenalean fungi in keratin-rich habitats of Maharashtra (India) and description of three novel taxa. *Mycopathologia* 185(1): 67-85.

Sheehan D, Casey JP. 1993. Evidence for Alpha and Mu class glutathione *S*-transferases in a number of fungal species. *Comparative Biochemistry and Physiology Part B: Comparative Biochemistry* 104(1): 7-13.

Shi T, Li XQ, Zheng L, Zhang YH, Dai JJ, Shang EL, Yu YY, Zhang YT, Hu WP, Shi DY. 2021c. Sesquiterpenoids from the Antarctic Fungus *Pseudogymnoascus* sp. HSX2#-11. *Frontiers in Microbiology* 12: 688202.

Shi T, Yu YY, Dai JJ, Zhang YT, Hu WP, Zheng L, Shi DY. 2021a. New polyketides from the antarctic fungus *Pseudogymnoascus* sp. HSX2#-11. *Marine Drugs* 19: 168.

Shi T, Zheng L, Li XQ, Dai JJ, Zhang YT, Yu YY, Hu WP, Shi DY. 2021b. Nitrogenous compounds from the antarctic fungus *Pseudogymnoascus* sp. HSX2#-11. *Molecules* 26: 2636.

Sigler L, Aneja KR, Kumar R, Maheshwari R, Shukla RV. 1998. New records from India and redescription of *Corynascus thermophilus* and its anamorph *Myceliophthora thermophila. Mycotaxon* 68: 185-192.

Sigler L, Carmichael JW. 1976. Taxonomy of *Malbranchea* and some other hyphomycetes with arthroconidia. *Mycotaxon* 4: 349-488.

Sigler L, Guarro J, Punsola L. 1986. New keratinophilic species of *Chrysosporium. Canadian Journal of Botany* 64: 1212-1215.

Sigler L, Whitney HS, Carmichael JW. 1982. *Chrysosporiium filiforme*, a new hyphomycete associated with the bark beetle dendroctonus ponderosae. *Mycotaxon* 14(1): 261-265.

Silva KV, Lima MI, Cardoso GN, Santos AS, Silva GS, Pereira FO. 2017. Inibitory effects of linalool on fungal pathogenicity of clinical isolates of *Microsporum canis* and *Microsporum gypseum. Mycoses* 60(6): 387-393.

Singh CJ. 1997. Characterization of an extracellular keratinase of *Trichophyton simii* and its role in keratin degradation. *Mycopathologia* 137: 13-16.

Singh I, Mishra A, Kushwaha R. 2009. Dermatophytes, related keratinophilic and opportunistic fungi in indoor dust of houses and hospitals. *Indian Journal of Medical Microbiology* 27(3): 242-246.

Singh K, Vézina C. 1971. An extracellular proteolytic enzyme from *Scopulariopsis brevicaulis*. 1. Purification and properties. *Canadian Journal of Microbiology* 17(8): 1029-1042.

Skou JP. 1992. A series of xerophilic *Chysosporium* species. *Mycotaxon* 43: 237-259.

Smith JMB, Jolly RD, Georg LK, Connole MD. 1968. *Trichophyton equinum* var. *autotrophicum*; its characteristics and geographical distribution. *Medical Mycology* 6(4): 296-304.

Smith JMB, Marples MJ. 1964. *Trichophyton mentagrophytes* var. *erinacei. Sabouraudia: Journal of Medical and Veterinary Mycology* 3(1): 1-10.

Song W, Yu JF, Zhang TY. 2008. Notes on soil dematiaceous hyphomycetes from Mount Taibai and its surrounding area, China I. *Mycosystema* 27(1): 16-28.

Staerck C, Landreau A, Herbette G, Roullier C, Bertrand S, Siegler B, Larcher G, Bouchara JP, Fleury MJ. 2017. The secreted polyketide boydone A is responsible for the anti-*Staphylococcus aureus* activity of *Scedosporium boydii*. *FEMS Microbiology Letters* 364(22): fnx223.

Stchigel A, Cano JF, Guarro J, Cormack WPM, Archubi DI. 2013. Antarctic Chrysosporia: *Chrysosporium magnasporum* Stchigel, Cano, Mac Cormack & Guarro, sp. nov. and *Chrysosporium oceanitesii* Stchigel, Cano, Archubi & Guarro, sp. nov. *Persoonia-Molecular Phylogeny and Evolution of Fungi* 31: 268-269.

Stockdale PM. 1963. The *Microsporum gypseum* complex (*Nannizzia incurvata* Stockd., *N. gypsea* (Nann.) comb. nov., *N. fulva* sp.nov.). *Sabouraudia* 3: 114-126.

Subramaniam SS, Nagalla SR, Renganathan V. 1999. Cloning and characterization of a thermostable cellobiose dehydrogenase from *Sporotrichum thermophile*. *Archives of Biochemistry and Biophysics* 365: 223-230.

Sugiyama M, Ohara A, Mikawa T. 1999. Molecular phylogeny of onygenalean fungi based on small subunit ribosomal DNA (SSU rDNA) sequences. *Mycoscience* 40: 251-258.

Sun PL, Ho HT. 2006. Concentric rings: an unusual presentation of tinea corporis caused by *Microsporum gypseum*. *Mycoses* 49: 150-151.

Sun PL, Mu CA, Fan CC, Fan YC, Hu JM, Ju YM. 2014. Cat favus caused by *Microsporum incurvatum* comb. nov.: the clinical and histopathological features and molecular phylogeny. *Medical Mycology* 52(3): 276-284.

Sundram S, Abdullah F, Ahmad ZAM, Yusuf UK. 2008. Efficacy of single and mixed treatments of *Trichoderma harzianum* as biocontrol agents of *Ganoderma* basal stem rot in oil palm. *Journal of Oil Palm Research* 20: 470-483.

Kim S, Silva C, Evtguin DV, Gamelas JAF, Cavaco-Paulo A. 2011. Polyoxometalate/laccase-mediated oxidative polymerization of catechol for textile dyeing. *Applied Microbiology and Biotechnology* 89: 981-987.

Szathmάry S. 1960. Die Cleistothecium-Bildung von *Ctenomyces trichophyticus—Trichophyton gypseum* var. *asteroides*—auf sterilisiertem Heu. *Mycoses* 3(3): 77-83.

Takeda H, Kinoshita S. 1995. Production of fructosylxylosides by *Scopulariopsis brevicaulis* sp. *Journal of Fermentation and Bioengineering* 79(3): 242-246.

Takeda H, Sato K, Kinoshita S, Sasaki H. 1994. Production of 1-kestose by *Scopulariopsis brevicaulis*. *Journal of Fermentation and Bioengineering* 77(4): 386-389.

Takiuchi I, Higuchi D, Sei Y, Koga M. 1982. Isolation of an extracellular proteinase (keratinase) from *Microsporum canis*. *Sabouraudia* 20(4): 281-288.

Tamminen A, Kramer A, Labes A, Wiebe MG. 2014. Production of scopularide A in submerged culture with *Scopulariopsis brevicaulis*. *Microbial Cell Factories* 13(1): 89.

Tawfik MM, Adnan HA. 2000. Partial purification and some biochemical characteristics of exocellular keratinase of *Trichophyton mentagrophytes* var. *erinacei*. *Mycopathologia* 150: 121-125.

Taylor JW, Jacobson DJ, Kroken S, Kasuga T, Geiser DM, Hibbett DS, Fisher MC. 2000. Phylogenetic species recognition and species concepts in Fungi. *Fungal Genetics and Biology* 31: 21-32.

Taylor SL, Peterson RE, Gray LE. 1985. Isolation of gregatin A from *Phialophora gregata* by preparative high-pressure liquid chromatography. *Applied and Environmental Microbiology* 50(5): 1328-1329.

Tekkok IH, Higgins MJ, Ventureyra EC. 1996. Posttraumatic gas-containing brain abscess caused by *Clostridium perfringens* with unique simultaneous fungal suppuration by *Myceliophthora thermophila*: case report. *Neurosurgery* 39: 1247-1251.

Thom C, Raper KB.1932. The arsenic fungi of Gosio. *Science* 76(1980): 548-550.

Thomas AD, Sigler L, Peucker S, Norton JH, Nielan A. 2002. *Chrysosporium* anamorph of *Nannizziopsis vriesii* associated with fatal cutaneous mycoses in the salt-water crocodile (*Crocodylus porosus*). *Medical Mycology* 40(2): 143-151.

Tigini V, Prigione V, Varese GC. 2014. Mycological and ecotoxicological characterisation of landfill leachate before and after traditional treatments. *Science of the Total Environment* 487: 335-341.

Tong X, Xu H, Zou L, Cai M, Xu X, Zhao Z, Xiao F, Li Y. 2017. High diversity of airborne fungi in the hospital environment as revealed by meta-sequencing-based microbiome analysis. *Scientific Reports* 7: 39606.

Topakas E, Moukouli M, Dimarogona M, Christakopoulos P. 2012. Expression, characterization and structural modelling of a feruloyl esterase from the thermophilic fungus *Myceliophthora thermophila*. *Applied Microbiology and Biotechnology* 94(2): 399-411.

Topakas E, Stamatis H, Biely P, Christakopoulos P. 2004. Purification and characterization of a type B feruloyl esterase (StFAE-A) from the thermophilic fungus *Sporotrichum thermophile*. *Applied Microbiology and Biotechnology* 63(6): 686-690.

Traaen AE. 1914. Untersuchungen über bodenpilze aus Norwegen. *Nytt Magasin for Naturvidenskapene* 52: 1-121.

Tripathi N, Kushwaha RKS. 2005. *Chrysosporium christchurchicum*: a new species from Indian. *Indian Phytopathology* 58(3): 305-307.

Tsipouras A, Goetz MA, Hensens OD, Liesch JM, Ostlind DA, Williamson JM, Dombrowski AW, Ball RG, Singh SB. 1997. Sporandol: a novel antiparsitic binaphthalene from *Chrysosporium meridarium*. *Bioorganic & Medicinal Chemistry Letters* 7(10): 1279-1282.

Turland NJ, Wiersema JH, Barrie FR, Greuter W, Hawksworth DL, Herendeen PS, Knapp S, Kusber WH, Li DZ, Marhold K, May TW, McNeill J, Monro AM, Prado J, Price MJ, Smith GF. 2018. International Code of Nomenclature for algae, fungi and plants (Shenzhen Code). Königstein, Germany: Koeltz Botanical Books.

Udagawa SI. 1997. Taxonomic studies on Plectomycetes (Cleistothecial ascomycetes). *Japanese Journal of Mycology* 38: 143-157.

Udagawa SI, Uchiyama S, Kamiya S. 1993. *Gymnostellatospora*, a new genus of the Myxotrichaceae. *Mycotaxon* 48(1): 157-164.

Ulfig K, Guarro J, Cano J, Gene J, Vidal P, Figueras MJ, Łukasik W. 1997. The occurrence of keratinolytic fungi in sediments of the river Tordera (Spain). *FEMS Microbiology Ecology* 22(2): 111-117.

Untereiner WA, Naveau FA. 1999. Molecular systematics of the Herpotrichiellaceae with an assessment of the phylogenetic positions of *Exophiala dermatitidis* and *Phialophora americana*. *Mycologia* 91(1): 67-83.

van den Brink J, Samson RA, Hagen F, Boekhout T, de Vries RP. 2012. Phylogeny of the industrial relevant, thermophilic genera *Myceliophthora* and *Corynascus*. *Fungal Diversity* 52: 197-207.

van Oorschot CAN. 1977. The genus *Myceliophthora*. *Persoonia* 9: 401-408.

van Oorschot CAN. 1980. A revision of *Chrysosporium* and allied genera. *Studies in Mycology* 20: 1-89.

van Oorschot CAN, Piontelli E. 1985. *Chrysosporium vallenarense* sp. nov. *Persoonia* 12: 487-488.

Vanbreuseghem R. 1950a. Contribution à l'étude des dermatophytes du Congo belge: Le *Sabouraudites* (*Microsporum*) *langeroni* nov. sp. *Annales de Parasitologie Humaine et Comparée* 25(5-6): 509-517.

Vanbreuseghem R. 1950b. Étude sur le *Trichophyton soudanense*: sa présence au Congo Belge création du genre *Langeronia*. *Annales de Parasitologie Humaine et Comparée* 25(5-6): 493-508.

Vanbreuseghem R. 1951. Essai de synthèsur les dermatophytes et te traitement des affections qu'ils déterminent. Rapport présenté à la séance du Cinquantenaire de la Societé beige de Derm. et Syphil. des

17 et 18 novembre 1951. Bruxelles: Impremerie Medicale et Sclentifique.
Vanbreuseghem R. 1952. Intérêt théorique et pratique d'un nouveau dermatophyte isolé du sol: *Keratinomyces ajelloi* gen. nov., sp. nov. *Bulletin de l'Académie Royale des Sciences de Belgique Classe des Sciences* 38: 1068-1077.
Vanbreuseghem R, de Vroey C, Takashio M. 1966. Guide pratique de mycologie médicale et vétérinaire. Masson, Paris, France.
Vander PYLD, Cans P, Debernard JJ, Herman F, Lelievre Y, Tahraoui L, Vuilhorgne M, Leboul J. 1995. RPR113228, a novel farnesyl-protein transferase inhibitor produced by *Chrysosporium lobatum*. *The Journal of Antibiotics* 48(7): 736-737.
Venkateshwarlu N, Reddy SM. 1993. Production of lipase by five thermophilic fungi. *Indian Journal of Microbiology* 33(2): 119-124.
Veselská T, Homutová K, Fraile PG, Kubátová A, Martínková N, Pikula J, Kolařík M. 2020. Comparative ecophysiology revealed extensive enzymatic curtailment, lipases production and strong conidial resilience of the bat pathogenic fungus *Pseudogymnoascus destructans*. *Scientific Reports* 10: 16530.
Vidal P, Guarro J, de Vroey C. 1996. Studies on keratinophilic fungi. VII. *Chrysosporium vespertilium* sp. nov. from Zaire. *Mycotaxon* 59: 189-196.
Vidal P, Ulfig K, Valmaseda M, Guarro J. 1999. Studies on keratinophilic fungi. XI. *Chrysosporium undulatum* sp. nov. *Antonie van Leeuwenhoek* 75: 171-182.
Vidal P, Valmaseda M, Vinuesa MÁ, Guarro J. 2002. Two new species of *Chrysosporium*. *Studies in Mycology* 47: 199-209.
Vidal P, Vinuesa M de los A, Sánchez-Puelles JM, Guarro J. 2000. Phylogeny of the anamorphic genus *Chrysosporium* and related taxa based on rDNA internal transcribed spacer sequences. *Rev Iberoamericana de Micologia* 17: 22-29.
Vidyasagar GM, Hosmani N, Shivkumar D. 2005. Keratinophilic fungi isolated from hospital dust and soils of public places at Gulbarga, India. *Mycopathologia* 159(1): 13-21.
Vikineswary S, Kuthubutheen AJ, Ravoof AA. 1997. Growth of *Trichoderma harzianum* and *Myceliophthora thermophila* in palm oil sludge. *World Journal of Microbiology and Biotechnology* 13: 189-194.
Villanueva P, Vásquez G, Gil-Durán C, Oliva V, Díaz A, Henríquez M, Álvarez E, Laich F, Chávez R, Vaca I. 2021. Description of the first four species of the genus *Pseudogymnoascus* from antarctica. *Frontiers in Microbiology* 12: 713189.
Vissiennon T. 1999. Fungal flora in chicken stalls and its etiopathogenic importance for humans and animals. *Berliner und Munchener Tierarztliche Wochenschrift* 112(3): 104-107.
Vohník M, Fendrych M, Albrechtová J, Vosátka M. 2007. Intracellular colonization of *Rhododendron* and *Vaccinium* roots by *Cenococcum geophilum*, *Geomyces pannorum* and *Meliniomyces variabilis*. *Folia Microbiologica* 52(4): 407-414.
Volesky B, Zajic JE. 1971. Batch production of protein from ethane and ethane-methane mixtures. *Applied Microbiology* 21(4): 614-622.
von Arx JA. 1971. Uber die Typudart. Zwei neue und einige weitere arten der Gattung *Sporotrichum*. *Persoonia* 6(2): 179-184.
von Arx JA. 1973. Further observations on *Sporotrichum* and some similar fungi. *Persoonia* 7(2): 127-130.
von Arx JA. 1975. On *Thielavia* and some similar genera of Ascomycetes. *Studies in Mycology* 8: 1-31.
von Arx JA. 1979. Ascomycetes as fungi imperfecti. In: Kendrick B. The Whole Fungus, The sexual-asexualsynthesis. 201-213. Ottawa: National Museums of Canada.
von Klopotek A. 1974. Revision der thermophilen *Sporotrichum*-Arten: *Chrysosporium thermophilum*

(Apinis) comb. nov. Und *Chrysosporium fergusii* spec. nov. = status conidialis von *Corynascus thermophilus* (Fergus und Sinden) comb. nov. *Archiv für Mikrobiologie* 98: 365-369.

Vroumsia T, Steiman R, Seigle-Murandi F, Benoit-Guyod JL. 2005. Fungal bioconversion of 2,4-dichlorophenoxyacetic acid (2,4-D) and 2,4-dichlorophenol (2,4-DCP). *Chemosphere* 60(10): 1471-1480.

Wagoner MD, Badr IA, Hidayat AA. 1999. *Chrysosporium parvum* keratomycosis. *Cornea* 18(5): 616-620.

Wang HF. 2009. Species diversity of dematiaceous hyphomycetes from soil in northern Qinghai Province. Tai'an: Master's Thesis of Shandong Agricultural University. [王洪凤. 2009. 青海北部地区土壤中暗色丝孢菌物种多样性研究. 泰安: 山东农业大学硕士学位论文.]

Wang XW, Han PJ, Bai FY, Luo A, Bensch K, Meijer M, Kraak B, Han DY, Sun BD, Crous PW, Houbraken J. 2022. Taxonomy, phylogeny and identification of Chaetomiaceae with emphasis on thermophilic species. *Studies in Mycology* 101: 121-243.

Westling R. 1911. Uber die grunen Species der Gattung *Penicillium*. *Arkiv för Botanik* 11: 1-156.

Wijayawardene N, Hyde KD, Al-Ani L, Tedersoo L, Haelewaters D, Rajeshkumar KC, Zhao RL, Aptroot A, Leontyev DV, Saxena RK, Tokarev YS, Dai DQ, Letcher PM, Stephenson SL, Ertz D, Lumbsch HT, Kukwa M, Issi IV, Madrid H, Phillips AJL, Selbmann L, Pfliegler WP, Horváth E, Bensch K, Kirk PM, Kolaříková K, Raja HA, Radek R, Papp V, Dima B, Ma J, Malosso E, Takamatsu S, Rambold G, Gannibal PB, Triebel D, Gautam AK, Avasthi S, Suetrong S, Timdal E, Fryar SC, Delgado G, Réblová M, Doilom M, Dolatabadi S, Pawłowska JZ, Humber RA, Kodsueb R, Sánchez-Castro I, Goto BT, Silva DKA, de Souza FA, Oehl F, da Silva GA, Silva IR, Błaszkowski J, Jobim K, Maia LC, Barbosa FR, Fiuza PO, Divakar PK, Shenoy BD, Castañeda-Ruiz RF, Somrithipol S, Lateef AA, Karunarathna SC, Tibpromma S, Mortimer PE, Wanasinghe DN, Phookamsak R, Xu J, Wang Y, Tian F, Alvarado P, Li DW, Kušan, Matočec N, Mešić A, Tkalčec Z, Maharachchikumbura SSN, Papizadeh M, Heredia G, Wartchow F, Bakhshi M, Boehm E, Youssef N, Hustad VP, Lawrey JD, Santiago ALCMA, Bezerra JDP, Souza-Motta CM, Firmino AL, Tian Q, Houbraken J, Hongsanan S, Tanaka K, Dissanayake AJ, Monteiro JS, Grossart HP, Suija A, Weerakoon G, Etayo J, Tsurykau A, Vázquez V, Mungai P, Damm U, Li QR, Zhang H, Boonmee S, Lu YZ, Becerra AG, Kendrick B, Brearley FQ, Motiejūnaitė J, Sharma B, Khare R, Gaikwad S, Wijesundara DSA, Tang LZ, He MQ, Flakus A, Rodriguez-Flakus P, Zhurbenko MP, McKenzie EHC, Stadler M, Bhat DJ, Liu JK, Raza M, Jeewon R, Nassonova ES, Prieto M, Jayalal RGU, Erdoğdu M, Yurkov A, Schnittler M, Shchepin ON, Novozhilov YK, Silva-Filho AGS, Gentekaki E, Liu P, Cavender JC, Kang Y, Mohammad S, Zhang LF, Xu RF, Li YM, Dayarathne MC, Ekanayaka AH, Wen TC, Deng CY, Pereira OL, Navathe S, Hawksworth DL, Fan XL, Dissanayake LS, Kuhnert E, Grossart HP, Thines M. 2020. Outline of Fungi and fungus-like taxa. *Mycosphere* 11(1): 1060-1456.

Wijayawardene NN, Hyde KD, Rajeshkumar KC, Hawksworth DL, Madrid H, Kirk PM, Braun U, Singh RV, Crous PW, Kukwa M, Lücking R, Kurtzman CP, Yurkov A, Haelewaters D, Aptroot A, Lumbsch HT, Timdal E, Ertz D, Etayo J, Phillips AJL, Groenewald JZ, Papizadeh M, Selbmann L, Dayarathne MC, Weerakoon G, Jones EBG, Suetrong S, Tian Q, Castañeda-Ruiz RF, Bahkali AH, Pang KL, Tanaka K, Dai DQ, Sakayaroj J, Hujslová M, Lombard L, Shenoy BD, Suija A, Maharachchikumbura SSN, Thambugala KM, Wanasinghe DN, Sharma BO, Gaikwad S, Pandit G, Zucconi L, Onofri S, Egidi E, Raja HA, Kodsueb R, Cáceres MES, Pérez-Ortega S, Fiuza PO, Monteiro JS, Vasilyeva LN, Shivas RG, Prieto M, Wedin M, Olariaga I, Lateef AA, Agrawal Y, Fazeli SAS, Amoozegar MA, Zhao GZ, Pfliegler WP, Sharma G, Oset M, Abdel MA, Takamatsu S, Bensch K, de Silva NI, de Kesel A, Karunarathna A, Boonmee S, Pfister DH, Lu YZ, Luo ZL, Boonyuen N, Daranagama DA, Senanayake IC, Jayasiri SC, Samarakoon MC, Zeng XY, Doilom M, Quijada L, Rampadarath S, Heredia G,

Dissanayake AJ, Jayawardana RS, Perera PH, Tang LZ, Phukhamsakda C, Hernández-Restrepo M, Ma XY, Tibpromma S, Gusmao LFP, Weerahewa D, Karunarathna SC. 2017. Notes for genera: Ascomycota. *Fungal Diversity* 86: 1-594.

Wirth JC, Beesley TE, Anand SR. 1965. The isolation of xanthomegnin from several strains of the dermatophyte, *Trichophyton rubrum*. *Phytochemistry* 4(3): 505-509.

Woudenberg JHC, Meijer M, Houbraken J, Samson RA. 2017. *Scopulariopsis* and *Scopulariopsis*-like species from indoor environments. *Studies in Mycology* 88: 1-35.

Wu YM. 2006. Taxonomic studies on soil dematiaceous Hyphomycetes from north of North China. Tai'an: Master's Thesis of Shandong Agricultural University. [吴悦明. 2006. 中国华北北部地区土壤中暗色丝孢菌分类研究. 泰安: 山东农业大学硕士学位论文.]

Wu YM, Zhang TY. 2008. Soil dematiaceous hyphomycetes from Dahingganling Mountain Range, Inner Mongolia, China I. *Mycosystema* 27(1): 5-15.

Wu YM, Zhang TY. 2011. Two new species of *Phialophora* from soil. *Mycotaxon* 115(1): 251-254.

Wyss M, Brugger R, Kronenberger A, Rémy R, Fimbel R, Oesterhelt G, van Loon AP. 1999. Biochemical characterization of fungal phytases (myo-inositol hexakisphosphate phosphohydrolases): catalytic properties. *Applied and Environmental Microbiology* 65(2): 367-373.

Xing Y, Cai L, Yin TP, Chen Y, Yu J, Wang YR, Ding ZT. 2016. Improving the antioxidant activity and enriching salvianolic acids by the fermentation of *Salvia miltiorrhizae* with *Geomyces luteus*. *Journal of Zhejiang University Science B* 17(5): 391-398.

Xu JJ. 2006. Taxonomic studies on soil dematiaceous Hyphomycetes from South-eastern Zones and Yunnan Province of China. Tai'an: Doctoral Dissertation of Shandong Agricultural University. [许俊杰. 2006. 中国东南部地区及云南省土壤暗色丝孢菌分类研究. 泰安: 山东农业大学博士学位论文.]

Xu X, Liu T, Ren XW, Liu B, Yang J, Chen LH, Wei CD, Zheng JH, Dong J, Sun LL, Zhu YF, Zhu Y. 2015. Proteogenomic analysis of *Trichophyton rubrum* aided by RNA sequencing. *Journal of Proteome Research* 14(5): 2207-2218.

Yamamoto H. 1994. Studies on taxonomy of the causal fungus of brown stem rot of adzuki bean and soyabean, *Phialophora gregata* (Allington et Chamberlain) Gams. *Memoirs of the Faculty of Agriculture Hokkaido University* 19(1): 57-98.

Yamashita M, Kawai Y, Uchida I, Komori T, Kohsaka M, Imanaka H, Sakane K, Setoi H, Teraii T. 1984. Sturcture and total synthesis of chryscandin, a new antifungal antibiotic. *Tetrahedron Letters* 25(41): 4689-4692.

Yamagishi Y, Kawasaki K, Ishizaki H. 1997. Mitochondrial DNA analysis of *Phialophora verrucosa*. *Mycoses* 40(9-10): 329-334.

Yang F, Chen GD, Gao H, Li XX, Wu Y, Guo LD, Yao XS. 2012. Two new naphthalene derivatives from an endolichenic fungal strain *Scopulariopsis* sp. *Journal of Asian Natural Products Research* 14(11): 1059-1063.

Yang SW, Buevich A, Chan TM, Terracciano J, Chen G, Loebenberg D, Patel M, Boehm E, Gullo V, Pramanik B, Chu M. 2003. A new antifungal sterol sulfate, sch 601324, from *Chrysosporium* sp. *The Journal of Antibiotics* 56(4): 419-422.

Yasmina MF, Alberto MS, Andrew NM, Josep G, Jose FCL. 2015. A re-evaluation of the genus *Myceliophthora* (Sordariales, Ascomycota): its segregation into four genera and description of *Corynascus fumimontanus* sp. nov. *Mycologia* 107(3): 619-632.

Ye F, Chen GD, He JW, Li XX, Sun X, Guo LD, Li Y, Gao H. 2013. Xinshengin, the first altenusin with tetracyclic skeleton core from *Phialophora* spp. *Tetrahedron Letters* 54(34): 4551-4554.

Yu RJ, Harmon SR, Blank F. 1968. Isolation and purification of an extracellular keratinase of *Trichophyton mentagrophytes*. *Journal of Bacteriology* 96(4): 1435.

Yu X, Seena K, Obadah A, Baden LR, Marty FM, Nathan W, Sophia K. 2016. Examining the *in vitro* volatile metabolite profile of pathogenic *Scedosporium* and *Lomentospora* species. *Open Forum Infectious Diseases* 3(Suppl.1): 1564.

Yu ZG, Lang G, Kajahn I, Schmaljohann R, Imhoff JF. 2008. Scopularides A and B, cyclodepsipeptides from a marine sponge-derived fungus, *Scopulariopsis brevicaulis*. *Journal of Natural Products* 71(6): 1052-1054.

Zajic JE, Volesky B, Wellman A. 1969. Growth of *Graphium* sp. on natural gas. *Canadian Journal of Microbiology* 15(10): 1231-1236.

Zanphorlin LM, Facchini FDA, Vasconcelos F, Bonugli-Santos RC, Rodrigues A, Sette LD, Gomes E, Bonilla-Rodriguez GO. 2010. Production, partial characterization, and immobilization in alginate beads of an alkaline protease from a new thermophilic fungus *Myceliophthora* sp. *Journal of Microbiology* 48(3): 331-336.

Zelenková H. 2006. *Geomyces pannorum* as a possible causative agent of dermatomycosis and onychomycosis in two patients. *Acta Dermatovenerologica Croatica* 14(1): 21-25.

Zhan P, Liu W. 2017. The Changing face of dermatophytic infections worldwide. *Mycopathologia* 182(1-2): 77-86.

Zhang Q, Qin L, Wang J, Li L. 2002. Report of two cases of tinea infection with scutula-like lesions caused by *Microsporum gypseum*. *International Journal of Dermatology* 41: 372-373.

Zhang W. 2006. Soil dematiaceous Hyphomycetes from cold temperate zone-temperate zone of Northeast China. Tai'an: Master's Thesis of Shandong Agricultural University. [张伟. 2006. 中国东北寒温带-温带地区土壤中的暗色丝孢菌. 泰安：山东农业大学硕士学位论文.]

Zhang X. 2020. The taxonomy of *Cephalotrichum* and *Scopulariopsis*. Guiyang: Master's Thesis of Guizhou University. [张欣. 2020. 头束霉属和帚霉属的分类研究. 贵阳：贵州大学硕士学位论文.]

Zhang Y, Wu WP, Hu DM, Su YY, Cai L. 2014. A new thermophilic species of *Myceliophthora*, from China. *Mycological Progress* 13(1): 165-170.

Zhang YL. 2007. Taxonomic studies of dematiaceous hyphomycetes from soil of middle-southern subtropical and tropical zones in China. Tai'an: Doctoral Dissertation of Shandong Agricultural University. [张悦丽. 2007. 中国中南部亚热带及热带地区土壤暗色丝孢菌分类研究. 泰安：山东农业大学博士学位论文.]

Zhang YW, Chen WH, Zeng GP, Wang YR, Zou X, Han YF, Qiu SY, Liang ZQ. 2016. Two new *Chrysosporium* (Onygenaceae, Onygenales) from China. *Phytotaxa* 270(3): 210-216.

Zhang YW, Han YF, Liang JD, Liang ZQ. 2013. A new species of the genus *Chrysosporium* from the rhizosphere soil of palm. *Mycosystema* 32(4): 612-616.

Zhang YW, Zeng GP, Zou X, Han YF, Liang ZQ, Qiu SY. 2017a. Two new keratinophilic fungal species. *Phytotaxa* 303(2): 173-180.

Zhang ZF, Liu F, Zhou X, Liu XZ, Liu SJ, Cai L. 2017b. Culturable mycobiota from Karst caves in China, with descriptions of 20 new species. *Persoonia* 39: 1-31.

Zhang ZF, Zhou SY, Eurwilaichitr L, Ingsriswang S, Raza M, Chen Q, Zhao P, Liu F, Cai L. 2021a. Culturable mycobiota from Karst caves in China II, with descriptions of 33 new species. *Fungal Diversity* 106: 1-108.

Zhang ZY, Dong CB, Chen WH, Mou QR, Lu XX, Han YF, Liang ZQ. 2020. The enigmatic Thelebolaceae (Thelebolales, Leotiomycetes): one new genus *Solomyces* and five new species. *Frontiers in Microbiology* 11: 572-596.

Zhang ZY, Han YF, Chen WH, Liang ZQ. 2019. Phylogeny and taxonomy of three new *Ctenomyces*

(Arthrodermataceae, Onygenales) species from China. *MycoKeys* 47: 1-16.
Zhang ZY, Han YF, Chen WH, Tao G. 2023a. Additions to Thelebolales (Leotiomycetes, Ascomycota): *Pseudogeomyces lindneri* gen. et sp. nov. and *Pseudogymnoascus campensis* sp. nov. *MycoKeys* 95: 47-60.
Zhang ZY, Li X, Chen WH, Liang JD, Han YF. 2023b. Culturable fungi from urban soils in China II, with the description of 18 novel species in Ascomycota (Dothideomycetes, Eurotiomycetes, Leotiomycetes and Sordariomycetes). *MycoKeys* 98: 167-220.
Zhang ZY, Pan H, Tao G, Li X, Han YF, Feng Y, Tong SQ, Ding CY. 2024. Culturable mycobiota from guizhou wildlife park in China. *Mycosphere* 15(1): 654-763.
Zhang ZY, Shao QY, Li X, Chen WH, Liang JD, Han YF, Huang JZ, Liang ZQ. 2021b. Culturable fungi from urban soils in China I: description of 10 new taxa. *Microbiology Spectrum* 9: e00867-21.
Zhao YZ, Zhang ZF, Cai L, Peng WJ, Liu F. 2018. Four new filamentous fungal species from newly-collected and hive-stored bee pollen. *Mycosphere* 9(6): 1089-1116.

索　引

真菌汉名索引

A

B

C

H

J

K

L

R

S

T

W

X

Y

Z

真菌学名索引

D

E

F

G

H

S

T

V

X

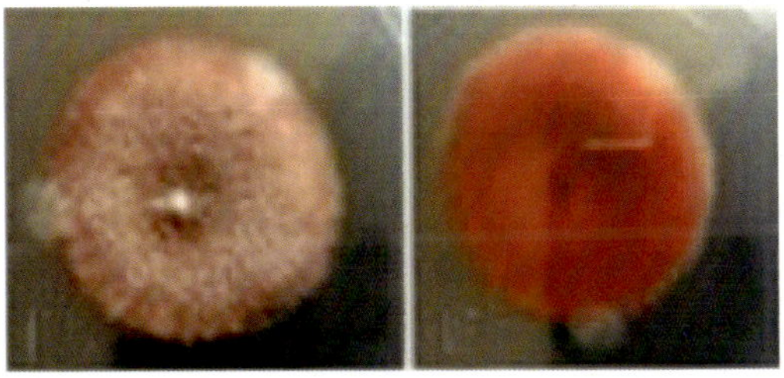

图 1　乔治金孢 *Chrysosporium georgiae*

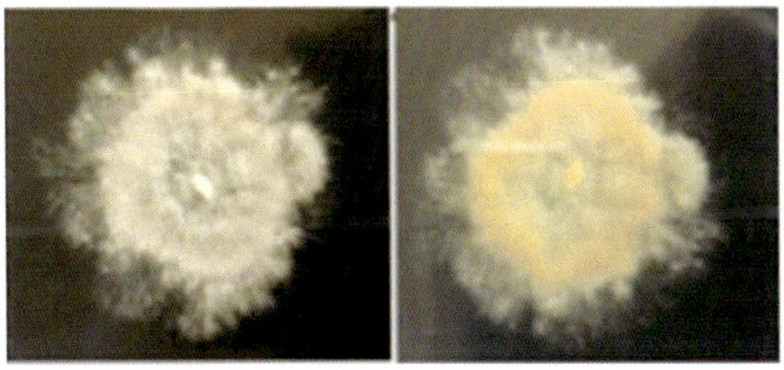

图 2　不规则金孢 *Chrysosporium irregularum*

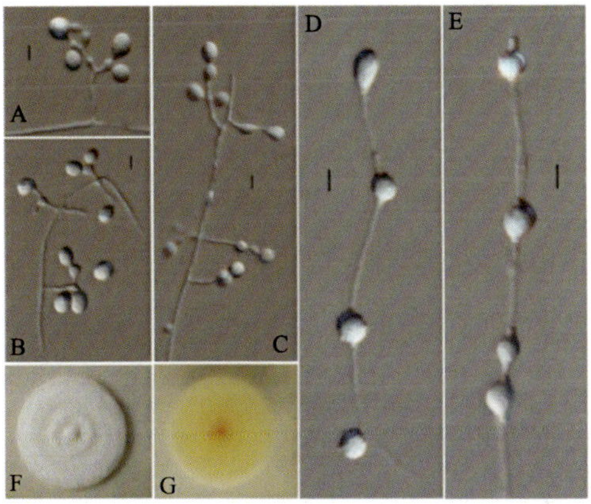

图 3　白色栉霉 *Ctenomyces albus*

A~C. 产孢结构与分生孢子；D、E. 间生孢子；F、G. PDA 培养基上的菌落。标尺：A~E = 10 μm

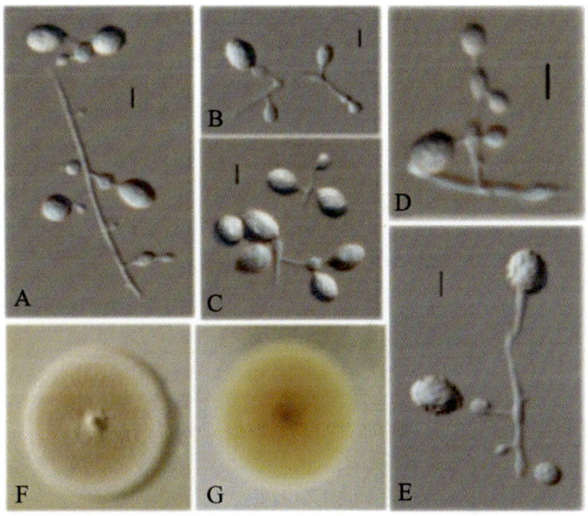

图 4　锯齿栉霉 *Ctenomyces serratus*

A~E. 产孢结构与分生孢子；F、G. PDA 培养基上的菌落。标尺：A~E = 10 μm

图版 II

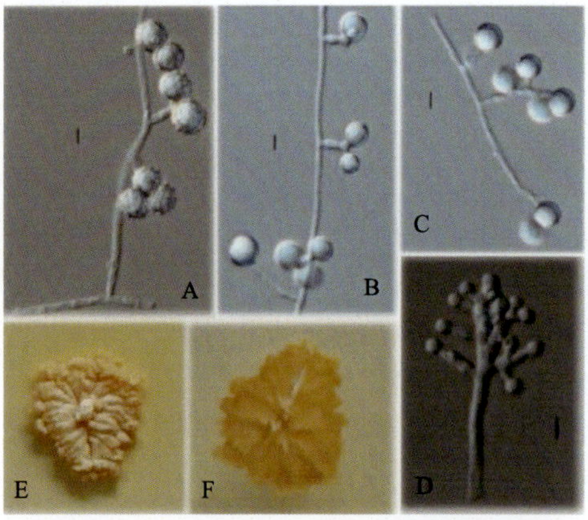

图 5　土黄毁丝霉 *Myceliophthora lutea*

A~D. 产孢结构和孢子；E、F. PDA 培养基上的菌落。标尺：A~D =10 μm

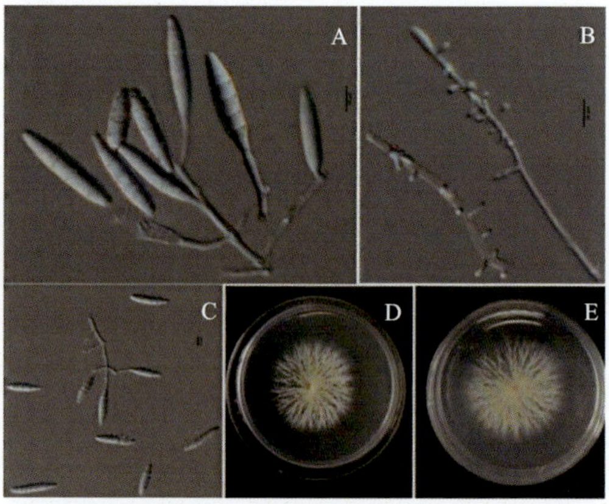

图 6　弯奈尼兹皮菌 *Nannizzia incurvata*

A、C. 大分生孢子；B. 小分生孢子；D、E. PDA 培养基上的菌落。标尺：A~C=20 μm

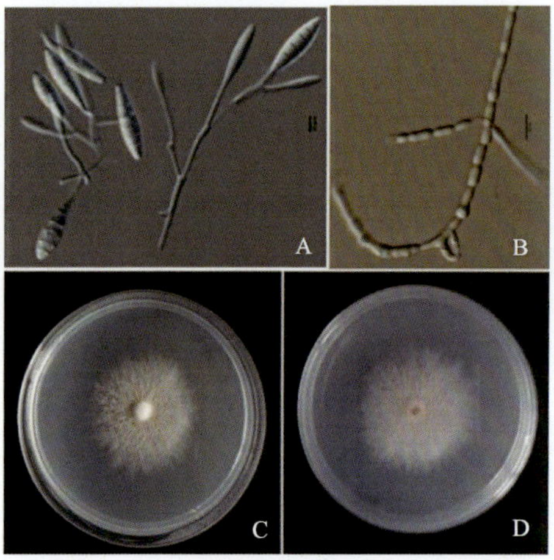

图 7　奇妙帕拉癣菌 *Paraphyton mirabile*

A. 大分生孢子；B. 节孢子状菌丝；C、D. PDA 培养基上的菌落。标尺：A、B=20 μm

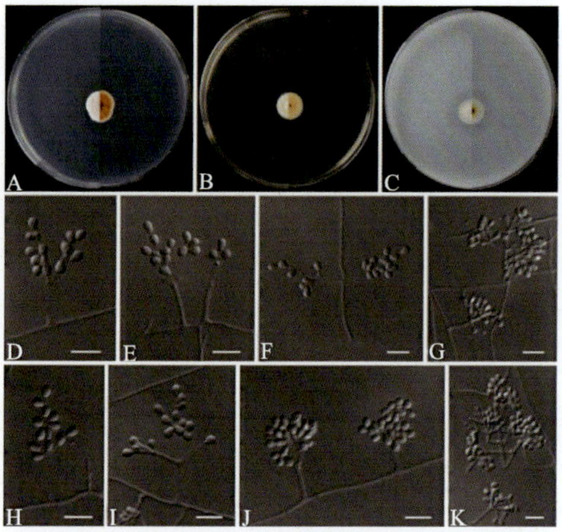

图 8　宗琦假裸囊菌 *Pseudogymnoascus zongqii*

A~C. PDA、MEA 和 OA 培养基上的菌落；D~K. 分生孢子梗、分生孢子和间生孢子。标尺：D~K =10 μm

(SCPC-BZBEZF21-0003)

ISBN 978-7-03-081146-2

定 价：**268.00** 元